Series on
Reproduction and Development in Aquatic Invertebrates

Volume 2

Reproduction and Development in Mollusca

Series on

Reproduction and Development in Aquatic Invertebrates

Volume 2

Reproduction and Development in Mollusca

T. J. Pandian

Central Marine Fisheries Research Institute
Kochi, Kerala
India

CRC Press
Taylor & Francis Group
Boca Raton London New York

CRC Press is an imprint of the
Taylor & Francis Group, an **informa** business

A SCIENCE PUBLISHERS BOOK

Cover page: Representative examples for aplacophora, monoplacophora, polyplacophora, shell-less opisthobranch, pteropod, prosobranchs, bivalves, scaphopod and cephalopods.

CRC Press
Taylor & Francis Group
6000 Broken Sound Parkway NW, Suite 300
Boca Raton, FL 33487-2742

First issued in paperback 2020

ISBN-13: 978-1-138-71045-0 (hbk)
ISBN-13: 978-0-367-78206-1 (pbk)

Library of Congress Cataloging-in-Publication Data

Names: Pandian, T. J.
Title: Reproduction and development in mollusca / T.J. Pandian, Central Marine Fisheries Research Institute, Kochi, Kerala, India.
Description: Boca Raton, FL : CRC Press, Taylor & Francis Group, 2017. | Series: Series on reproduction and development in aquatic invertebrates ; volume 2 | "A Science Publishers book."
| Includes bibliographical references and indexes.
Identifiers: LCCN 2017007407| ISBN 9781138710450 (hardback) | ISBN 9781315200637 (e-book)
Subjects: LCSH: Mollusks--Reproduction. | Mollusks--Development.
Classification: LCC QL403 .P36 2017 | DDC 594--dc23
LC record available at https://lccn.loc.gov/2017007407

Visit the Taylor & Francis Web site at
http://www.taylorandfrancis.com

and the CRC Press Web site at
http://www.crcpress.com

Preface to the Series

Invertebrates surpass vertebrates not only in species number but also in diversity of sexuality, modes of reproduction and development. Yet, we know much less of them than we know of vertebrates. During 1950s, the multi-volume series by L. E. Hyman accumulated bits and pieces of information on reproduction and development of aquatic invertebrates. Through a few volumes published during 1960s, A. C. Giese and A. S. Pearse provided a shape to the subject of Aquatic Invertebrate Reproduction. Approached from the angle of structure and function in their multi-volume series on Reproductive Biology of Invertebrates during 1990s, K. G. Adiyodi and R. G. Adiyodi elevated the subject to a visible and recognizable status.

Reproduction is central to all biological events. The life cycle of most aquatic invertebrates involves one or more larval stage(s). Hence, an account on reproduction without considering development shall remain incomplete. With passage of time, publications in large numbers are appearing in newly established journals on invertebrate reproduction and development. The time is ripe to update the subject. This treatise series proposes to (i) update and comprehensively elucidate the subject in the context of cytogenetics and molecular biology, (ii) view modes of reproduction in relation to Embryonic Stem Cells (ESCs) and Primordial Germ Cells (PGCs) and (iii) consider cysts and vectors as biological resources.

Hence, the first chapter on Reproduction and Development of Crustacea opens with a survey of sexuality and modes of reproduction in aquatic invertebrates and bridges the gaps between zoological and stem cell research. With capacity for no or slow motility, the aquatic invertebrates have opted for hermaphroditism or parthenogenesis/polyembryony. In many of them, asexual reproduction is interspersed within sexual reproduction. Acoelomates and eucoelomates have retained ESCs and reproduce asexually also. However, pseudocoelomates and hemocoelomates seem not to have retained ESCs and are unable to reproduce asexually. This series provides possible explanation for the exceptional pseudocoelomates and hemocoelomates that reproduce asexually. For posterity, this series intends to bring out six volumes.

August, 2015   T. J. Pandian
Madurai-625 014

Preface

Molluscs are unique for the presence of a protective external shell, defensive inking, distribution from the depth of 9,050 m to an altitude of 4,300 m, gamete diversity, the use of nurse eggs/embryos to accelerate the first few mitotic divisions in embryos, the occurrence of natural androgenics, gigantism induced by elevated polyploidy, the complementary role of mitochondrial genome in sex determination by nuclear genes and the uptake of steroids from surrounding water. Not surprisingly, many books were authored or edited on one or other aspect of molluscs. This book represents perhaps the first attempt to comprehensively project the uniqueness of molluscs, covering almost all aspects of reproduction and development from aplacophorans to vampyromorphic cephalopods.

The book is organized in nine chapters. Considering their uniqueness, three chapters are devoted to shell and reproduction, shell and acidification, and aestivation. In view of its importance, regeneration is woven with the claimed asexual reproduction. Molluscan strategy is to carry shell(s) to escape from predators, limit energy expense on motility, and allocate the thus saved energy on reproductive output. Being proliferative breeders, molluscs inject a huge volume of fodder into the trophic dynamics, contribute 22 mt to the global food basket, colonize thermal vents and cold seeps by successfully engaging endosymbionts; whereas the adults move at 'a snail pace', their larvae embark on trans-Atlantic dispersal; they also serve as intermediary hosts to the dreadful disease causing schistosomes. For the first time, the second chapter attempts to relate shell thickness to reproductive output, and chirality to mating and insemination between dextrals and sinistrals. With increasing size/age, the allocation for shell growth is progressively diminished but with increasing reproductive output. At low temperatures and higher altitudes, energy cost of calcification of shell increases. The shell size, shape and thickness are phenotypic traits. Like other animals, the 2% shell-less molluscs depend on chemical defense; aplysiaids and cephalopods are known for inking and escape.

From the fairly long third chapter, it has become apparent that, the presence of shell in molluscs has afforded iteroparity and comparatively longer life span in prosobranchs and bivalves but its absence semelparity and short life span in opisthobranchs and cephalopods. Notably, >25%

of molluscs is hermaphordites. In shell-less molluscs, gonochorism has facilitated faster growth and relatively larger body size in cephalopods but hermaphroditism small body size in opisthobranchs. Gastropods display great powers of regeneration inclusive of 'brain' and nervous system. The claimed asexual reproduction in the pteropod *Clio pyramidata* is more of regeneration involving multi-potent stem cells. The fifth chapter recognizes the occurrence of seasonal aestivation in snails and tidal aestivation in bivalves. For the first time, the effect of varying durations of aestivation on growth and reproduction in a freshwater snail and supratidal inhabitant bivalves are comprehensively described. For the first time, the scope for employing aestivating, schistosome-infected snails as 'biological weapon' is also explored.

The sixth chapter commences with description of spawning induction and gamete preservation, indicates the presence of heterogamety in less than a dozen gastropods and bivalves, and reports the rare occurrence of androgenics in bivalves; the induced triploid and tetraploid bivalves grow many times faster than their counterpart diploids, produce more and tastier meat with higher glycogen content and are available in markets during off season. The unique but perhaps 'vestigial' presence of Doubly Uniparental Inheritance (DUI) in bivalves and its possible causative complementary role of sperm mitochondrial genes in sex determination by nuclear genes is brought to light. Irrespective of sexuality in molluscs, sex is irrevocably determined by a few genes harbored on autosomes. Sex differentiation is solely accomplished by neurohormones. The genes responsible for biosynthesis of vertebrate-type steroid hormones are indicated to be missing in molluscan genome. Surprisingly, molluscs absorb and accumulate steroid hormones from surrounding waters. The ongoing debate on the role played by the steroid hormones in regulation of reproductive cycle is discussed. While tributyltin induces sex reversal from female to male in fishes, it only superimposes masculine traits on female neogastropods. Trematode parasites exploit many molluscs as intermediate host and partially or completely sterilize them.

Sex ratio represents the cumulative end product of sex determination and differentiation processes. In molluscs, sex ratio of $1♀ : 1♂$ is maintained at population level. Among different families of a molluscan species, sex ratio varies from nearly all females to nearly all males. But the ratio remains constant between successive broods in a family. Hence, the ratio is more a family trait than that of species. The primary sex differentiation is determinate and not amenable to environmental influence. However, the secondary differentiation from male to female in protandrics is labile, protracted and amenable to environmental factors like food supply and temperature.

The eighth chapter highlights the effects of acidification on molluscan shell. Since the advent of industrial era, the oceans have absorbed 127 billion metric tons of CO_2, resulting in decreased pH, which renders the acquisition of calcium costlier. Consequently, the molluscan shells suffer from pitting, dissolution and loosening of the hinge to valve mechanism. However, with abundant food supply, many molluscs do adapt and not suffer from shell dissolution.

This book is a comprehensive synthesis of over 907 publications carefully selected from widely scattered information from 249 journals and 87 other literature sources. The holistic and incisive analyses have led to harvest several new findings related to reproduction and development in molluscs and to project the uniqueness of mollusca. Hopefully, this book serves as a launching pad to further advance our knowledge on reproduction and development of molluscs.

September, 2016   **T. J. Pandian**
Madurai, 625014

Acknowledgements

It is with pleasure that I thank Drs. T. Balasubrmanian, S. Mohamed and E. Vivekanandan for critically reviewing parts of the manuscript of this book and offering valuable suggestions. I am thankful to my students Drs. M. A. Haniffa and J. P. Arockiam, whose findings in my laboratory and my earlier editorial service on molluscan energetics (Pandian, 1987, *Animal Energetics*, Academic Press) emboldened me to author this book. The Central Marine Fisheries Research Institute (CMFRI), Kochi provides the best library and excellent service to visitors. I wish to place on records my sincere thanks to the Director, CMFRI and his library staff especially Mr. V. Mohan. Thanks are due to the American College, Madurai for providing valuable books. I also thank Drs. R. Jeyabaskaran, G. Kumaresan, N. Munuswamy, P. Murugesan, B. Senthilkumaran and K. K. Vijayan for helping with many publications. I gratefully appreciate Dr A. Gopalakrishnan, Director, CMFRI, who has very thoughtfully engaged me to author this book serial. The manuscript of this book was prepared by Mr. T. S. Balaji, B.Sc. and I wish to thank him for his good work.

Firstly, I wish to thank many authors/publishers, whose published figures are simplified/modified/compiled for an easier understanding. To reproduce original figures from published domain, I gratefully appreciate the permissions issued by *Biological Bulletin*, CMFRI, *Fisheries Bulletin*, *Genetics*, *Indian Journal of Fisheries*, *Indian Journal of Geo-Marine Science*, International Ecology Research Institute (*Marine Ecology Progress Series*), *International Journal of Developmental Biology*, New Zealand Marine Studies Centre, *Proceedings of Natural Academy of Science*, USA and Rajeev Gandhi Centre for Biotechnology. For permission issued to reproduce their original figures, I remain thankful to Drs. M. A. Haniffa, C. S. Hickman, O. Kah, E. L. Kenchington, P. Krug and F. Melzner. Special thanks are due to Dr. F. Ghiselli and Dr. L. M. Roger for providing their figures and permission to reproduce them. For advancing our knowledge in this area by their rich contributions, I thank all my fellow scientists, whose publications are cited in this book.

September, 2016   T. J. Pandian
Madurai, 625014

Contents

1

Introduction

1.1 Molluscan Science

The molluscs comprise edible abalones and snails, musssels and oysters as well as cuttlefish and squids. The molluscan fisheries contribute to the global food basket annually 22 million tons (mt) worth US$ 26.7 million (FAO, 2014). Since ages, their shells have served as raw material for construction and painting. They are also used as ornamentals, curios and decoratives. Some are considered as sacred by Hindus and Buddhists. Not surprisingly, their importance has commanded frequent references in Greek and Sanskrit literature. The 2000-year old Tamil Academy at Madurai (India) speaks volumes on the flourishing pearl trade between Tamils and Romans. Being a costly gem, the marine pearls have been traditionally a very valuable commodity. In terms of color, luster and shape, the pearl produced by freshwater mussels *Hyriopsis cumingii* has also become a valuable gem. About 1.2 million ton (mt) of *H. cumingii* is annually produced in China (Bai et al., 2011). The less pretty *Conus gloria maris* is the rarest and costliest shell in the world, as it is perhaps completely lost from the Philippines coral reef after an earthquake (Hyman, 1967, p 342).

Known for proliferative breeding, the molluscs inject a huge volume of 'fodder' into the trophodynamics of aquatic system. In fact, some neretids, limpets and nudibranch (gastropods) annually release so many gametes that weigh more than their own body weight (Hughes, 1971b, Parry, 1982, Todd, 1979). For example, the gastropod *Patella peroni* releases gametes equivalent to 2.2 times of its own body weight (Parry, 1982) and life time reproductive investment ranges from 2.5 times in the dumpling squid *Euprymna tasmanica* to 5.5 times in the prosobranch *Lacuna pallidula*, of their respective body mass (Squries et al., 2013, Grahame, 1977); these values may be compared with the most prodigious pufferfish *Canthigaster valentine*, producing eggs equivalent to 1.5 times of its body weight (see Pandian, 2010, p 14). In terms of number, *Haliotis midae* produces up to 25 million eggs (Poore, 1973). In the Coosa River below Jordan Dam, USA, the viviparous snail *Tulotoma magnifica* produces 163 million young ones (Christman et al., 1996). Amazingly, a 2.6 kg weighing

Aplysia californicus produces 478 million eggs in 18 weeks (MacGinitie, 1934). Near La Jolla, California, *Loligo opalescens* population spawns 1.76×10^{12} eggs covering half the total 1,600,000 m^2 of spawning area with egg capsules (212 eggs per capsule) at the density of 10,400 capsules per m^2 (Okutani and McGowan, 1969). In terms of standing biomass, the quantum of gametes released ranges from 12% of assimilated energy in *Choromytilus meridionalis* (Griffiths, 1977, 1981b) to 80% in *Mytilus edulis* (Kautsky, 1982). The quantity of gametes released by bivalves is on average an impressive 76% of their weight (Griffiths and Griffiths, 1987). With an estimated production of 8×10^7 eggs per m^2 over 160 km^2 area (up to 30 m depth) of the Baltic Sea, *M. edulis* alone contributes more than 50% of the zooplankton (Kautsky, 1982). About 90% of the filtered energy by *C. meridionalis* is released to other consumers in the form of gametes, feces and pseudofeces (Griffiths, 1981b). The pelagic larval duration of many gastropods lasts for 107, 242 and 320 days in *Phalium granulatum, Tonna galea* and *Cymatium nicobaricum*, respectively (Scheltema, 1971). Hence, these larvae serve as food for longer durations to the consumers. Annually, an estimated 100 mt squids serve as feed for spermwhales (Clarke, 1980). Serving as the host, the bivalve *Scrobicularia plana* provides adequate resources to its trematode parasite *Gymnophallus fossarum* to release 73,163 cercariae per day (Bartoli, 1974). Whereas molluscan reproductive biology offers considerable scope for ecologists, the amazing tenacity of some snails to successfully survive over long periods from 3 months to 23 years on aestivation (see Hyman, 1967, p 613) provides astounding opportunity to physiologists.

Some unique bivalves thrive in the hydrothermal vents and deep-sea cold seeps (Goffredi et al., 2003). Hosting prokaryotic endosymbionts in their gills, they survive as chemoautotrophs in the vents and seeps (Trask and Van Dover, 1999). These bivalves pose super challenges to explorers. Further, amino acids are reported to enter the gills through the large surface area of bivalves (*M. edulis*, Pandian, 1975, p 73). In tropical Pacific waters, bacterial production ranges between 0.2 and 0.5 g Carbon (C)/m^2/d. These exceed the primary production of phytoplankton (0.2–0.3 g C/m^2/d). Wright et al. (1982) have also reported that bacterioplankton accounts for 112 µg C/l, compared to 150 µg C/l for phytoplankton. About 30% of the bacterioplankton populations compose of aggregates larger than 4 µm (see Pandian, 1975, p 63). The filter system of many bivalves, say, *M. edulis* has a mesh size of 0.6 x 2.7 µm (see Pandian, 1975, p 89). Hence, water-borne amino acids and bacterioplankton may serve as important nutrient sources to many bivalves. Briefly, the exploration of different nutrient sources to these chemoautotrophic bivalves in the vent and seeps may prove rewarding.

Known to have originated some 16.5 million years ago (Cunha et al., 2005), the family Conidae includes 329 species belonging to four genera *Conus* (with 85% of known species), *Californiconus, Conasprella* and *Profundiconus* (Puilliandre et al., 2015). Many of them are known for endemism. For example, 41 out of 49 valid species are endemic in the Cape Verde Islands (Cunha et al., 2005). The venom of *Conus* called conotoxin contains a large number of peptides. These peptides are small and inherently stable, making them ideal means of peptides therapeutics. There is considerable scope for modifying these peptides for effective oral delivery (Omar, 2013). However, the venom remains to be explored for its medicinal use. The Central Marine Fisheries Research Institute (CMFRI, 2012) has for the first time commercialized an unidentified drug contained in the mussel powder as a cure for knee pain in aged (post-menopaused) women (CMFRI News Letter, 2012). Describing medicinal attributes of the sacred chank *Turbinella pyrum*, Lipton et al. (2013) list a dozen traditional (Siddha) drugs consisting of finely powered chank shell mixed with one or other medicinal herb products. These drugs are known to restrict peptic ulcer and other gastric disorders. A sleep-inducing peptide from the venom of the Indian cone snail *Conus araneosus* has been reported by Franklin and Rajesh (2005). Besides, freshwater snails are a potential agent for the control of submerged aquatic weeds (e.g. *Pomacea canaliculata*, Estebenet and Gazzanniga, 1992).

On the other hand, some planorbid snails host the blood-flukes; these flukes cause the dreadful disease schistomiasis in 200 million people and another 500 million are at the risk of infection, especially in Africa (Brown, 2002). Hence, containing the disease and controlling the vector snails provide an array of opportunities for medical biologists in the service of Africans. Interestingly, the infected but aestivating snails may serve as a weapon in biological warfare. The pholadid (Angel wings) and teredinid (shipworms) bivalves bore and consume wood (Paalvast and van der Velde, 2013). Hence, the teredinids have attracted much attention. It is an ongoing debate, whether some of these teredinids feed exclusively on wood or on symbiotic bacteria feeding the wood (Nair and Saraswathy, 1971). To wade off the fouling and boring molluscs, tributyltin containing paints are used. But they have led to 'imposex' involving superimposition of maleness on females of many gastropods (cf Pandian, 2015, p 105, 135). As foulers in harbors, on the ships and inside the coolant pipelines of nuclear power plant, the bivalves are serious pests. For example, the mytilid *Septifer virgatus* contributes 89% of the foulers in the coolant pipeline of Hong Kong (Morton, 1991). In rice fields, ampullarid gastropod *P. canaliculata* is a major pest (Yusa, 2004b).

Being an important component of the benthic community, the non-targeted gastropods suffer the most by trawl fishing. From every haul of trawl fishing, hundreds and thousands of gastropods are discarded, displaced, injured and/or killed (Thomas et al., 2014). Conservation biologists have raised serious concern over the discarded gastropods and benthic disturbances by fishing gears.

This curtain raiser has just introduced the relationship between man and molluscs. So widespread is the relationship that there are many molluscan academic and professional societies, and a large number of subject journals publish molluscan research work. They are too many but Table 1.1 represents an attempt to list some of them. Incidentally, the list may be compared with

TABLE 1.1

List of some societies and subject journals of Mollusca

Societies	Journals
American Malacological Society	American Journal of Conchology
Belgische Vereniging voor Conchyliologie (Dutch)	American Malacological Bulletin
Conchological Society of Great Britain and Ireland	Bollitteno Malacologico
Conchologists of America	Bull of Russian Far East Malacological Soc
Deutsche Malakozoologische Gesellschaft	Bull of the Institute of Malacology
Eesti Malakoloogia Ühing	Cour Forsch Inst Senenberg Frankfort
European Quaternary Malacologists	Fish & Shellfish Immunology
Freshwater Mollusk Conservation Society	Folia Conchyliologica
Magyar Malakológiai Társaság	Folia Malacologica
Malacological Society of Australasia	International Journal of Malacology
Malacological Society of London	Japanese Journal of Malacology
Malacological Society of the Philippines	Journal de Conchyliologie
Nederlandse Malacologische Vereniging	Journal of Conchology
Sociedad Española de Malacología	Journal of Medical and Applied Malacology
Sociedad Mexicana de Malacología y Conquiliología	Journal of Molluscan Studies
Sociedade Brasileira de Malacologia	Malacologia
Società Italiana di Malacologia	Malacologica Bohemoslovaca
Société Belge de Malacologie (French)	Malacological Review
Stowarzyszenie Malakologów Polskich	Molluscan Research
Western Society of Malacologists	The Nautilus The Conchologists' Exchange
	The Veliger
	Z Malakozool/Malakozoologische Blätter

just two subject journals (Crustaceana, Journal of Crustacean Biology) of another important taxon Crustacea. The molluscan and other journals have accumulated a voluminous literature on reproduction and development of molluscs. Hence, this book provides a 'snap-shot' of living molluscs rather than an in depth or exhaustive account on each item listed in 'Contents'.

1.2 Taxonomy and Distribution

Among metazoans, the phylum Mollusca is second only to arthropods in number of living species (see Ghiselli et al., 2012). The number of described molluscan species ranges from 85,000 (Chapman, 2009) to 100,000 (Strong et al., 2008) and to ~ 130,000 (Geiger, 2006). The estimate by Appeltans et al. (2011) ranges between 135,870 and 164,107 species inclusive of ~ 50,000 and odd species remaining to be described. Of the seven molluscan classes, the 110,000 known gastropod species contribute 85% of all molluscs and the 60,000 species of prosobranchs contribute a significant part of this group (Gruner et al., 1993). A massive sampling in the west coast of New Caledonia, south-west Pacific has revealed the presence of 2,733 molluscan species within 295-km^2 site; occurrence of considerable proportion of rare species in the collected samples has suggested that actual species richness of marine molluscs is probably underestimated (Bouchet et al., 2002). Understandably, many malacological journals continue to devote some pages for taxonomy and description of new species. To maintain continuity with the earlier volume in this series (Pandian, 2016), the conventional but condensed version of systematic resume of Mollusca is listed in Table 1.2. Strikingly, the gastropods, arguably the most diverse molluscan group, encompass 78% of living molluscan species. Within gastropods, prosobranchs contribute 53% followed by 43 and 4% by pulmonates and opisthobranchs, respectively (Venkatesan and Mohamed, 2015).

All the living taxons belonging to Aplacophora, Polyplacophora, Monoplacophora, Opisthobranchia, Scaphopoda and Cephalopoda are strictly marine habitants. To gastropods and bivalves, the availability of calcium carbonate, a major constituent of shell, is a critical limiting factor in colonizing freshwater (cf Dillion, 2000). Consequently, their presence in freshwater is limited to 9.5% of bivalves (856 of ~ 9,000 species) and 5.0% of gastropods (4,000 of 78,000 species) (Table 1.3). Of 4,000 freshwater gastropods, Hydrobiidae (1,250 species, 400 genera, see Wilke et al., 2001) Planorbidae (250), Cochlipidae (246), Amnicolidae (200), Pleurocaridae (200) and Pachichilidae (165–225) are the largest families. Among 856 freshwater bivalve species, the most speciose family Unionidae alone comprises 674 species (Graf and Cummings, 2007).

Table 1.4 shows the geographical distribution of these freshwater gastropods and bivalves. The pH of waters progressively decreases with increasing latitudes (Feely et al., 2009). A major consequence is the decreased availability of carbonate ion (CO_3^{2-}) making the acquisition of calcium carbonate ($CaCO_3$) costlier for the molluscs with calcifying shell (Wood et al., 2008a). It is difficult to comprehend how freshwaters of Arctic support 53% (1,993/3,795) gastropods and 41% (347/856) bivalves. Shell thickness and hence calcium carbonate requirement for formation of shell progressively decreases with increasing latitudes in gastropods but not in bivalves (Watson et al., 2012). This may be the reason why freshwater gastropods are relatively more abundant in Arctics than freshwater bivalves. Employing hermaphroditism and lecithotrophy/viviparity, 24,000 pulmonate species have colonized the *terra firma*. These snails may either selectively feed on calcium-rich plants (e.g. lettuce) or forego the shell; for example, veronicellids shed their shell

TABLE 1.2

Condensed version of systematic resume of Mollusca

Phylum: Mollusca (85,000 species, Chapman, 2009, ~ 1,30,000 species, Geiger, 2006)
Class: Aplacophora, *Chaetoderma* (263 species, Appeltans et al., 2011)
Class: Polyplacophora, *Acanthochiton* (930 species, Appeltans et al., 2011)
Class: Monoplacophora, *Neopilina* (30 species, Appeltans et al., 2011)
Class: Gastropoda (78,000–100,000, Geiger, 2006, Strong et al., 2008)
Subclass: Prosobranchia
Order: Archaeogastropoda, *Haliotis, Trochus*
Order: Mesogastropoda, *Littorina*
Order: Neogastropoda, *Murex, Buccinum*
Subclass: Opisthobranchia
Order: Tectibranchia, *Bulla, Aplysia*
Order: Pteropoda, *Clio*
Order: Nudibranchia, *Doris, Aeolida*
Subclass: Pulmonata
Order: Basommatophora, *Lymnae, Planorbis*
Order: Systellommatophora, *Vaginulus, Laevicaulis*
Order: Stylommatophora, *Helix*
Class: Scaphopoda, *Dentalium* (572 species, Appeltans et al., 2011)
Class: Pelecypoda (Bivalvia) (9,000 species, Appeltans et al., 2011)
Order: Protobranchia, *Nucula, Yoldia*
Order: Filibranchia, *Mytilus, Ostrea*
Order: Eulamellibranchia, *Mercenaria, Teredo*
Order: Septibranchia, *Poromya, Cuspidaria*
Class: Cephalopoda (761 species, Appeltans et al., 2011)
Subclass: Nautiloidea, *Nautilus*
Subclass: Coeloidea
Order: Decapoda, *Loligo, Illex*
Order: Octopoda, *Octopus, Argonauta*

before hatching or retain a vestigial shell, as in *Testacella* (Hyman, 1967, p 449–450).

A vast majority (90%) of molluscs are predominantly benthic. However, pteropods and some argonaut cephalopods are holopelagic. Being small, the pelagic pteropods are known to thrive in hundreds and thousands in sub-arctics (Orr et al., 2005) and up to a depth of 300 m (Urban-Rich et al., 2001). The nektonic cephalopods can be either bathypelagic (e.g. *Opisthoteuthis massyae* moving between 877 and 1,378 m) or bathybenthic like *Cirrothauma murrayi* found between 2,430 and 4,850 m depth (Collins et al., 2001). Being large and carnivorous, the density ranges from $10/km^2$ in *Benthoctopus ergasticus* to $60/km^2$ in *Sepiola atlantica* (Table 1.5). The filter-feeders such as

TABLE 1.3

Distribution of molluscs in aquatic and terrestrial habitats
(compiled from Hyman, 1967, Graf and Cummings, 2007,
Strong et al., 2008, Appeltans et al., 2011)

Taxon	Habitats and Species (no)		
	Marine	Freshwater	Terrestrial
Aplacophora	263	0	0
Polyplacophora	930	0	0
Monoplacophora	30	0	0
Gastropoda	78,000	4,000	24,000
Bivalvia	9,000	856	0
Scaphopoda	572	0	0
Cephalopoda	800	0	0

TABLE 1.4

Geographical distributions of freshwater gastropods and bivalves
(compiled from Graf and Cummings, 2007, Strong et al., 2008)

Zone	Bivalvia	Gastropoda
Palaearctic	45	1408–1711
Neoarctic	302	585
Neotropic	172	440–533
Afrotropic	85	366
Indotropic	219	509–606
Australiasian	33	490–514
Pacific Oceanics	?	154–169
Total	856	3795–3972

mussels and oysters form crowded and multilayered matrics (Briones et al., 2014). The density of benthic molluscs runs in dozens and hundreds/m^2 and increases with 'increased packing'. For example, the number of *Mytilus edulis* on a single, long term rope increases from 387 at 0.5–1.0 m to 704 at 2–3 m and then progressively decreases to 55 at 14–15 m depth. Correspondingly, their biomass in the Black Sea also increases from 34 kg dry weight including shell at 0–1 m depth to 1,836 t at 1–2 m depth and subsequently decreases to 50 kg at 15 m depth (Kautsky, 1982).

Grassle and Maciolek (1992) reported interesting observations on the species richness from their survey at 1,500–2,500 m depth off New Jersey, USA. Their samples contained 798 species belonging to 171 families and 14 phyla. This diversity is maintained by a combination of biogenic microhabitat heterogeneity in a system with a few barriers to dispersal. Notably, the number of species decreased from ~ 350 at 1,400 m depth to

TABLE 1.5

Density and biomass for some molluscs

Species	Density	Biomass	References
	Pelagic pteropods		
Limacina sp	~ 8,000/m²	~ 8 g/m²	Mackas and Galbraith (2012)
	Nektonic cephalopods		
Sepiola atlantica	60/km² at 0–100 m		Collins et al. (2001)
Benthoctopus ergasticus	1–10/km² at 500–1,500 m		
Grimpoteuthis sp	5–10/km² at 3000–5000 m		
	Benthic marine gastropod		
Conus milarus	0.02/m²		Frank (1969)
Nerita albinica	1.6/m²		Frank (1969)
N. peloronta	8.7/m²		Hughes (1971a)
N. versicolor	21/m²		
N. tessellata	93/m²		
Patella vulgata	32/m²	353 kJ/m²	Wright and Hartnoll (1981)
P. vulgata	69–82/m²		Baxter (1983)
Fissurella barbadensis	46/m²	300 g dry flesh/m²	Hughes (1971b)
Spiratella inflata	3.0/m³		Lalli and Wells (1973)
	Benthic freshwater gastropod		
Pila globosa	3–16/m²	26–104 g/m²	Haniffa (1978a)
Elliptio complanata	28/m²		Downing et al. (1989)
Tulotoma magnifica	58.2/m²	389–3501 mgC/m²	Chirstman et al. (1996)
Lymnaea palustris	120–5940/m²		Hunter (1975)
	Benthic bivalves		
Guenkensia demissa	10/m²	818390 Kj/m²	Kuenzler (1961)
Perumytilus purpuratus	43–180/m²		Briones et al. (2014)
Choromytilus meridionalis		7.4 kg/m²	Griffiths (1981b)
Tridacna crocea	0.16/m²	115.6 g/m²	Hardy and Hardy (1969)
Mytilus edulis	414500/m²		
	387 in rope at 1 m	183 kg/m³	Kautsky (1982)
	704 in rope at 2 m		
	55 in rope at 15 m		

~ 180 at 4,700 m and density of individuals within a species also decreased from ~ 2,300 at 1,400 m to < 400 at 4,700 m. Table 1.6 summarizes the maximum depths, at which selected molluscs are known to occur and thrive. Amazingly, representative of aplacophoran *Pachymenia abyssorum* is known from the depth of 4,000 m, polyplacophoran *Lepidopleurus benthus* from 4,200

TABLE 1.6

Vertical distribution of molluscan taxons

Species	Collection Area	Depth (m)	Reference
Aplacophora			
Pachymenia abyssorum	Off California	3,900–4,000	Hyman (1967, p 68)
Chaetoderma nitidulum	Gulf of Maine	45–3,200	Hyman (1967, p 67)
Polyplacophora			
Lepidopleurus benthus		1,000–4,200	Hyman (1967, p 142)
L. pelgicae		4,200	Hyman (1967, p 142)
Monoplacophora			
Neopilina galatheae	Costa Rica	3,570	Hyman (1967, p 143)
N. ewingi	California & Peru-Chile Trenches	2,700–5,700	Hyman (1967, p 143)
Gastropoda			
Procymbullia, pteropod	Indian ocean	2,000	Hyman (1967, p 511)
Opisthobranchs		2,000	Hyman (1967, p 523)
Deminucula atacellana	North Atlantic	1,100–3,800	Etter et al. (1999)
Malletia abyssorum	North Atlantic	2,800–5,000	Etter et al. (1999)
Buccinids	Kuril Trench	6,800 7,200	Bruun (1957)
Prosobranchs	Kuril Trench	8,300–9,050	Bruun (1957)
Valvata sp, non-pulmonate	Tibetan Plateau	4,300***	Clewing et al. (2014)
Radix, Gyraulus, pulmonates	Tibetan Plateau		Clewing et al. (2014)
Bivalvia			
*Idas modiolaeformis***	Cheops Mud Valvano	3,000	Gaudron et al. (2012)
Tindaria callistiformis	North Atlantic	3,806	Turekian et al. (1975)
*Calyptogena pacifica**	45° N: 130° W, 47° N: 129° W	1,549–3,313	Cary and Giovannoni (1993) and Peek et al. (1998)
*C. phaseoliformis**	36° N: 122° W, 40° N: 144° E	3,399–6,370	
Pisidium, Musculiam	Tibetan Plateau	> 3,000 ?	Clewing et al. (2014)
Cephalopoda			
Cirrothauma murrayi	Northeast Atlantic	2,430–4,850	Collins et al. (2001)
Grimpoteuthis sp	Northeast Atlantic	4,802–4,848	Collins et al. (2001)

* from thermal vents, **deep-sea cold seeps, ***above sea level

m, cephalopod *Grimpoteuthis* sp from 4,848 m, monoplacophoran *Neopilina ewingi* from 5,700 m, and bivalve *Calyptogena phaseoliformis* from as deep as 6,370 m. Some prosobranchs have also been collected from 9,050 m depth from the Kuril Trench. Curiosity to know how these creatures inhabit at these cold depths has made this a hot area of research. Notably, a prosobranch gastropod *Valvata* sp. is reported to inhabit on the 'roof of the world' at an altitude of 4,300 m in the Tibetan Plateau.

1.3 Thermal Vents and Cold Seeps

The most startling discovery is the surprisingly speciose benthic assemblages, especially in the hydrothermal vents and cold seeps (Etter et al., 1999). Many bivalves belonging to two families Vesicomytilidae (~ 50 species) and Mytilidae (~ 250 species) reside mostly in sulfide-rich, oxygen-poor, deep-sea habitats at depths below 500 m. In recent years, their adaptive success in such challenging habitat with no photosynthetic primary production has received much attention (Duperron et al., 2006).

Vesicomyids are regarded as monophyletic. Their phenotypic morphological characters like body shape, hinge structure and ligament are unreliable for confirming systematics. However, molecular taxonomic studies have assigned the vesicomyids into three genera: *Vesicomysis, Calyptogena* and *Ectenagena*. Mitochondrial COI sequences of the clams of North America have revealed five discrete evolutionary lineages (Goffredi et al., 2003). The vesicomyids dominate the dense invertebrate assemblages in vents and seeps throughout world oceans. They are highly specialized and rely on chemoautotrophic bacteria. Their productivity can be high and much of the chemical energy fixed by the symbiotic association is transferred to surrounding heterotrophic species. Hence these clams play a vital role in the deep-sea communities through nutrient recycling and cleaning of water column and sediments (Goffredi et al., 2003).

The vesicomyids have retained the usual complement of digestive organs but with reduced, almost non-functional structures associated with filter-feeding (Kennish and Lutz, 1992). The vesicomyids such as *C. elongata, C. magnifica, C. pacifica* and *C. soyoae* harbor sulfur-oxidizing endosymbionts in their gills (Endow and Ohta, 1990). As a result, these clams must rely primarily on sulfur-oxidizing thiotrophic endosymbionts-mediated chemoautotrophic nutrition. Hence, they are restricted to the sulfide enriched interface between ambient oxygenated sea water and hydrothermal fluids or sediment pore waters delivering hydrogen sulfide readily (Cary and Giovannoni, 1993). Co-adaptation between these clams and their sulfur-oxidizing bacteria is supported by the following evidences: 1. The endosymbionts are not found free living in these habitats. 2. They are not amenable to culture out of the clams. 3. The clams have specialized morphological and biochemical adaptations to support and sustain these thiotrophic symbionts. 4. Isotopic signatures from the clam tissues have shown chemoautotrophic carbon fixation. 5. The vertical symbiont transmission from ovary to eggs has been confirmed by cytological evidence and molecular markers (see below). The estimated age of co-adaptation between the clams and bacterial symbiont lineages is probably > 100 million y (Peek et al., 1998).

C. *magnifica* produces large yolk-laden eggs suggesting lecithothrophic larval strategy. However, buoyancy of these eggs indicates broadcast type dispersal (Cary and Giovannoni, 1993). Presumably, the life cycle of C. *magnifica* includes a short non-feeding pelagic larval stage. TEM picture has shown the consistent presence of gram-negative sulfur oxidizing endosymbiotic bacteria in the primary oocytes of C. *soyoae* (Fig. 1.1), indicating the possible vertical transmission of these symbionts from mother to progeny (Endow and Ohta, 1990). At least, a single bacterium may be required to initiate the symbiosis in developing larva. Cary and Giovannoni (1993) employed an oligonucleotide probe complementing to a unique domain of the vesicomyid

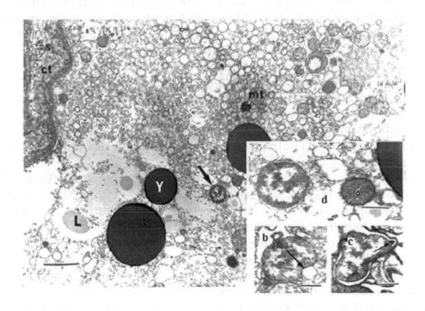

FIGURE 1.1

Transelectron micrograph showing the presence of symbiotic bacteria in the oocytes of *Calyptogena soyoae*. a. Arrow showing the bacterium near the basement of vitellogenic oocyte (scale: bar = 2 μm), d. Shows the enlarged bacterium (scale: bar = 0.5 μm), b. Arrow shows the bacterium with electron-transparent vacuole in periplasm, c. Arrow showing the electron dense material deposited in periplasm (scale: bar = 0.5 μm) (from Endow, K. and S. Ohta. 1990. Occurrence of bacteria in the primary oocytes of vesicomyid clam *Calyptogena soyoae*. Mar Ecol Prog Ser, 64: 309–311, modified, reproduced with kind permission by Inter-Research).

clam symbionts *16S rDNA* assay capable of detecting extremely low copy numbers of the symbiont's rDNA from mixed nucleic acid sample. They found that 1. The symbiont *16S rRNA* genes were consistently amplified from the visceral tissues of *C. magnifica, C. phaseoliformis* and *C. pacifica*. 2. The nucleotide sequences of the genes amplified from the ovaries were identical to those of the respective host symbionts. 3. Vesicomyid clams inoculated the host specific symbionts into all progenies using transovarial mechanism of symbiont transmission.

Mytilids: Within the family Mytilidae, a single clade Bathymodiolinae comprises all species inhabiting deep sea cold seeps, hydrothermal vents and organic (wooden) falls (Lorion, 2012). In these mussels, two phytogenetically and physiologically distinct Gammaproteobacteria thrive as symbionts. Some of these mussels harbor only thiotrophic (e.g. *Idas* sp, Duperron et al., 2008) or methanotrophic symbionts alone (e.g. *Bathymodiolus childressi*, Eckelbarger and Young, 1999). In others, both these bacteria co-occur in gill bacteriocytes (e.g. *B. azoricus* and *B. puteoserpentis*, Duperron et al., 2006). Methanotrophic bacteria use methane both as electron and a carbon source (Distel and Cavanaugh, 1994). Unlike the thiotrophic-dependent vesicomyid clams with reduced, non-functional filter-feeding structures, these mussels are capable of filter feeding (Page et al., 1991). As a result, they may simultaneously employ thiotrophic, methanotrophic and phytosynthetic energy sources. They filter feed on chemotrophs from water (Peek et al., 1998). The relative importance of these nutritional energy sources may change from site to site in the vents as well as during ontogenetic development. Wide differences in the carbon and nitrogen isotopic compositions were observed in *Bathymodiolus* sp collected from two chemically distinct vents. These differences indicate that the mussel expressed a nutritional adaptive response to chemical variations in the vents (Trask and Van Dover, 1999).

The presence of yolkless eggs and feeding larvae suggests that *B. thermophilus* pass through planktotrophic pelagic larval stage. Carbon isotopic composition of larvae and post-larvae indicates greater dependence on the methanotrophic symbionts. Hence, Trask and Van Dover (1999) suspected that only methanotrophic symbiont is transformed from mother to larvae and the functional symbiosis develops subsequently. However, recent ultrastructure studies have shown that both methanotrophic and thiotrophic symbionts are not transmitted from mother to progeny. From their ultra-structure studies, Le Pennec and Beninger (1997) reported the absence of prokaryotes from the male gametes of *B. elongatus, B. puteoserpentis* and *B. thermophilus*. Eckelbarger and Young (1999) also found no evidence for the presence of any prokaryotes in female and male gametes of *B. childressi*. In *Bathymodiolus*, Won et al. (2003) too found no similarity between the host and symbionts. *Idas modiolaeformis* hosts six distinct bacterial symbionts including sulfur- and methane-oxidizers (Duperron et al., 2008). On application of either general bacteria specific probe or methanotrophic bacteria specific probe,

FISH experiments showed no positive signal for the presence of bacterial symbionts in the reproductive organs of *I. modiolaeformis* (Gaudron et al., 2012). However, ultra-structure studies of Salerno et al. (2005) revealed the presence of methane- and sulfur-oxidizing symbionts in the gills of *B. azoricus* and *B. heckerae*. Evidently, the mytilid mussels environmentally acquire both the symbionts during the long period (> 5 week in *I. modiolaeformis*) of larval/post-larval stages. A comparative account on contrasting features of the two bivalve families is summarized in Table 1.7.

TABLE 1.7

Contrasting characteristic features of vesicomyid clams and mytilid mussels of the thermal vents

Characteristics	Vesicomyid Clams	Mytilid Mussels
Symbionts	Thiotrophics only	Thiotrophic or Methanotrophic or both of them
Filter feeding structure	Reduced, non-functional	Functional
Eggs/larvae	Yolk-laden eggs, lecithotrophic short non-feeding pelagic larvae	Yolk-less eggs, planktotrophic long (5 weeks) pelagic larvae
Acquisition of symbionts	Transovarian transmission from eggs to all progenies	Environmental acquisition during larval/post-larval stage

1.4 Energy Budgets

In animals, energy budget is assessed by estimations of C = F + U + R + P, where C is the food energy consumed, F, U and R, the energy lost on feces, urine and metabolism, respectively and P, the energy gained due to growth (e.g. Pandian, 1987). In ecological energetics, one or more of the energetics components have been estimated at population level in many gastropods (e.g. Parry, 1982) and bivalves (e.g. Griffiths and Griffiths, 1987). The physiological energetics is associated with relatively more precise estimates of these components (e.g. *Pila globosa*, Haniffa, 1982, *Octopus maya*, see Van Heukelem, 1983b). The available information on energetics remains incomplete for one or other of the following: 1. In many investigations, one or more of these components were not estimated and assumed to be negligible (see Table VII, Griffiths and Griffiths, 1987), 2. Barring perhaps Van Heukelem, none has undertaken long term study to cover the entire life span of molluscs. 2a. Understandably, such study may not be possible, as the life span of snail like *Patella vulgata* lasts for 14–17 years (Wright and Hartnoll, 1981), 2b. Different proportions of populations of many gastropod species (e.g. intertidal limpets, Parry, 1982) and bivalve species (e.g. *Choromytilus meridionalis*, Griffiths and Buffenstein, 1981) are submerged only for 12 hours/day limiting the duration of access to food. 2c. Some gastropods aestivate for short or longer duration, when they do not feed (e.g. *P. globosa,* Haniffa, 1978b). 2d. Ripening ovary reduces the space available within the hemocoelom and appetite, as well. Many cephalopods do not feed while brooding. For example, the brooding squat octopod *Bathypolypus arcticus* does not eat for one year (O'Dor and Macalaster, 1983).

Like most aquatic animals, molluscs profusely secrete mucus to (i) chemically clean themselves, (ii) facilitate smooth gliding on substratum (snails; Davies et al., 1990) and (iii) selectively bind the retained food particles in the gills into strings (bivalves; Foster-Smith, 1975) (Fig. 1.2). In *P. vulgata,* the ingested food energy lost on mucus secretion ranges from 4% (Wright and Hartnoll, 1981) to 23% (Davies et al., 1990). In freshwater gastropods > 20% of the ingested energy is lost on mucus production (Calow, 1977). Using ¹⁴Carbon labeled feed, Kofoed (1975) made a more precise estimate and

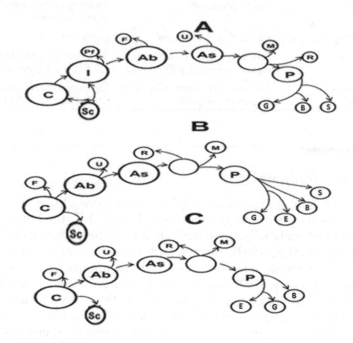

FIGURE 1.2

Energetics components of A. Bivalvia, B. Gastropoda and C. Cephalopoda. C = consumption, F = feces, Pf = pseudofeces, I = ingested food after the loss of Pf, U = urine, Ab = absorption, R = respiration, M = mucus, P = production, S = shell, B = bodygrowth, G = gonadal production, E = egg capsules and the like, Sc = microbial contribution to C, A = energetic components of bivalves. Note the contribution of microbes to C and the loss of C via Pf, B = energetic components of gastropods. Note the contribution of microbes to C through 'aufwuch' and the loss of G through egg capsules and the like. Note the absence of S and the contribution to C by Sc in cephalopods.

reported that 9% of the assimilated carbon is excreted as mucus in the deposit-feeding prosobranch *Hydrobia ventrosa*. With 70% assimilation efficiency, the prosobranch invests 13% of the ingested food on mucus secretion, a value that falls between 4 and 23% reported for *P. vulgata*. Many authors, who have reported data on energy budget, have not quantified the mucus component.

Food acquisition by radular scraping in gastropods (e.g. Haniffa and Pandian, 1974), especially in drillers (see p 54–55) and filter-feeding in bivalves last for longer durations. The fact that oxygen uptake doubles during feeding in comparison to the starvation period suggests that there is a cost of food acquisition in filter feeders (Bayne and Scullard, 1977). The cost of food acquisition by radular scraping in gastropods is not yet estimated. In filter-feeding bivalves, particles retained on the gills are bound into mucus strings and transported to the mouth by labial palps. When density of suspended food particle increases, some of the material is rejected as pseudofeces (Fig. 1.2). In *Mytilis edulis*, for example, pseudofecal production rate are increased from 2 to 6 mg/g body weight/d with increasing particle density from 20 to 80 mg/l, while defecation rate remains constant at ~ 0.8 mg/g/d up to the density of 80 mg/l (Tsuchiya, 1980). Hence, pseudofeces production represents a significant loss of the C in filter feeding lamellibranchs.

Bivalves have many sources to satisfy their energy and nitrogen requirements. To acquire the required food energy in the thermal vents, the vesicomyid bivalves almost entirely depend on sulfur-oxidizing endosymbiotic bacteria, while the mytilids partially depend on thiotrophic and/or methanotrophic endosymbiotic bacteria. Freshwater gastropods like the lymnaeid acquire valuable nutrients from 'aufwuchs' (Hunter, 1975); it may not be able to acquire some of these nutrients from the usual algal foods. Available information on wood boring teredinids reveals the importance of microbial nutrients. For example, wood containing 0.2% nitrogen cannot satisfy the nitrogen requirement of the teredinid shipworm with 5.8% nitrogen in its body and 4.4% nitrogen in its pediveliger larvae. Table 1.8 represents a simplified version of the data reported by Gallager et al. (1981) on the energy (dry weight basis), carbon and nitrogen budgets of the larviparous shipworm *Lyrodus pedicellatus*. About 8% of the ingested food originates from bacterial source. Despite this small proportion of the C from the wood (Fig. 1.2), the ingested bacteria supply 82% of the nitrogen required for growth (52%) and reproduction (19%) of the shipworm. Using isotope ratio of $\delta^{13}C$, Paalvast and van der Velde (2013) demonstrated that the symbiotic cellulolytic nitrogen fixing bacteria supplied not only the required nitrogen but also carbon to *Teredo navalis*. In fact, the teredinids use the wood more for protection than for food. Hence, microbes may constitute an important component of C in these bivalves.

Many marine gastropods provide their embryos with extra-embryonic nutrition through capsular fluids and/or nurse eggs. This aspect is elaborated later. Lecithotrophic eggs of gastropods and cephalopods are protectively laid within a strand, string, jelly coat and/or capsule. However,

TABLE 1.8

Dry weight, carbon and nitrogen budgets of the larviparous shipworm *Lyrodus pedicellatus* fed on wood for a period of 4 m (compiled from Gallager et al., 1981)

Parameter	Consumption	Feces + Urine	Respiration	Growth	Reproduction
	Dry weight basis (mg dry weight)				
Wood	294 + 24* = 318†	212	63	29	14
Carbon	134.5 + 49.7* = 184†	106.1	63.2	11.7	3.15
Nitrogen	0.6 + 2.6* = 3.6†	0.9	–	1.7	0.6
	Percentage basis (%)				
Wood	92 + 8* = 100†	66.7	19.8	9.0	4.3
Carbon	73 + 27* = 100†	57.6	34.3	6.3	1.7
Nitrogen	18 + 82* = 100†	29.2	–	51.9	18.8

*from other sources, † total

no attempt has yet been made to estimate the weight or cost of the chorion, jelly, string or strand that is used to protect the cephalopod eggs (Fig. 1.2). For a few gastropods, some data are available on the cost of capsule protecting the eggs. The capsule wall represents 9% of the total egg mass in *Crepidula dilatata* (Chaparro et al., 1999). This value is 37% for *Conus pennaceus* (Perron and Corpuz, 1982), 39–45% in *Thais lamellosa* (Sticke, 1973), and 50% in *C. vexillum* (Perron, 1981). Notably, most marine bivalves simply broadcast their gametes to be fertilized externally. Hence, they invest no energy on protection of their embryos.

Reproductive Effort (RE) is the material/energy invested on reproduction as a fraction of Production (P), i.e. Somatic growth including shell (S), Egg capsules (E) and Reproduction (R). Population, which is repeatedly depressed to low densities (r-selection), invests more on RE. Conversely, stable population remaining close to its carrying capacity in a habitat (K-selection) invests relatively less on RE. RE is calculated in three ways: (i) R/P, (ii) $R^{-1}/R^{-1} + W^{-1}$ and (iii) R/B. The first is currently used and cumulative, for which relatively more information is available. The weight specific RE^{-1} is one, in which R^{-1} refers to the reproduction per unit time and W^{-1}, the somatic weight/energy content just prior to spawning, over which R^{-1} is measured (Hughes and Roberts, 1980, Perron, 1982). The biomass specific RE relates to R to standing Biomass (B), for which some data are available for bivalves (Griffiths and Griffiths, 1987). The (currently used) cumulative RE increases with age (Fig. 1.3A); it increases linearly during rapid growth (juvenile) phase but begins to level off gradually towards its asymptotic value of unity, as somatic growth declines towards zero, while reproductive growth continues (see also Fig. 2.12). Considering the cumulative and weight specific RE, the investment on RE decreases in the following order: Planktotrophic *Littorina neritoides* > the oviparous *L. rudus* > lecithotrophic (directly developing) *L. nigrolineatus*

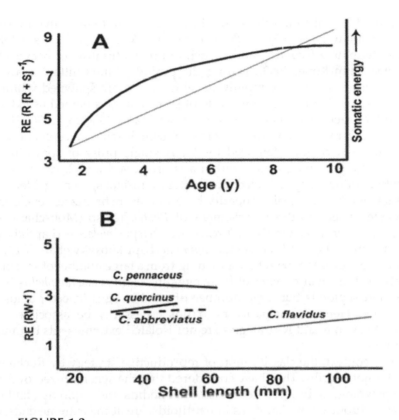

FIGURE 1.3

A. *Conus pennaceus*: age specific reproductive effort as measured by R (R + S)$^{-1}$. The trend for relationship between age and post-spawning body energy is shown by thick line. B. Relationship between reproductive effort (RW^{-1}) and shell length in selected *Conus* spp (redrawn from Perron, 1982). In Fig. 1.2, R represents respiration. But R represents reproduction in this figure.

(Huges and Roberts, 1980). Browne and Russel-Hunter (1978) have also found that in freshwaters, the investment on RE is less in ovoviviparous snails than oviparous snails. Considering weight specific RE alone, the decreasing order of RE investment is the planktotrophic *Conus pennaceus* < lecithotrophic *C. quercinus* < *C. abbreviatus* < *C. flavidus* (Fig. 1.3B).

Considering the life history strategies such as planktotrophy (indirect development involving one or more larval stages) and lecithotrophy (direct development including ovoviviparity, viviparity) as well as semelparity and iteroparity, Calow (1978, 1987) has developed theoretical model. Accordingly, a combination of lecithotrophy and iteroparity requires less investment on RE than planktotrophy (r-strategy) and semelparity (K-strategy). For example, the RE of the semelparous cephalopods ranges from 14.2 in *Octopus cyanea* (Van Heukelem, 1983) to 30 in *Loligo opalescens*. Incidentally, it is little different from others, in which the intermittent spawning pulses last from a few days in *Sepia officinalis* to 4 weeks in *L. pealei* characterized by indeterminate fecundity and asynchronous maturation of oocytes (also *L. forbesi*, Lum-Kong, 1993). During the period of intermittent spawning, the immature oocytes are rapidly developed and are spawned within the last few spawning pulses. This mode of semelparity is named as extended and intermittent terminal spawning (Rocha and Guerra, 1996). Barratt et al. (2007) recognized two patterns of semelparity in cephalopods: 1. Synchronous ovulation followed by 1a. A single pulse of spawning and 1b. Multiple spawning pulses within a short period and 2. Asynchronous ovulation followed by extended intermittent terminal spawning. Hence, the claims made by potential iteroparity based on asynchronized development of oocytes in one to three speciemens of *Photololigo* sp (Moltschaniwskyj, 1995), *Argonauta argo, Ocythoe tubercula, Tremoctopus violaceus* (Laptikhovsky and Salman, 2003) and *Kondakovia longimana* (Laptikhosvky et al., 2013) may require more evidence on actual spawning from a large number of specimens. Notable is the small number of large eggs in *Nautilus macromphalus* (Ward, 1983) and sepiolids but large number of smaller eggs in octopodids and teuthoides. Hence, semelparity need not necessarily be associated with fecundity. Also r- and K-strategies are not isolated extreme ends but more a continuum.

In this context, the classification of reproductive strategy by Rocha et al. (2001) is noteworthy. They have considered that the synchronized ovulation and spawning, as in a few teuthoids, octopodids and sepioids (Table 1.9) and asynchronized ovulation, as in nautiloids, are at the extreme ends; they represent clear semelparity and iteroparity, respectively. Others displaying synchronized ovulation and monocyclic eggs released in batches within a single spawning (e.g. *S. officinalis*) or with intervening somatic growth (e.g. *Dosidicus gigas*) are all within the boundary of semelparity. However, the polycyclic *Nautilus*, which lays eggs in batches at different seasons, may have to be considered as iteroparous. On applying GT (as a fraction of LS) values over the classification of Rocha et al., it becomes apparent that all the cephalopods under the headings I, Ia and Ib (Table 1.9), in which GT ranges from > 0.5 to 1.0 are all semelparous. However, GT values seem to fall below 0.5 in iteroparous nautiloids.

RE is also affected by factors other than simple planktotrophy/lecithotrophy and semelparity/iteroparity. Firstly, food quality, a factor hitherto not considered, has profound effect on RE, especially in herbivorous molluscs. RE ranges from 16–17 in *Aplysia dactylomela* and *A. punctata* fed on less preferred algae *Cladophora* sp and *Laminaria digitata*, respectively (Table 1.10). But RE is high as 32 for *A. dactylomela* and 72 for *A. juliana*, when they are fed on more preferred algae *Enteromorpha* sp. Secondly, many bivalves broadcast their larvae (e.g. oysters, Walne, 1964) into the pelagic realms, whereas most gastropods do it as gametes. Consequently, the investment on RE is much greater for bivalves (Table 1.11) than those

TABLE 1.9

Classification of reproductive strategy of cephalopods (compiled from Boyle, 1983, Moltschaniwskyj, 1995, Rocha et al., 2001) * no growth or † growth between spawnings. GT = generation time as percentage of life span (LS)

Groups	Spawning Patterns	Examples	Life Span	GT
I Synchronized ovulation and spawning*	Synchronized terminal spawning	**Teuthoidea:** *Todarodes pacificus* **Octopoda:** *O. vulgaris,* *O. cyanea* **Sepiodea:** *Loligo opalescens,* *Gonatus fabricii*	> 12 m ~ 1 y > 3 y < 3 y	 1.0
II Synchronized ovulation a. Monocyclic: eggs laid in batches*	Intermittent terminal spawning	**Sepiodea:** *Sepia officinalis,* *Euprymna scoleps, Loligo paelei,* *L. forbesi, L. vulgaris, Sepietta* *oweniana, Photololigo* sp	< 1 y	0.8
b. Monocyclic: eggs laid in batches within a spawning season†	Multiple spawning	**Sepiodea:** *Dosidicus gigas*	> 2 y	0.5
c. Polycyclic: eggs laid in batches during different spawning seasons†	Polycyclic spawning	**Nautilidae:** *Nautilus* spp	> 20 y	< 0.3 ?
III Asynchronized ovulation Eggs laid through an extended and continuous period†	Continuous spawning	*Idiosepius pygmaeus* *Argonauta hians*		

TABLE 1.10

Effect of food quality on cumulative reproductive effort (RE) of *Aplysia* spp (compiled from Carefoot, 1967a, b, 1970)

Aplysia Species	Food Species	RE
A. dactylomela	*Cladophora* sp	16
	Lawrencia papillosa	21
	Ulva fasciata	23
	Enteromorpha sp	32
A. juliana	*U. fasciata*	60
	Enteromorpha sp	72
A. punctata	*Laminaria digitata*	17
	Placomium cocci	50
	Enteromorpha intestinalis	50

TABLE 1.11

Reproductive effort of some pelecypods

Species		RE = R/P	RE = R/B	Reference
Guekensis demissa		17	13	Kuenzler (1961)
Crassostrea virginica		16	32	Dame (1976)
Perna perna		16	57	Berry (1978)
Mercenaria mercenaria		46	33	Hibbert (1977)
Ostrea edulis		47	–	Rodhouse (1978)
Lyrodus pedicellatus		47	–	Gallager et al. (1981)
Scrubicularia plana		48	46	Hughes (1970)
Aulacomya alter		63	70	Griffiths and Griffiths (1987)
Choromytilus meridionalis		73	12	Griffiths (1981a)
Exposed for 50% duration		37	–	Griffiths (1981b)
C. gigas		98	43	Bernard (1974)
Tellina tenuis	young	8.8	–	Trevallion (1971)
	Mature	54.0	–	Trevallion (1971)

of gastropods (Table 1.12). Within bivalves biomass specific RE is greater in mussels and oysters, whose densities are higher than burrowing clams, which are not densely populated. Quinn (1988) has brought evidence for density affecting reproductive output and thereby RE in intertidal pulmonate gastropod *Siphonaria diemenensis*. The snail produces less (2) number of egg masses at higher density (24 snails/cage) than at lower density (four egg masses in a cage with six snails). Thirdly, age/size have also have profound effect on RE. For example, energy allocated for reproduction increases from 22% in 2-year old *Pinctada margaritifera* to 50% in the 5-year old pearl oyster (Pouvreau et al., 2000a, b) and RE increases from 8.8 in young *Tellina tenuis* to 54.0 in mature ones (Table 1.11) from 0 in 11 mm size of the drill *Natica maculosa* to 41.5 in 19 mm size (Table 1.12) and from 0.35 in 2-year old *C. pennaceus* to 0.85 in 10 year old ones (Perron, 1982). Interestingly, another gastropod drill *Polinices alderi* is reported to invest 90% of the assimilated energy on reproduction suggesting an overall reproductive output of 10-times of its body weight (Ansell, 1982). In the southern dumpling squid *Euprymna tasmanica*, the life time reproductive investment is 2.53 times of its body mass (Squires et al., 2013).

Food availability or accessibility for different durations in the intertidal zones or total deprivation of food for different durations of aestivation may also affect RE. For example, *Nerita peloronta*, an inhabitant of highly stressed upper intertidal zone with much reduced duration of accessibility to food matures at larger size and invests more on RE (Table 1.12). However, with abundant food supply, the gastropods may opt to increase fecundity or

TABLE 1.12

Effect of planktotrophy and lecithotropy on reproductive effort of gastropods

Species	RE	Remarks	References
Patelloida alticostata*	10.3	Mid littoral	Parry (1982)
Notoacmea petterdi*	11.5	Supra littoral	Parry (1982)
Cellana tramoserica*	12.1	Mid littoral	Parry (1982)
Patella peroni†	26.6	Sub littoral	Parry (1982)
Tegula funebralis	7.7		Paine (1971)
Nerita peloronta (18 mm)**	21.0	Supra littoral	Hughes (1971a)
N. versicola (16 mm)**	12.0	Mid littoral	Hughes (1971a)
N. tessellata (14 mm)**	30.0	Sub littoral	Hughes (1971a)
Conus pennaceus RE	> 7.0	Nutrient fluid, but	Perron (1982)
C. pennaceus RE[1]	~ 3.5	Planktotrophic	Perron (1982)
C. quercinus RE[1]	~ 2.5	Lecithotrophic	Perron (1982)
C. abbreviatus RE[1]	~ 2.4	Lecithotrophic	Perron (1982)
C. flavidus RE[1]	< 2.0	Lecithotrophic	Perron (1982)
Fissurella barbadensis	10.3	Planktotrophic ?	Hughes (1971a)
Natica maculosa (11 mm)	0.0		Berry (1983)
(19 mm)	41.5		
Polinices alderi	> 90		Ansell (1982)

* annual spawner, † spawns twice a year, ** capsular eggs containing no nursing eggs

spawning frequency (Spight and Emlen, 1976). Yet, the presently available data on RE are not complete, as many authors have not estimated one or more components of bioenergetics. Hence, it is difficult to develop a theoretical model or to make generalizations.

1.5 Life Span and Generation Time

In consonance with the broad morphological diversity, the size and weight of molluscs also range widely from 1 mm in an aplacophoran to 18 m in the cephalopod *Architeuthis* sp. Table 1.13 lists the smallest and largest sizes and body weights in each of the seven classes of aquatic molluscs. Lack of shell and high motility seem to have facilitated the cephalopods like *Octopus dofleini* (9.6 m arm span, 272 kg) and *Architeuthis* sp (18 m, 450 kg) to achieve the largest size and body weight.

TABLE 1.13

Size and weight of selected molluscs

Species	Size	Weight	References
Aplacophora	~ 1 mm		Hyman (1967, p 13)
Polyplacophora			
Cryptochiton stelleri	0.4 m		see Boyle (1987b)
Monoplacophora			
Neopilina galatheae	25–37 mm		Hyman (1967, p 143)
Gastropoda			
Prosobranchia			
Ammonicera minortalis	0.4 mm		see Gofas and Waren
Tokina tetraquettra	> 0.3 m	1.4 kg	(1988)
			see Boyle (1987b)
Opisthobranchia			
Limacina retroversa	< 1.3 mm		Dadon and deCidre (1992)
Aplysia vaccaria	1 m	14 kg	see Boyle (1987b)
Pulmonata			
Pelecypoda			
Condylo nucula	0.6 mm		see Gofas and Waren
Hipponus hipponus	1.4 m	21 kg	(1998)
			see Boyle (1987b)
Cephalopoda			
Euprymna tasmanica	3 mm*	10 g	Hanlon et al. (1997)
Architeuthis sp	18 m	450 kg	see Boyle (1987b)

* mantle length

In crustaceans, larval duration is regulated by temperature-sensitive juvenile hormone (see Pandian, 2016, p 34, 184–187). Hence, the investment on Generation Time (GT) as percentage of Life Span (LS) averages to 28, 48 and 78% across all taxons in tropical, temperate and arctic crustaceans, respectively. A similar trend is not apparent for mollusca (Table 1.14), suggesting the absence of an analogue of juvenile hormone in molluscs. In general, information on LS and GT of molluscs is not only limited but also widely scattered in literature. Most opisthobranchs and cephalopods are annuals and semelparous. For example, the GT as percentage of LS is 96 and 98% in the pteropods *Spiratella inflata* and *S. trochiformis*, respectively (Wells, 1976). Another opisthobranch *Philine aperta* with LS of 3–4 years passes through two winters (> 1.5 years) in veliger stage (Hyman, 1967, p 502). However, there are exceptions: the opisthobranch *Cadlina* lives for 2 years and is iteroparous (see Boyle, 1987a, p 324). Similarly, especially the nautiloids have longer LS (> 20 years) and are iteroparous (see Table 1.9). The

TABLE 1.14

Life span (LS) and generation time (GT) in selected molluscs

Species	Location	GT	LS	GT/LS	References
Polyplacophora					
Chaetopleura tuberculatus	Bermuda	3–4 years	9–12 y	0.25	Hyman (1967, p 123)
C. apiculata	Cape Cod	3 years	>4 y	<0.75	Hyman (1967, p 123)
Prosobranchia					
Tulotoma magnifica	USA	♂ 1 year	>2 y	0.50	Christman (1996)
		♀ 48 weeks	>102 w	0.47	Christman (1996)
Viviparus georgianus	USA	36 months+	12 m+	0.33	Browne (1978)
Pomacea canaliculata		25 months	10 m	~0.5	Estebenet and Gazzanica (1992)
Patella vulgata	England	3–4 years	>14 y	~0.3	Wright and Hartnool (1981)
Protandric sex change at 4–5 y					
Opisthobranchia					
Limacina helicina (protandric)	Arctic	Spring 75 days	~150 d	0.5	Gannefors et al. (2005)
		Autumn 105 days	210 d	0.7	Gannefors et al. (2005)
Pteria penguin (protandric)	Austarlia	7 months	24 m	0.29	Milione et al. (2011)
Pelecypoda					
Elliptio arca, hermaphrodite	USA	2 years	20 y	0.1	Haag and Staton (2003)
Lampsilis arnata, female	USA	1 year	18 y	0.06	Haag and Staton (2003)
Potandric hermaphrodite passing through simultaneous hermaphroditism					
Tindaria callistiformis	North Atlantic, 3806 m depth	100 years	50 y	0.5	Turekian et al. (1975)
Pinctada margaritifera (protandric)	Polynesia	3 years	12 y	0.25	Chávez-Villalba et al. (2011)

LS of molluscs ranges from ~ 100 days in freshwater pulmonates belonging to the genera *Bulinus* and *Biomphalaria* (see p 153) to > 100 years in *Tindaria callistiformis*, a deep sea inhabiting bivalve (Table 1.14). Similarly, the GT as percentage of LS also ranges from 0.06 in freshwater bivalve *Lampsilis arnata* to > 0.75 in the polyplacophora *Chaetopleura apiculata*. However, this simple relationship between GT and LS is considerably altered by season, in which the offspring are hatched. In subartic pteropod *Limacina helicina*, it is 0.5 for offspring hatched in spring but 0.7 for those of autumn. In an another subtropical freshwater pulmonate *Bulinus tuberculata*, the snail hatched in August requires 200 days to complete GT (intervened by aestivation?) but those of May just 50 days (see Hyman, 1967, p 611). The presence of protandric and serial hermaphrodites also makes the estimate of GT difficult. Notably, simultaneous hermaphrodites like *Bulinus* and *Biomphalaria* invest less (~ 0.2) on GT and undergo 2–3 generations within a year (Brown, 2002). Hence, these snails produce new progeny and make uninfected juveniles available almost throughout the year for infection by miracidia discharged as eggs through urine or feces by *Schistosoma* almost throughout the year.

1.6 Spermatogenesis and Fertilization

Within the animal kingdom, the molluscs exhibit perhaps the greatest diversity in structural organization and morphology of spermatozoa (Maxwell, 1983). Two types of spermatozoa are produced; fertilization is external in marine archaeogastropods but internal in meso and neo-gastropods, opisthobranchs, pulmonates, freshwater gastropods, pelecypods and cephalopods (see also Fig. 1.5). The two types of sperm have been named as (i) eupyrene or fertilizing eusperm and (ii) apyrene/oligopyrene or non-fertilizing parasperm of the gastropod (Hyman, 1967, p 287, Hodgson and Heller, 2000). Production of two types of spermatozoa is widespread in prosobranchs, especially in neogastropods and some mesogastropods (Fig. 1.4). In fact, morphological differences in spermatozoa and spermatogenesis are used as key for systematic and phylogenetic studies (Zabala et al., 2009). Many publications provide detailed description on structure and genesis of these dimorphic sperm. Considering the volutid *Adelomelon ancilla* (Zabala et al., 2009, 2012) and the cerithioids *Melanopsis* (Hodgson and Heller, 2000) as representative, spermatogenesis and spermiogenesis are briefly described. In these gastropods, spermatogenesis commences with spermatocytes undergoing the first and second mitotic divisions in quick succession (Fig. 1.4). The spermatids are interconnected by cytoplasmic bridges. Briefly, spermiogenesis includes the lateral flattening of the head, giving it a spatulate shape, positioning of the conical acrosome anteriorly, housing the centriole complex within the short posterior fossa, placing of

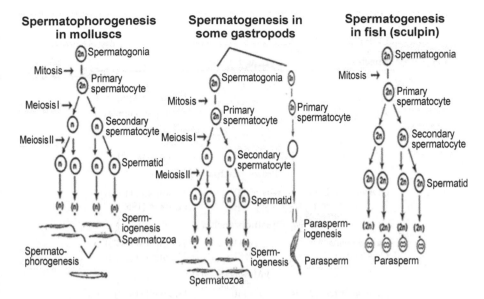

FIGURE 1.4

Spermatogenesis in Mollusca. Note the occurrence of typical spermiogenesis (not shown in figure) and spermatophoregenesis in Mollusca. But spermatogenesis simultaneously includes typical spermatogenesis and paraspermiogenesis in some gastropods. Note the absence of nucleus in the parasperm. For comparison, paraspermiogenesis in fish sculpin is also given. Note the presence of diploid nuclei.

equal (e.g. Modulidiae) or unequal (Cerithiidae) 2 + 2 mitochondria around the axonema and a glycogen piece in the mid piece (Fig. 1.5B). Essentially, the process of spermiogenesis remains the same, although species specific dimensions may vary (Table 1.15). Finally, the filiform eusperm consists of (i) an anterior conical laterally flattened acrosome, (ii) a rod-shaped, solid and highly electron-dense nucleus with centriolar complex, (iii) an elongated mid piece consisting of axonema, (iv) an elongated glycogen piece, (v) reduced thin layer of cytoplasm surrounding the nucleus and (vi) short free tail (e.g. marine volutid snail *A. ancilla*). But the terrestrial stylommatophoran snails have veritable long tail (e.g. 800 μm in *Arianta arbustorum*).

Conversely, the vermiform paraspermatozoan is relatively large and consists of a head with no trace of nucleus and tail tuft composed of numerous flagella (Fig. 1.5C, D). At the early stage, it can be identified by its chromatin pattern and a small round dark nucleus. Essentially, the process of spermatogenesis involves the nuclear break down, reduction in mitochondrial size, increased activity of Golgi complex and formation of axonemes. Nevertheless, a few littorinids retain an intact DNA remnant. Besides being a source of nutrient to eusperm, parasperm are suggested to serve in sperm transfer and inhibition of sperm competition (Zabala et al., 2012). Incidentally, parasperm are also reported to occur in fishes like the sculpin *Hemilepidotus gilberti*. But the

TABLE 1.15

Spermatozoa length of some molluscs

Species	Length (µm)	Reference
Pelecypoda		
Unio pictorum	39	Franzen (1955)
Transennella tantilla	66	Thompson (1973)
Aplacophora		
Chaetoderma nitilus	90	Franzen (1955)
Prosobranchia		
Gibbula umbilicalis	50	Thompson (1973)
Triphora perversa	140	Franzen (1955)
Opisthobranchia		
Bulla ampulla	~ 110	Thompson (1973)
Odostomia sp	< 876	Thompson (1973)
Pulmonata		
Agriolimax reticulatus	140	Maxwell (1977)
Pleurodonta acuta	1750	Maxwell (1983)
Cephalopoda		
Loligo pealei	50	Austin et al. (1964)
Eledone cirrhosa	550	Maxwell (1974)

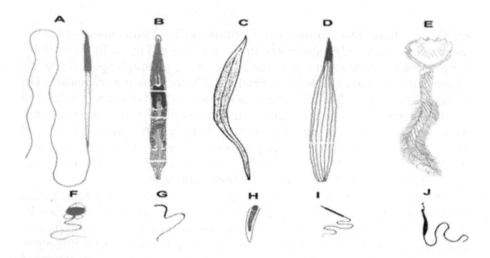

FIGURE 1.5

Free hand drawings of A. Eusperm, B. its structure, C. Parasperm and D. its structure, E. Spermatozeugma (Hyman, 1967, Zabala et al., 2009), F. Eusperm of a primitive type and G. modified type (from Maxwell, 1983), H. spermatozoa of clam *Nutricola tantilla* (from Lutzen et al., 2015), I. *Nautilus* and J. *Octopus* (from Maxwell, 1983).

piscine parasperm are binuclear, due to incomplete second mitotic division (Fig. 1.4) and hence cannot fertilize eggs. They are 10-times larger than the eusperm. Lumps of parasperm hinter the sperm of sneakers from reaching the eggs (see Pandian, 2010, p 146–147).

Fertilization is external in many pelecypods and archaeogastropods, which broadcast their gametes. Chance plays an important role in actual meeting of the egg and sperm. Hence most broadcast spawners are prodigious. For example, a single *Mytilus edulis* female release 8×10^7 million eggs. However, a few palaeogastropods have adopted suitable strategies to minimize the wasteful production of gametes. The Antarctic patellid limpet *Nucella concinna* forms 'stacks' prior to spawning, which ensures the release of gametes in close proximity (Picken and Allan, 1983). Incidentally, most Antarctic limpets do not develop in pelagic realm. For example, *Acmae rubella* and *Problacmea* brood their eggs (Picken, 1980). The other modes of external fertilization involve spermatozeugma or spermatophore. In scalid prosobranchs, the remarkably modified parasperm consists of a fibrous plate and a long tail (Fig. 1.4E) to which myriads of eusperm are attached. With long and many flagella, the spermatozeugma transport the eusperm (Hyman, 1967, p 287). In cephalopods too, dimorphic sperm and transfer strategies are known (e.g. *Loligo bleekeri*, Hirohashi and Iwata, 2013, Iwata et al., 2015). They are explained elsewhere (p 102). Interestingly, the number of spermatophores inside the Needham's sac of *Octopus insularis* ranges from 15 (0.02% body weight) in a small mature male (80 mm mantle length [ML], 50 g) to 66 (0.15% body weight) in the largest male (150 mm ML, 1700 g). The Spermatophore Length (SL) relation to either ML or Body Weight (BW) followed a logarithmic model, indicating that spermatophore size is stabilized at certain range of ML and BW (de Lima et al., 2014). Incidentally, the presence of spermatophore may provide an easier means to assess the number of spermatozoa in spermatophore-barring molluscs.

1.7 Ontogenetic Development

Being phylogenetically closely related, the indirect development in annelids and molluscs is characterized by the presence of trochopore larva. In Aplacophora, Polyplacophora and possibly Monoplacophora, the development is characterized by external fertilization and planktonic trochophore larval stage. Only a very few aplacophorans such as *Halomenia gravida* and *Pruvotina providens* cloacally brood their embryos (Hyman, 1967, p 51).

In gastropods and pelecypods, the development is indirect and passed through a short period of trochophore and an extended period of veliger larval stages. Figure 1.6 shows representative examples for the ontogenetic pathways in molluscs. In archeogastropods, the indirect development

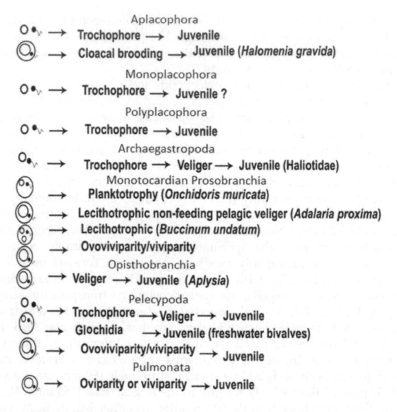

FIGURE 1.6

Ontogenetic development in mollusca. O ●ᵥ = external fertilization, ⊙⁙ = internal fertilization involving brooding, ovoviviparity or viviparity ⊛ = internal fertilization with nursing eggs.

includes external fertilization, and planktotrophic trochophore and veliger stages, as in Patellidae and Haliotidae. In monotocardian prosobranchs, and opisthobranchs, the development is also indirect and includes planktotrophic larval stages (e.g. *Onchidoris muricata*, Todd, 1979, *Nassarius burchardi*, Borysko and Ross, 2014). In nudibranchs, lecithotrophy ranges from a short (2 days) non-feeding planktonic late veliger stage (e.g. *Adalaria proxima*, Todd, 1979) to lecithotrophic egg producing fully developed benthic juvenile, which feeds as trochophore and veliger on the nurse eggs within the egg capsule (e.g. *Buccinum undatum*, Smith and Thatje, 2013).

In oysters, the early (trochophore) larvae are brooded for 7–10 days (e.g. *Ostrea edulis*, Walne, 1964). In some freshwater bivalves, development is indirect with glochidium larva. In these bivalves, glochidia are symbiotics in cyprinid fishes like the rose bitterling. Due to their bitterness, these cyprinids are avoided by predators. The bitterling spawns her eggs on the ctenidia of some freshwater bivalves. Following the completion of incubation within the interlamellar space, the young bitterling, attached with glochidia larvae of its host, disperse the glochodia (see Pandian, 2010, p 28–30). Using isotope ratios of $\delta^{15}N$ and $\delta^{13}C$, Fritts et al. (2013) have shown that glochidia of the unionoid bivalve *Lampsilis cardium* is indeed parasitic on the host fish *Micropterus salmoides* and acquires the required 57% of the $\delta^{15}N$ from the host tissues.

In both marine and freshwater gastropods and bivalves, brood protection, ovoviviparity or viviparity is not uncommon. The pelagic pteropods provide examples for brood protection and viviparity. Except for a few pulmonates (*Melampus* spp, *Amphibola*, *Siphonaria japonica*) characterized by very short veliger stage, all pulmonates directly develop involving lecithotrophy or ovoviviparity. In almost all cephalopods, development is direct, although a few teuthoids pass through a planktonic predatory rhynchoteuchum larva.

1.8 Locomotion and Dispersal

The molluscs are proverbial examples for slow motility at 'a snail pace' (Kappes and Haase, 2012). They exhibit three modes of locomotion (i) crawling (e.g. snails), (ii) swimming (e.g. scallops, cephalopods) and burrowing (e.g. clams) (Trueman, 1983). Marked and recaptured cowry has shown that *Monetaria annulus* does not move even 5 meter out of its 'home' even after 4 months (Frank, 1969). The estimated movement by *Aplysia californica* is 2.1 m/d (Angeloni et al., 2003). The snails glide at the speed of 3.3 mm/second (s) (Lai et al., 2010). But the scallops swim employing jet propulsion generated by pressure pulses by opening and closing the valves. For example, *Chlamys opercularis* swims at the speed of 30 cm/s (Moore and Trueman, 1971). The cephalopods swim at high speed employing mantle muscular propulsion to perform powerful and precisely directed jet thrusts, which afford its swimming ability far greater than any other molluscs. For example, *Loligo vulgaris* swims 65 m/s (Campos et al., 2012). *Todarodes pacificus* travels 2,000 km at a speed of 25.9 km/d (Shevtsov, 1973/1974), which may be compared with spawning migration of *Penaeus indicus* covering a distance of 380 km at a speed of 5.8 km/d (CMFRI, 2013). Brown (1979) assessed the energy cost of burrowing in the whelk *Bullia digitalis* as 5×10^{-4} J/digging cycle. Burrowing may be costlier than other forms of locomotion. In the snails, crawling costs an energy expenditure of 0.90 j per g/m (Denny, 1980). In bivalves, the widely scattered estimates suggest that the motility (speed) is accelerated with increasing body size (Fig. 1.7). In nudibranchs like *Melibe leonina*, serotonin (10^{-5} M/l) increases the duration of swimming and swimming speed (Lewis et al., 2011). Many other molluscs such as mussels and oysters are sedentary. Apparently, the strategy of gastropods and bivalves is to save energy on locomotion and use it for reproduction.

With regard to larval dispersal, two distinct planktotrophic and lecithotrophic modes of development are recognized. Planktotrophic

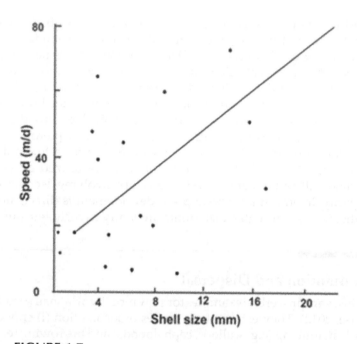

FIGURE 1.7

Speed of locomotion of some bivalves as function of cell size (drawn using data reported by Kappes and Haase, 2012).

development is characterized by a large number of small eggs, prolonged pelagic feeding larval stage and higher dispersal abilities, which allow them to have a wide and continuous geographic, range but with low genetic divergence. Conversely, the lecithotrophic development involves large but small number of eggs involving either no or short, non-feeding larval stage with very little dispersing ability that results in more restricted distribution and geographical isolation condusive for rapid speciation (cf Cunha et al., 2005, Puilliandre et al., 2015). To exemplify these contrasting features, complementary information on two gonochoric prosobranchs *Lacuna vincta* and *L. pallidula* as well as two hermaphroditic nudibranchs *Adalaria proxima* and *Onchidoris muricata* is summarized in Table 1.16. Both these prosobranchs are annuals but *L. pallidula* has a larger body biomass (15.9 mg dry weight) than *L. vincta* (8.3 mg). The former feeds exclusively on *Fucus serratus*, while the latter on a variety of algae including *F. serratus* (Grahame, 1977). The annual semelparous *A. proxima* can ill-afford reproductive failure and hence opt for lecithotrophy, as energy available for reproduction is unpredictable (Todd, 1979).

Briefly, bivalves invest 76% of their body weight on gamete production (Griffiths and Griffiths, 1987). Gastropod families that have many representatives with long duration of pelagic development include

TABLE 1.16

Comparative features of planktotrophic and lecithotrophic molluscs

Reference/Features	Planktotrophic	Lecithotrophic
Grahame (1977)	*Lacuna vincta*	*L. pallidula*
Fecundity (no./female)	53,432	1,365
Egg diameter (μm)	0.88	30.78
Reproductive effort	127	119
Ratio of egg weight to body weight	3.18	5.80
Todd (1979)	*Onchidoris muricata*	*Adularia proxima*
Fecundity (no./female)	43,364	4,224
Egg diameter (μm)	90	180
Body weight (mg/dry)	20	62
Eggs/mg body weight	4795	2209
Incubation period (d)	20	47
Larval period (d)	60	1–2
Reproductive effort	47.7	55.6

Architectonicidae, Bursidae, Cassidae, Coralliophilidae, Cymatidae, Cypraeidae, Lamellariidae, Muricidae, Naticidae, Neritidae, Ovulidae and Tonnidae (Scheltema, 1966). In bivalves and gastropods, planktotrophic development provides considerable scope for dispersal. The high frequency of gene flow in the French Polynesian amphidromous neretid *Clithon spinosus*, the most likely long-lived planktotroph, readily survives the 140-km trip between the Leeward and Windward Islands (Myers et al., 2000). The detailed investigation by Scheltema (1971) is chosen as a representative to describe dispersal. The dispersal of gastropod larvae serves not only as a means of colonizing new regions but also as agents of gene flow between widely scattered populations and thereby minimizes the scope for geographic isolation. The direction and routes of larval dispersal are determined by the principal ocean current. The major warm shallow current in the tropical Atlantic extends between 50 and 100 m depth. Scheltema (1971) has recognized four routes involving warm water currents in the Atlantic, through which the planktotrophic gastropod larvae may be dispersed (Table 1.17). It is through these routes > 70% of the teleplanic larvae of these dozen gastropod families disperse in the tropical Atlantic. These 'teleplans' (i) originate from continental shelf-inhabiting gastropods, (ii) involve pelagic larval development for long durations and (iii) are regularly dispersed over long distances. The estimated pelagic larval durations of some of these gastropods range between 42 days in *Pedicularia sicula* and 207 days in *Cymatium nicobaricum* (Table 1.18). However, many gastropod larvae do not metamorphose in the absence of their preferred food or substratum. For example, the onset of metamorphosis of *Pyramidellid*

TABLE 1.17

Recognized dispersal routes and estimated duration of Trans-Atlantic drift (from Scheltema, 1971, modified)

Routes	Estimated Current Velocity (km/h)	Estimated Duration for Trans-Atlantic drift (d)
North Atlantic Drift from Bahamas to N.W. Africa	0.5–1.3	128–400
North Equatorial Current from N.W. Africa to West Indies	0.9–1.2	128–171
South Equatorial Current from Guinea Gulf to Brazil	1.0–3.2	60–154
Equatorial Undercurrent from Brazil to Guinea Gulf	2.0–96	96

TABLE 1.18

Estimated durations of pelagic larval stage of some gastropods (from Scheltema, 1971)

Species	Larval Period (d)	Delayed Period (d)	Total Pelagic Duration (d)
Cymatium nicobaricum	207	113	320
C. parthenopeum	155	138	293
Charnonia variegata	219	57	276
Tonna galea	148	94	248
T. maculosa	198	–	198
Phalium granulatum	107	–	107
Thais haemastoma	62	28	90
Philippia krebsii	67	7	74
Smaragdia viridis	25	30	55
Pedicularia sicula	42	–	42

odostomia is accelerated by the presence of the polychaete worm *Pomatoceros* but delayed in its absence. The prosobranchs *Rissoa splendida* and *Bittium reticulatum* settle preferentially on the alga *Cystoseira*. The deposit-feeder *Nassarius obsoletus* delays settlement until finding suitable sediment (see Scheltema, 1971). Larvae of the naticid gastropod *Polinices pulchellus* extend their larval stage for ~ 6 months in the absence of suitable settlement cue (Kingsley-Smith et al., 2005). The estimated pelagic larval duration inclusive of the delayed metamorphosis lasts from 55 days in *Smaragdia viridis* to 320 days in *C. nicobaricum*. Based on fecundity estimates, frequency of veliger presence in selected stations across the Atlantic and geographical distribution of the selected gastropods in the western and eastern Atlantic shelf, Scheltema concluded: 1. The duration required for the transport of a gastropod larva across the Atlantic is ~ 170 days; incidentally, the trans-Atlantic voyage of Columbus to discover America took 86 days, 2. The probability for a larval transportation across the Atlantic is 1 in 80 billion and 3. *C. parthenopeum, Charnonia variegata* and *T. galea* hold the highest probability of Trans-Atlantic dispersal. But *C. nicobaricum* and *Tonna maculosa* may only very rarely be dispersed across the Atlantic.

Not surprisingly, most gastropods have opted for lecithotrophic, as the strategy represents a supreme compromise in terms of total energy demand for reproduction in the unpredictable conditions, duration of the pre-benthic larval stage, number of eggs and individual probabilities of survival to metamorphosis but with limited probabilities of dispersal. Interestingly, notable is the dispersal strategy of bivalves inhabiting the thermal vents and cold seeps. The vesicomyids, which depend primarily on thiotrophic symbionts alone, have opted for relatively less risky lecithotrophy but the mytilids with options for thiotrophic, methanotrophic and photosynthetic energy sources have opted for planktotrophy. For example, *Bathymodiolus thermophilus* has the highly dispersive planktotrophic larval stage.

The slow motile/sessile gastropods and bivalves have developed ingenious strategies for dispersal over long distances. For example, many cyprinid bitterling oviposit their eggs on the gills of mussel, the symbiotic partner, which irrigate the developing embryos of bitterling in secured habitat. The mussel attaches glochidia to the bitterling larvae, which disperse the glochidia. Thus a half a dozen bitterling are known to have established symbiotic relation with mussels: *Rhodeus sericeus* is hosted by *Unio, R. ocellatus* by *Anodonta woodiana, Acheilognathus rhombeus* by *Inversidens brandti, A. tabiara tabiara* by *Obovalis omiensis* and *Tanaka lanceolata* by *Inversiunio jokohamensis* (see Pandian, 2010, p 28–30). Table 1.19 lists selected gastropods and bivalves employing floating plants, fishes and other terrestrial and aerial vectors as well as ballast water for dispersal. The slow motile bivalves and gastropods, with their own dispersal capacity of 0.03–0.7 m/d, have employed strategies for dispersal from 75 km by the glochidium larva of *Amblema plicata* through its host *Micropterus dolomeiu* to as long as 18,200 km through ballast water by *Potamopyrgus antipodarum* from New Zealand to England.

1.9 Molluscan Fisheries

The fisheries comprise edible and ornamental molluscs. The global molluscan capture production has increased from < 100 t during the 1950s to 6.7 mt in 2012. Though known from the days of Romans, viable aquaculture production has commenced during 1980s with 200 t production. It has increased to 14.9 mt in 2012. The present world status of major capture and cultured molluscan

TABLE 1.19

Dispersal of some gastropods and bivalves

Species/References	Description
I Potential motility	
Physidae Lymnaeidae see Kappes and Haase (2012)	12 m/d by *Physa fontinalis* 72 m/d by *Radix bathyica*
II Active dispersal	
Gastropods Bivalves see Kappes and Haase (2012)	0.7 m/d by *Potamopyrgus antipodarum* 0.06 m/d by *Elliptio complanata*
III Passive dispersal: 1. Via floating plant/log	
Dreissena (bivalve) Horvath and Lamberti (1997)	*Vallisneria americana* rafted with adults transported to 260–430 km
Thais haemastma see Scheltema (1971)	Floating log transported the snail over 450 km
2. Via host fish	
Amblema plicata Lyons and Kanehl (2002)	Glochidium transported to 75 km by *Micropterus dolomeiu*
3. Via fish gut	
Pisidium idahoensis *Valvata sincera* (snail) Brown (2007)	Survived post-ingestion by fishes like *Corgonus pidschian*
Corbicula fluminea Voelz et al. (1998)	Defecated bivalve transported to 1.2 km/y from the place of its ingestion
4. Via attaching byssus threads	
Dreissena polymorpha *Corbicula fluminea* Leuven et al. (2009)	Transported to 199 km/y Transported to 276 km/y
5. Terrestrial vectors	
Sphaeriid clams Wood et al. (2008b)	Clinging to amphibian toes, the clams are transported between local ponds
6. Aerial vectors	
D. polymorpha Ricciardi et al. (1995)	Half air-exposed bivalve survive 10–14 d at 15°C
Lymnaea stagnalis *Helisoma trivolvis* Boag (1986)	Attached to wings of birds, the snails withstood 12–20 h aerial transportation and dispersed to a distance of ~ 10 km
7. Ballast water	
Potamopyrgus antipodarum Stadler et al. (2005)	Transported from New Zealand to England over a distance of 18,200 km

and other fisheries is briefly summarized in Table 1.20. The following may be inferred: 1. Molluscan capture and culture fisheries contributes 22 mt worth US$ 26.7 million to the food production sector. 2. Culture production of especially oystreids and venerids (see Ghiselli et al., 2012) (15 mt) is two times more than capture production. Hence, molluscs especially bivalves are highly amenable to culture than capture. 3. Captured molluscan taxons decrease in the following order: cephalopods < scallops < clams/cockles < oysters < mussels, whereas the cultured molluscs decrease in the order of: clams/cockles < oysters < scallops < mussels. Within cephalopods, capture of octopods is nearly 2- and 4-times more than that of cuttlefish and squids, respectively (Mangold, 1983). 4. The value of culture production of mussels is several times more than that of capture production. Barring mussels, however, the values for capture and culture of almost all the molluscs are almost equal in terms of a unit (one mt). Hence, countries like India have to encourage culture of mussels than other molluscs. 5. Calcium carbonate being a critical limiting factor, the inland capture and cultured molluscs are very low, i.e. 0.7 mt, in comparison to 22 mt marine production. Hence, priority may be given to marine molluscan aquaculture rather than freshwater molluscs

TABLE 1.20

The world status of capture and culture molluscan as well as other fisheries in 2012 and 2013 (modified, mean values compiled from FAO, 2014)

Taxon	Capture (mt)	Culture (mt)	Value (in US$)		Production Value (m US$/mt)	
			Capture	Culture	Capture	Culture
Mussels	0.1	1.83	43	2053	430	1122
Oysters	0.2	4.74	176	3900	880	823
Clams, cockles	0.6	5.00	598	4952	997	990
Scallops	0.8	1.65	1439	2849	1799	1727
Cephalopods	4.0	0	8055	0	2014	0
Other molluscs	0.7	1.35	529	939	766	696
Inland molluscs	0.4	0.3	–	–	–	–
Sub total	6.7	14.87	10,840	14,693	1148+	1072+
Total	21.7		26,697		–	
Inland molluscs	0.4	0.3	0.7			
Marine molluscs	6.6	14.9	21.5			
Fishes	78.6	44.2	79,036	87,499	1006	1980
Crustaceans	6.4	6.5	24,121	30,864	3711	4823
Molluscs	7.0	14.9	10,840	14,693	1148	1072
Grand total	92.0	65.6	113,997	133,056		
Global total	157.6		247053			

+ mean values, m US$/mt = million US$/metric ton

including pearl culture. 6. The capture and culture fisheries contributes 160 mt worth US$ 250 billion into the global food basket. In terms of unit value, the values for the finfish and crustacean fisheries are 1.0 and ~ 2.0 times more valuable than that for molluscan fisheries. It is suggested that the culture of molluscs merits priority support (than capture molluscan fisheries), as they may require no feeding cost, when cultured in natural habitats.

In view of the fact that the economic value of cephalopods is four times higher than that of fishes, considerable efforts have been made to culture on a larger scale, especially the cuttlefish *Sepia officinalis*, the loligonid squid *Spioteuthis lessoniana* and the octopodids *Octopus maya* and *O. vulgaris* (Vidal et al., 2014). Short life span (< 1 year) and semelparity are attractive features of cephalopods. Roughly 70 of 700 known cephalopods have been reared and/or cultured in captivity for four and seven generations of *Octopus joubini* (Thomas and Opresko, 1973) and *S. officinalis* (Forsythe et al., 1984), respectively. Octopodids feed voraciously (~ 10% body weight/d) and converts food at an efficiency of 50% (Van Heukelem, 1983, Vidal et al., 2014). Hence, very high growth rates are a hall mark of them. Their mean annual weight increase, expressed as ratio of the maximum attainable size, ranges from 0.40 to 0.99, which may be compared with those of fishes 0.07–0.24 (Hanlon, 1987, see also Table 3.21). They produce sufficient eggs and develop directly. They are hardy and adapt even to the 'crude' rearing facility and can withstand high levels of NH_4, NO_3 and NO_2, provided pH and oxygen levels are maintained. The greatest challenge is to feed them, especially on artificial feeds. Some observations and remarks on the scope for aquaculture of *O. vulgaris* are summarized in Table 1.21.

TABLE 1.21

Octopus vulgaris as a candidate species for aquaculture (compiled from Mangold, 1983, Vidal et al., 2014)

Features	Observations	Remarks
Habitat	Occurs in the coast of Kerala and Tamil Nadu	Spawns at 700 g size at 20–30°C in Western subtropical Africa
Behavior	Solitary, territorial swims fast at 60 m/s to distances of 3 and ~ 40 km/d but remains in its den for long, nocturnal, have to be fed during nights	Explore possibility of reducing swimming speed and duration by administering serotonin-antogonist. Also reduce aggression and territoriality by administering dopamine or its analogue (see Pandian, 2013, p 227)
Feeding	Gastropods and bivalves are important prey. Converts 50% of the consumed food for body growth	Explore feeding with molluscs bycatch or artificial feed from bycatch. Preferable to culture males, as they do not lose much body weight on milting
Growth	Easily grows to a kilo within 8 mo	
Reproduction	Spawning throughout the year. A female produces up to 500, 000 eggs. Egg size is not dependent on female body size	Explore induced spawning by temperature changes. Stock eggs on substratum. Can be readily incubated for 4 mo

Ornamental molluscan fisheries: A few publications on the ornamental molluscan shells indicate the need to culture them. In India, ornamental shells belonging to 21 gastropod families are collected. Of them Turritellidae, Muricidae and Trochidae contributed the lion share (Babu et al., 2011). However, the most valuable is the sacred chank *Turbinella pyrum*. In their book, Lipton et al. (2013) have collected scattered information on the fishery of four varieties of the chank along the coast of Gulf of Mannar (southeast coast of India). The chank requires 20 years to attain live body weight of 2.5 kg. There are adequate hints that the chank fisheries have declined to < 50%; for example, it has decreased from 33,289 pieces in 1952–53 to 13,816 in Thanjavur district and from 1,185 in 1975–76 to 0 in 1996–97 in Gujarat, India (Fig. 1.8A). Notably, the Gulf of Mannar has sustained prolonged harvest of

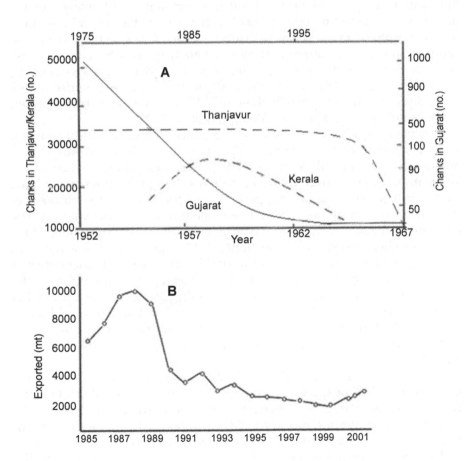

FIGURE 1.8

Declining trends of A. the chank harvest in Thanjavur, Gujarat and Kerala coasts (drawn using data reported by Lipton et al., 2013) and B. Exported quantity of shells from the Philippine (from Fisheries Statistics of the Philippine, 1985, 2001).

the chank, while the harvest in Gujarat coast drastically has fallen to zero. The Philippines have exported raw capis shell *Placuma placenta* alone to 41 countries (Floren, 2003). The exported shells and shell craft has drastically decreased (Fig. 1.8B). Thanks to Central Marine Fisheries Institute (CMFRI, India), Lipton et al. (2013) have standardized techniques to incubate the eggs of the chank and rearing its veliger from stage I to V and up to 360 days, when the chank attains a body length of 70 mm.

The disturbed benthos: With progressively increasing use of more and more efficient gears like trawls as well as fossil fuel for their operation during the last 60 years (cf Vivekanandan et al. 2013), the global capture fisheries have impressively increased (Pandian, 2015, p 31) but at the heavy cost of disturbing the structure and functioning of benthic communities. Being the largest components of the benthic communities, (i) many benthic molluscs are crushed by bottom trawl in the path of the net, (ii) are killed on being captured in the net and (iii) are discarded at sea, as they are of little commercial value. The impacts are more severe with beam trawling, because of its deeper penetration into bottom sediments. Being more vulnerable, larger filter feeding bivalves in the path of the beam trawl suffer < 20% mortality. Consequently, some of them like *Arctica isandica* have disappeared in the heavily trawled areas (see Malaquias et al., 2006).

Trawling affects both substratum and benthos. "Direct effects of trawling include scraping and ploughing of the substrate, sediment re-suspension and destruction of benthos" (Jones, 1992). Depending on the nature of the substrate and water movements over the track, an injury caused by the trawl track remains visible for varying periods. As little as a 1 mm silt layer over the settlement surface may prevent settlement of *Ostrea virginica* spat. The churning of the soft bottom may create an anaerobic turbid condition, which may even kill *Platinopecten* sp. (see Jones, 1992). Specimens at low densities (~ 0.1/m^2) suffer relatively less mortality (23%) but those at higher densities (> 24/m^2) incur up to 47% mortality. Many bivalves and gastropods suffer 4–20% mortality immediately following a single passage on the track of 4 or 20 m beam trawl (Table 1.22). Experimental trawling with standard bottom trawl gear in the east coast of India reveals the disappearance of one bivalve and two gastropod species immediately following trawling (Muthuvelu et al., 2013).

More information is available on bycatch and discarded molluscs. For example, of > 50% of species captured and discarded by southern Portugal trawling in the European waters, the molluscan share was 19%. Forty-four molluscan species including 28 gastropods, 15 bivalves and one polyplacophore were identified from the discarded specimens. The number and proportion of gastropods and bivalves differ considerably from fish trawls to crustacean trawls. Only 20 species were caught and discarded by crustacean trawls. But it was just 10 species for fish trawls. For both trawls 14 species were common. Together the crustacean trawls accounted

TABLE 1.22

Direct mortality of bivalves and gastropods caused by a single pass of 12 or 4 m beam trawl (compiled from Bergman and van Santbrink, 2000)

Species	Mortality (%)
Bivalvia	
Arctica islandica	20
Spisula sp	20
Tellimya ferrugionga	19
Donax vittatus	10
Corbula gibba	9
Nuclula nitidosa	4
Mysella bidentata	4
Gastropoda	
Turritella communis	20
Cylichna cydrindacaea	14

for 34 molluscan species but the fish trawls for 24 species only. Furrowing deeply into the 'biogenic benthic habitats' (see Collie et al., 1997) crustacean trawls capture more molluscs than fish trawls. The discarded molluscs from crustacean trawls increased in the following order: *Ranella alearium* > *Ampulla priamus* (gastropods) > *Venus nux* (bivalve). The order for the fish trawl was *V. nux* > *Calliostoma granulatum* (gastropod) > *Anadra divulli* (bivalve) (Malaquias et al., 2006).

In the Indian waters, considerable efforts were made to study the trawling effects on benthic gastropods. Being an important component of benthic community, gastropods constitute a significant bycatch that are eventually discarded. From their studies on the effects of trawling off the Mangalore coast from 2007 to 2010, Thomas et al. (2014) reported that 1. Thirty five species belonging to 18 families and 4 orders of gastropods were identified among the discards at 0–50 m, 51–100 m and 101–200 m depths. At all the depth zones, Muricidae (17%) dominated the bycatch followed by Bursidae, Casidae and Naticidae (Fig. 1.9). However, *Tipia* sp. (45%), *Turris* sp. (14%) and *Toma* sp. (14 species) dominated the upper, middle and deeper investigated zones, respectively. Diversity index (H) in the discarded gastropods decrease from 3.4–4.2 at 50 m depth to 1.4–3.0 at 200 m depth. Shannon-Weiner diversity also decreased almost parallel with depth. Being the mega fauna, the discarded gastropod diversity may serve as an index of the negative effects of trawl fishing on the benthic community. Notably, no study is yet available on the time required for the recovery of these discarded gastropods.

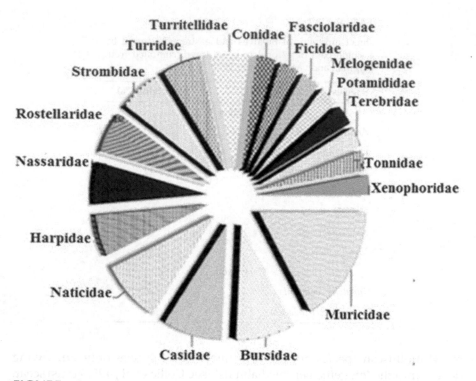

FIGURE 1.9

Bycatch of gastropods families that suffer by trawling in the west coast of India (from Thomas et al., 2014).

2

Shell and Reproduction

Introduction

To escape from predators, the development and maintenance of strong skeleton/shell(s) in slowly motile/sessile invertebrates became inevitable during the checkered history of evolution. The skeleton/shells provide structural protection and are often essential components of morphological defense in the ever-escalating evolutionary 'arm-race' between prey and predator (Nienhaus et al., 2010). Not surprisingly, most aquatic invertebrates have developed an array of calcified 'dermis' (e.g. corals, crustaceans, echinoderms), calcareous tubes (e.g. polychaetes) or shells (molluscs). These defensive external skeleton/shells are further strengthened by the development of armor, spikes, spines and the like at a cost of energy, which otherwise would be used for growth and/or reproduction (Dodson, 1984). Even the relatively fast moving vertebrates like the coffer fishes, chelonians and crocodiles do have external skeleton.

Molluscan shells are abundant, persistent, ubiquitous structure in aquatic habitats (Gutierrez et al., 2003). However, the development and maintenance of shell is not unique to molluscs. Crustaceans like the spinicaudates and ostracods do have shells (see Pandian, 2016, p 91, 105). Uniquely, the molluscan shells form the substrates for settlement and attachment of epibionts and provide refuges from predators (e.g. hermit crabs, Hazlett, 1981, fish embryo, Pandian, 2011, p 28–30). Changes in availability of molluscan shells have important consequences for these animals. Colonization of shelled habitats depends on individual shell traits and spatial arrangements of shells. Briefly, molluscan shells enhance species richness at habitat level (Gutierrez et al., 2003).

2.1 Structure and Diversity

A vast majority (98%) of molluscs have shell(s). However, the shell is absent in aplacophorans and coeloid cephalopods (Table 2.1). In opisthobranchs, it is

TABLE 2.1

Presence/absence of shell in mollusca (compiled from Hyman, 1967 and others)

Taxon	Observation
Aplacophora	Shell absent
Polyplacophora	Shell consists of a longitudinal succession of 8 valves embedded in the mid-dorsal surface. First and last valves are different in shape
Gastropoda	Shell is present in most snails but absent in a few
Prosobranchia	Shell is present with maximal diversity in size and shape
Opisobranchia	Shell may be present or absent in Cephalaspidea, Pteropoda, Notaspidea
	Shell is absent in Onchidiacea, Acochlidacea, Philinoglossacea, Saccoglossa, Nudibranchia, Parasita
Pulmonata	Shell is present. It is lost at hatching in Veronicellidae
Scaphopoda	Tusk-like shell present
Pelecypoda (bivalvia)	Equal sized and shaped bivalve shell present
Septibranchia	
Cephalopoda	
Nautiloidea	Single valved thin shell present
Coeloidea	An internal shell is present in decapods but not in octopods. The subclass includes 4 orders of living cephalopods: Sepioidea, Teuthoidea, Octopoda and Vampyromorpha. In these orders, the size of internal shell is progressively reduced to chitinous pen, the gladius and two rods

totally absent in groups like nudibranchs. In others like pteropods, shells may be present in thecosomates but not in gymnosomates. The thecosomatous shells vary widely in shape; they may be tusk-like (Fig. 2.2E), fan-like, triangular or even quadriangular. A shell is present in most pulmonates. But it is lost prior to hatching in veronicellids.

Typically, the synthesis of shell by molluscs is a highly organized process, nucleating from the inner mantle and radiating uni-directionally in systematic brick-like layers to the outer periostracum (Checa et al., 2007). Shell thickness ranges from 7–12 μm in the holopelagic pteropod *Creseis acicula* (Roger et al., 2012) to a few centimeters in the benthic bivalve *Tridacna gigas* (Hardy and Hardy, 1969). During its one year life span, shell length of the south Atlantic (40–50° S) pteropod *Limacina retroversa* grows to a maximum size of nearly 1.33 mm (Dadon and de Cidre, 1992). Growing at the rate of 5–8 cm length/y, *T. gigas* attains a maximum shell length of 1.4 m during its ~ 20 years life (Hardy and Hardy, 1969). There are also ontogenic changes in shell morphology with reduction in size and number of knobs and spines.

Gratefully acknowledging L. H. Hyman, it must be stated that her description is so precise that it is difficult to re-do it (however, see Geiger, 2006).

A typical gastropod shell is an elongated cone wound into a *spiral* around the *columella*, which is the main site of attachment of the snail in its shell (Fig. 2.1). The turns of the spiral are the *whorls*, and are demarcated by smooth, wavy, irregular or intended *sutures*. The whorls are usually in close contact but may rarely be disconnected to produce a wormi-form tube-like structure (Fig. 2.2C). The largest, terminal body whorl is bound by an *aperture*, through which the snail protrudes or retracts. The aperture ranges from a simple rounded opening to an elongated slit, which may be reduced by the presence of teeth, ridges and the like. The body wall and apertures may be drawn anteriorily to form a *siphonal canal* (Fig. 2.2B). The upper, rather posterior whorls are collectively named as the *spire*. The edge of the aperture is the outer lip and its opposite is the inner lip, which is continuous with columella (Hyman, 1967, p 152–153, 171).

Every conceivable variation from this typical conical shape occurs. At individual level, shell shape and thickness is flexible and adaptive (e.g. Nagarajan et al., 2006). The shell shapes of most gastropods maintain self-similarity during ontogeny. Considering body surface, ornamentation and orientation of the aperture plane, comprehensive mathematical models have been developed (Cortie, 1989). The shell may be long and slender with many whorls (e.g. Turitellidae, Terbridae, Fig. 2.2D) or short and globose with large body whorl and small, reduced spire (e.g. *Pila globosa*) (Fig. 2.2F). The spire may be condensed into the body whorl or even disappear with entire shell being flattened with all whorls remaining in a single plane, as in planorbids

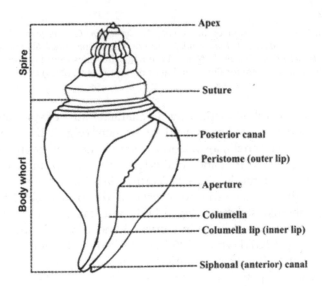

FIGURE 2.1

Morphology of a typical conical gastropod (from Lipton et al., 2013, modified, courtesy, CMFRI).

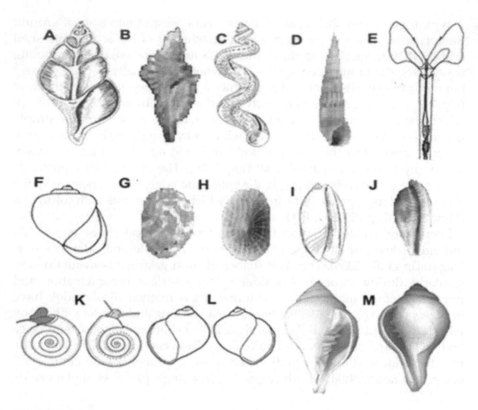

FIGURE 2.2

A. Longitudinal median section of *Buccinum*, B. *Cymatium*, C. *Tenagodus*, D. *Terebralia*, E. Pteropod, F. *Pila*, G. *Haliotis*, H. *Emarginula*, I. Olive, J. *Mauritia*, Sinistrals and dextrals of K. *Planorbis*, L. *Pila globosa*, M. *Turbinella pyrum*. Free hand drawings from Hyman (1967), Species Album of Rajiv Gandhi Centre for Biotechnology and Lipton et al. (2013).

(Fig. 2.2K). A conical shell displaying no characteristics of spire and aperture occupying the entire shell is typical of limpets (Fig. 2.2H). Loss of spire entails the loss of columella also. Consequently, the snail cannot retract into its shell but it clamps its foot and shell edge to the substratum. Strikingly, the assumption of sessile mode of life in Vermetidae, the spiral shell of juvenile is attached to the substratum and subsequently formed shell turns into a long curved tube-like structure (Fig. 2.2C). Unusually, the growth of shell in abalones leaves a series of holes in a curved line, of which only the last few remain open (Fig. 2.2G) (Hyman, 1967, p 155). With reduction of aperture into an elongated slit by the approximation of the lips brought about by inward rolling of the inner lip and spire, the olives (Fig. 2.2I) and cowries (Fig. 2.2J) look radically different from other gastropods.

In a vast majority of gastropods, the dextral shell opens on the left side of the axis right-handedly but a few are normally sinistral with an aperture opening on the right side (Fig. 2.2L). The occurrence of sinistral gastropods has been reported from a few marine and many freshwater prosobranchs. Sinistral shell occurs in many prosobranch taxons but are more prevalent in pulmonates. Notably, it occurs once for every 20,000 in the sacred chank *Turbinella pyrum* (Lipton et al., 2013). It is revered by Hindus and Buddhists. At the cost of US$10/g, a single sinistral sacred chank may fetch $1,000. It is also reported from two terrestrial prosobranchs, *Palaina taeniolatus hyaline* and *Schistoloma alta sibuyanica* and as many as 72 terrestrial pulmonate species belonging to 17 families (http://www.jaxshells.org/reverse1.htm).

The gastropod shell consists of an outermost thin layer, the periostracum and three calcareous layers namely an outer prismatic or palisade layer, a middle lamellar layer and an inner nacreous layer called hypostracum interleaved with organic layers. The periostracum primarily consists of organic matrix called conchiolin or conchin and protects the shell from corrosion. It is easily eroded and is totally wanting in cypraeids. The quantity of conchin ranges from 0.02 to 7% of shell weight (Hyman, 1967, p 161). In gastropods, "the prismatic and lamellae layers consist of calcium carbonate either as calcite or aragonite. These two forms of calcium are chemically identical but belong to different systems of crystallization. Aragonite is less stable form of calcium carbonate than calcite (Hyman, 1967, p 157)". In *Patella vulgata* (Hyman, 1967, p 159) and *Nucella lamellosa* (Nienhaus et al., 2010), the shells are made entirely of calcite but it is completely made by aragonite in the neritids. *Purpura* shells are externally calcitic and internally aragonitic. Calcium carbonate content of the shells ranges from 90% (e.g. *Helix*) to 100% (e.g. *Cypraea*). The shells may not have all three inner layers as in the trochid *Gibbula* but the number of layers also ranges from four to six in terrestrial pulmonates (Hyman, 1967). Whereas many traits of shell are phenotypic (Reed et al., 2013), shell breadth in the land snail *Arianta arbustorum* is a heritable trait. Interestingly, its heritability, estimated by the father-offspring regression, is five times smaller than that of mother-offspring regression. Understandably, the mother seems to have a greater role in heritability in shell breadth (Minoretti et al., 2013, Dillon and Jacquemen, 2015). Rarely, a gastropod like *Julia japonica* may be bivalved (Hyman, 1967, p 404).

Bivalves have a pair of equal (but some with unequal) sized and shaped calcareous valves connected by a hinge ligament that acts as a spring and allows the animal to gape and thereby permits exchange between the internal and external environment. Morphological variations in bivalves are relatively limited (Fig. 2.3). The variation ranges from the conically flat motile clams (Fig. 2.3A) and scallops (Fig. 2.3B) to flat sessile oysters (Fig. 2.3C) and mussels (Fig. 2.3D). The clams are burrowers (e.g. *Tapes aureus*) but the pectinid scallops are swimmers by opening and closing the valves (e.g. *Placopecten megallanicus*). Both internal and external surface of their valves are not iridescent or sometimes the inner surface alone is iridescent but not brightly. In the sessile oysters, the oval-shaped valves are broad both by length and breadth; the inner surface of both valves is iridescent. The

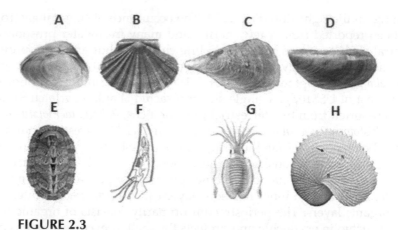

FIGURE 2.3

A. Clam *Perna viridis*, B. Scallop *Chlamys tranquebaricus*, C. Oyster *Pinna bicolor*, D. Mussel *Perna viridis* (from Album of Rajiv Gandhi Centre for Biotechnology), E. *Acanthochiton* (permission by New Zealand Marine Studies Centre), F. Scaphopod (redrawn from Boyle, 1987), G. *Sepia officinalis* (from Herklots, 1859), H. *Argonauta nodosa* (from Wolfe et al., 2012).

shape of the mussel, with attachment strength of byssus threads (e.g. 2.21 N [newton] in *Mytilus galloprovincialis*, Babarro et al., 2008, see also O'Donnell et al., 2013) to substratum, is increased by length wise alone; both surfaces of the valves are iridescent.

The valves of bivalves such as oysters (e.g. *Crassostrea virginica*) consist of only three layers, i.e. the periostacum, prismatracum and hypostracum (Lombardi et al., 2013). The prismatic layers contain curved, wedge-shaped prisms of calcite crystals (ostracons) under the conchion matrix (Fig. 2.4). The innermost hypostracum or calcite foliated layer is the major constituent of the shell and comprises sheets of calcites that are associated with providing shell strength. Composed of parallel sheets of calcite crystals, the microscopic structure is similar to that of 'nacre' layer of aragonite structures (Lombardi et al., 2013). To have an access to the bivalve tissue, a predator will have to penetrate the thick foliated layer within the valves.

Cephalopods are unique in that the coeloid octopods and decapods (cuttlefish, squids) may not have external argonaut shell. However, the nautiloids do produce external shells (Fig. 2.3H). A description of the argonaut shell is reported by Wolfe et al. (2012). In tune with the holoplanktonic mode of life, the females construct the thinner most (225 μm), brittle white paper nautilus shell, which is used to brood the embryo until hatching. The dwarf male lacks a shell (Finn and Norman, 2010) and may be 600 times smaller in weight than a female (Finn, 2009). The "shell has a central core as the site of bidirectional nucleation toward the inner end and outer shell surface, which lack an organic cover" (Wolfe et al., 2012). Hence, there is no evidence for the presence of periostracum. A hundred percent calcification

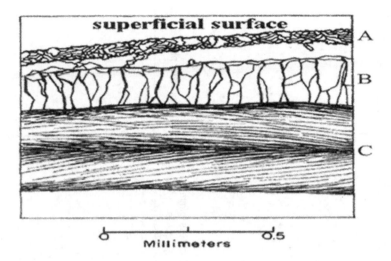

FIGURE 2.4

Cross section of a valve of *Crasostrea virginica* to show A. Periostracum, B. Prismatic layer, C. Folliated calcite layer (from Galtsoff, 1964, modified).

occurs by periodic mineralization. In the apparent bilaterally symmetrical shell, the disjointed and inconsistent nodules and ribs render it asymmetrical (Fig. 2.3H). Disordered crystalline grains radiate outward from the central core in fan-like units perpendicular to the inner and outer surfaces.

Argonautes are a group of pelagic octopus with benthic ancestry. Apart from serving as brood chamber, the thin brood shell also functions as a hydrostatic structure to precisely control buoyancy at varying depths. Pelagic female octopus employs the aragonaut shell to 'gulp' air at the sea surface, seal it using flanked arms and dive forcibly to different depths, where the captured sealed gas buoyancy counteracts body weight and thereby become neutrally bouyant (Finn and Norman, 2010).

In polyplacophorans, eight arched shell valves are arranged in longitudinal succession and medially embedded on dorsal surface (Fig. 2.3E). For more details, Hyman (1967) may be consulted.

2.2 Latitudes, Predators and Parasites

Latitudes: At the broad geographic scales, the magnitude of variations in shell thickness is modulated by latitudes through decreases in Sea Surface Temperature (SST) and pH (Fig. 2.5A, B) and calcium carbonate availability. The progressive reduction in availability of $CaCO_3$ renders the acquisition of biogenic $CaCO_3$ by molluscs more difficult and costlier (Wood et al., 2008a).

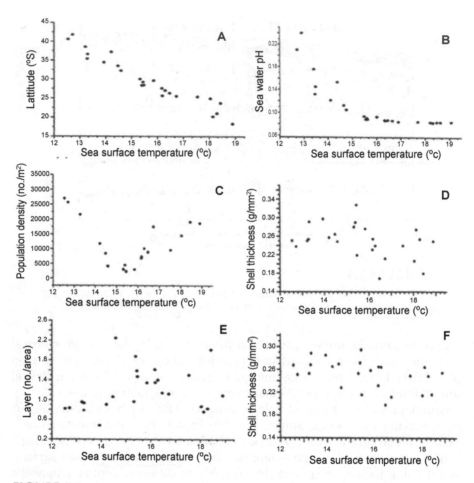

FIGURE 2.5

Decreasing sea surface temperature (SST) A. and B. pH with increasing latitude, as well as number of layer E. and shell thickness for corrected F. and uncorrected D. for population density C. in *Perumytilus purpuratus*. Note the differences in the scatter in Fig. 2.4D and F (from Briones et al., 2014, modified).

The energy cost of deposition and maintenance of calcified shell increases at low temperatures and hence at increasing latitudes. As $CaCO_3$ is less saturated and more soluble with decreasing temperature (Fig. 2.5A, B), the number of layers in the shell (Fig. 2.5E) and consequently the shell thickness (Fig. 2.5D, F) of molluscs inhabiting cold waters is expected to progressively decrease with increasing latitude (Vermeij, 1993). Notably, the solubility of aragonite is higher in colder waters than that of calcite. But, the expected relationship may be modulated by other environmental factors at levels of mini- and -mesoscales. For example, population density (Fig. 2.5C) is one such factor, which profoundly influences the shell thickness-latitude relation (Fig. 2.5D).

However, Valladares et al. (2010), who have not taken population density of *Mytilus chilensis* into consideration, have found that the expected relation was not apparent. Another mussel *Perumytilus purpuratus* is distributed between 18° and 42° S latitudes in the Southeast Pacific coast. The mussel form multi-layered 'beds' and display an intense intraspecific competition with increased 'packing'. Crowding and consequent increase in population density lead to self-thinning processes (Guinez, 2005). Hence, the expected negative relation between latitude-shell thicknesses may be modified. Not surprisingly, shell thickness of *P. purpuratus* does not show a clear negative relation of shell thickness with decreasing temperature (Fig. 2.5D). However, a clear negative relationship became apparent, when the relationship is corrected for population density (Fig. 2.5F). Hence, the relation between shell thickness-latitude is also dependent on population density (Briones et al., 2014).

In the intertidal snail *Littorina obtusata*, Trussell (2000) made an interesting investigation. This snail is distributed over 400-km gradient from north Manchester, New England coasts, Massachussetts (42.5° N) to south Lubec, Maine (45.0° N). Trussell's observation may briefly be summarized: 1. Northern (immature) snail shells were thinner, weaker and weighed less than those of southern snails (Table 2.2). 2. Shells became thicker and weighed more, when northern snails were transplanted to southern habitats

TABLE 2.2

Shell characteristics and body mass of *Littorina obtusata* in its natural northern and southern habitats and on transplantation from northern to southern habitat as well as southern to northern habitats (compiled from Trussell, 2000)

Parameter	North 45.0° N	South 42.5° N
Shell thickness (mm)	0.74	1.27
Shell strength required breaking force (N)	~ 150	~ 300
Shell weight (mg)	70	240
Body mass (mg)	54	35
Before and after transplantation from south to north		
Growth in shell length (mm)	2.55	2.75
Shell thickness (mm)	1.12	~ 1.27
Shell weight (mg)	170	240
Body mass (mg)	38	35
Before and after transplantation from north to south		
Growth in shell length (mm)	2.10	2.20
Shell thickness (mm)	0.74	1.07
Shell weight (mg)	160	170
Body mass (mg)	54	44

after a period of 90 days. The reverse was true for the southern snails after transplantation to northern habitats. 3. One hundred year-old museum specimens collected from the Gulf of Maine were thinner than those of the present day snails. The habitat of the predatory green crab *Carcinus maenas* has been expanding towards northern latitudes during the last 100 years, i.e. the southern snails were co-inhabiting with the crab for over 100 years but those of Maine for < 50 years. Briefly, shell thickness, strength and weight of southern snails, exposed to relatively higher temperature and longer period of co-habitation with the predatory crab, are significantly more than those of northern snails, which inhabit relatively cooler waters, in which they co-inhabit with the crab for a shorter period.

The directly-developing muricid snail *Acanthiana monodon* characterized by holobenthic life is distributed in the southeastern Pacific coast between 28° S and 36° S. The snail is highly polymorphic and exhibits variations in shell shape, color and thickness. Individuals from wave sheltered sites have robust and heavy shells, reduced aperture and short spire. However, those from wave exposed sites have thin shells with wider aperture and longer spire (Sepulveda et al., 2012). In *L. obtusata* too, snails inhabiting wave-exposed sites have a thin and light shell with a large foot to avoid dislodgement by waves (Trussell et al., 1993). On exposure to crab cue, *A. monodon* developed thicker shells, which do not significantly affect body weight (Sepulveda et al., 2012).

Predators: The predators of molluscs may be classified into (i) borers (ii) crackers (iii) openers and (iv) nippers. Borers and nippers may not kill the host/prey but the shell or siphon is regenerated at the cost of body growth and/or reproduction. Borers may further be divided into (ia) burrowers and (ib) drillers. Whereas the drillers employ a combination of chemical and mechanical force to penetrate through shell, the burrowers apply chemicals alone to inflict blisters over the shell surface or pores in the shell. Crackers such as chelonians, fishes and starfishes use 'biting force' but crabs employ the chelar force to crack the shell. Openers have to overcome the contracting adductor muscular power, which tightly holds the two valves together.

Burrowers cause an array of injuries on molluscan shell like etching, erosion, lesion, blister and hole. Lauckner (1983a, b, c) published an extensive summary of animal parasitic role on bivalves, amphineurans and scaphopods. This account is limited to selected examples of burrowers and their effects on shell, body growth and reproduction in 'host' molluscs.

Protista: Incertae sedis causes lesions on the inner surface of *Crassostrea gigas* (Fig. 2.6A). Desmopongids of the family Clionidae burrow into the calcareous substrate including live and dead molluscan shells. Among 69 species belonging to the genus *Cliona*, *C. celata*, *C. lobata*, *C. spirilla*, *C. truitti* and *C. vastifica* are known as shell-invading species. In India, the sacred chank shells are bored by *C. celata*, *C. carpenteri*, *C. lobata* and *C. vastifica* (Sunil Kumar and Thomas, 2011, Thomas et al., 1993, see also Lipton et al., 2013,

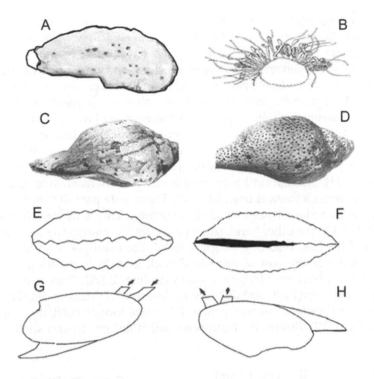

FIGURE 2.6

A. Left valve of *Crassostrea gigas* showing lesions caused by *Protista incertae sedis* (redrawn from Quale, 1961), B. *Cuna gambiensis* showing excessive polyp growth of *Monobrachium drachi* (redrawn from Marche and Marchad, 1963). C. *Turbinella pyrum* pored by *Cliona carpenteri*, D. *T. pyrum* but heavily infected by *C. carpenteri* (courtesy–Dr. P. A. Thomas, see Lipton et al., 2013). The effect of *Gymnophallus fossarum* metaceriae on the gap of healthy E. and infested F. *Cardium glaucum*. Note the widely opened gap in the infested cockle (free hand drawing from Bartoli, 1974). G. and H. Effect of *G. fossarum* infestation on the posture of *Tapes aureus* (free hand drawing from Bartoli, 1974).

p 157). Applying phosphatase and carbonic anhydrase, the Clio etch and bore shells. Perforations caused by clione species on the shell of *Turbinella pyrum* are numerous (Fig. 2.6C, D) and the width of each perforation on the shells caused by the Clio ranges from 0.3 to 1.6 mm (Thomas et al., 1983). Excessive growth of Clio impairs motility of scallops and causes the 'dark meat', when infestation is extended to soft body. Cnidarians mostly use the molluscan shell as a substrate. However, excessively growing *Monobrachium drachi* polyps (Fig. 2.6B) reduce soft body weight of *Macoma calcarea* from 335 to 146 mg (see Lauckner, 1983a).

Trematodes are more parasites than burrowers. As intermediary hosts, bivalves harbor sporocysts, cercariae and metacercariae of tremetodes. In the cockle *Cardium edule*, the number of cercariae increases from 20/host in a 2-year old (25 mm shell length) cockle to 220/host in a 4-year old (33 mm) cockle, i.e. for every mm increase in shell length, 25 cercariae are added. The trematode *Gymnophallus rebequi* produces protuberance on the outer mantle of *Abra ovata*. In the banded wedge-shell *Donax vittatus* infested by *G. strigatus*, these protuberances on the inner valves subsequently cause pits. *G. tokiensis* larvae erode shells, deplete body reserves and reduce longevity of the Pacific oyster *C. gigas*. *G. fossarum* infestation may cause different deformities in clams. Its cercarial infestation causes a wide gap between the valves in *C. glaucum* (Fig. 2.6E, F) and dramatic alteration in the posture of *Tapes aureus* (Fig. 2.6G, H). Infestation by *Himasthla elongata* metacercarae reduces the number of byssus threads from 61 to 40. Trematode parasites inflict greatest damage to gonadal tissues including castration. This is discussed elsewhere (Chapter 7). On the other hand, some parasites are known to accelerate body growth of the host resulting in 'gigantism'. For example, increased growth of body mass at the cost of decreased shell growth is reported in *C. edule* infested by *Labratrema minimus* sporocysts (Fig. 2.7A). This is also true in *C. virginica* infested by *Bucephalus cuculus*. Infested by protistan *Haplosporidium chitonis*, *Lepidochitona cinereus* grows 1½ times longer than the uninfested chiton (Fig. 2.7B). Briefly, the burrowers inflict deformities on shell/valve of

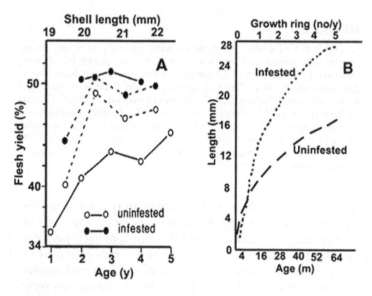

FIGURE 2.7

A. Effect of *Labratrema minimus* infestation on flesh yield of parasitized and unparasitized cockle *Cardium edule* of different age and size classes (redrawn from Bowers, 1969), B. Effect of *Haplospiridium chitonis* on shell growth of infested and uninfested *Lepidochitona cinereus* of different age and size classes (modified and redrawn from Baxter and Jones, 1978).

molluscs, decrease body size in many host species and reduce reproductive performance, whether the hosts are substratum seekers, pest or parasites. Bivalves also suffer from decreased body weight and castration by hosting pests and parasites.

Annelids: The cosmopolitan spionids *Polydora* and *Boccardia* compose a large number of burrowing (by boring) species of 'mudworms'. In India, the frequency of *Polydora* spp infestation ranges from 7 to 40% in *Crassostrea madrasensis* (Ghode and Kripa, 2001, see also Lipton et al., 2013, p 162). *P. websteri* damages oysters by burrowing into the shell through boring aided by a viscous fluid, which dissolves interperismatic and interlamella organic matrices and subsequently dissolves the exposed calcite crystals. With its thin valves *M. edulis* suffers more severely from *P. ciliata* infestation than oysters. The force required to pull apart the tightly closed valves of *M. edulis* has been determined under constant (2.55 kg) and linearly increasing load of 0.5 cm/s and found that the infected mussels exhibit a significant reduction in shell strength (Fig. 2.8A). Similarly, the closures of the mussel valves are also affected by *P. ciliata* due to presumably reduced posterior adductor muscle. In either case, the infested mussels have become openable more readily and easily. With larger adductors, Danish mussels are capable of resisting sea star attack than British mussels with smaller adductor muscle (see Lauckner, 1983a). Condition index (body dry weight divided by shell cavity volume) is recognized as a marker of health of molluscs. In *M. edulis*, the index is decreased with increasing parasitic load of *P. ciliata* (Fig. 2.8B).

Crustaceans: In comparison to any other invertebrates, marine bivalves are more infected by copepod parasites. Yet, no freshwater mollusc is so far reported for the incidence of copepod parasites. *Mytilicola intestinalis* adults are parasites of oysters like *M. galloprovincialis* and *M. edulis*. With increasing parasitic load of *M. intestinalis*, the live body weight of *M. edulis* decreases at the rate of one gram for every increase of 10 parasites/host (Fig. 2.8C). The pinnothorid pea crabs are the smallest crustaceans and usually inhabit the mantle cavity of bivalves (cf Pandian, 2016, p 212). They are more irritant gill pests than parasites. In the presence of *Pinnotheres maculatus*, both growth rate and body weight of *Argopecten irradians* decrease with increasing shell height (Fig. 2.8D).

Clams: Burrowing clams excavate an array of substrates including other molluscs. The thin valves and burrowing mode of life render smaller clams unsuitable for large burrowers. But, the relatively larger valved gastropods, mussels and thick valved oysters are suitable for a variety of burrowers. For example, the pholadid *Penitella conradi* burrows into the valves of *M. californianus* and the other pholadid *Pododesmus* species into the shell of *Haliotis* spp. *Rocellaria* spp infect the valves of silver-lip pearl oyster *Pinctada maxima* (see Lauckner, 1983a). The Indian pearl oyster suffers from fouling and boring by burrowing clams (Dharmaraj et al., 1987). Some bivalve pest

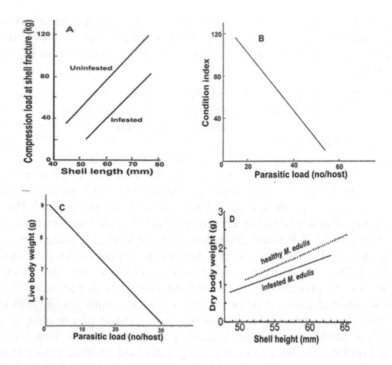

FIGURE 2.8

A. Effect of *Polydora ciliata* infestation on shell strength of *Mytilus edulis* (redrawn from Kent, 1981) B. Effect of parasitic load of *P. ciliata* on condition index of *M. edulis* (redrawn from Kent, 1979), C. Effect of parasitic load of *Mytilicola intestinalis* on live body weight of *M. edulis* (redrawn from Andreu, 1960), D. Effect of *Pinnotheres maculatus* infestation on dry body weight of *Argopecten irradians* (redrawn from Kruczynski, 1972).

may settle as larvae on the left valve of scallop. On growing, the pest induces thick layer of conchin and added mineralization to cause deformities.

Drillers: Sessile (e.g. oysters) and sedentary (e.g. *Nucella emarginata*) molluscs are predated by boring/drilling snails. Shell drilling gastropods are reported from prosobranch families Capulidae, Naticacae, Tonnacae, Muricacae and opisthobranch family Vayssiereidae. Their distribution is from intertidal zone to at least 2,700 m depth. But their number decreases with depth. The naticacids and muricacids possess an Accessory Boring Organ (ABO) and are capable of detecting prey from distances; their osphradia play a primary role in distance chemoreception (Carriker, 1981). Their prey includes mostly oysters and snails but rarely crayfish (Covich et al., 1994) and crabs (Huelsken, 2011). Unusually, the muricid snail *Reishia clavigera* is reported to drill the reinforced egg capsules of *Nerita japonica* (Fukumori et al., 2013). The whelk and oyster drillers combine the physical action of radular drilling with shell-dissolving secretion from an eversible ABO located on the foot to enable them to penetrate hard shelled prey (Carriker and Williams, 1978). Within 3 days of metamorphosis, the naticid *Polinices pulchellus* drills the bivalve *Lasaea adansoni* (~ 2 mm), later it drills *Cerastoderma edule* (~ 4 mm) and subsequently displays cannibalistic behavior (Kingsley-Smith et al., 2005). *Ocirebrina hispidula* is one of the smallest mussel drillers (Barco et al., 2013). Time budget estimates of *Acanthina punctulata* indicate that the drill spent 5% of its time in searching for its snail prey, 48–70% of time for drilling into the prey, and the remaining time on eating (Menge, 1974). *Nucella lapillus* requires 2.5 days to drill completely its prey, ingests it and moves to another prey (Bayne and Scullard, 1978). *Murex virgireus* requires 3 days to drill a round hole of 7–10 mm deep and 3–5 mm diameter on the shells of *Turbinella pyrum* (see Lipton et al., 2013, p 162). The pearl oyster *Pinctada fucata* is also drilled and predated by some gastropods (Chellam et al., 1983).

Crackers: Chelonians, fishes and starfishes more commonly crack the shells of bivalves but rarely gastropods. Crabs employ the chelar force to crack the shells of gastropods and bivalves. Lombardi et al. (2013) measured the hardness (in gigapascal, GPa) and compressive force (in newton, N) required to produce a crack in the shells of *Crassostrea virginica* and *C. ariakensis* as function of age (Table 2.3). They found no significant difference in the hardness between innermost and outermost layers of the shells of both oysters. It is likely that the foliated layers have similar properties and are composed of the same material in these oysters. Neither there was a significant difference in shell thickness between 4-year and 6 year old *C. virginica*, though the hardness was less in 1-year and 9-year old oysters. Notably, shells of *C. virginica* were thicker and denser than those of *C. ariakensis*. There are more numbers of publications

TABLE 2.3

Shell hardness and compressive force required to break shells as a function of age of oysters (compiled from Lombardi et al., 2013).

Age	Shell			
	Hardness (GPa)	Thickness (mm)	Density (g/ml)	Compressive Force (N)
Crassostrea virginica				
1-year-old shell	1.17	–	–	854
4-year-old shell	1.55	2.4	2.2	1090
6-year-old shell	1.47	4.5	–	3896
9-year-old shell	1.20	–	–	3513
Crassostrea ariakensis				
4-year-old shell	1.64	2.0	1.4	971
6-year-old shell	1.57	1.9	–	1552

on the molluscan response to predation by crabs but they are required more for chelonian and piscine predators. Further, the 6-year and 9-year old *C. virginica*, and 6-year old *C. ariakensis* withstood higher compressive force than the younger oysters in both species. Consequently, compressive force (N) required cracking the shells increased with shell thickness (Fig. 2.9). This finding explains why most molluscs opt to increase shell thickness against predation and why smaller predators prefer smaller prey. The chelar crushing forces may reach 27 N in the blue crab *Callinectes sapidus* (Singh et al., 2000) and 75 N in the rock crab *Cancer irroratus* (Block and Rebach, 1998). Hence, these crabs may predate only the spats of *C. virginica* (18 N) and *C. ariakensis* (25 N) (Newell et al., 2007).

Openers include Chelonians, fishes, starfishes and crabs. Expectedly, the chemical cue arising from predatory turtle *Chinemys reevesii* is detected earlier and responded more strongly by females than males of the freshwater snail *Pomacea canaliculata* (Xu et al., 2014). Adductor muscle weight increases with increasing body size (e.g. *Aulacomya alter*, Griffiths, 1977). Correspondingly, the adductor muscle strength may also increase, though no measurements have yet been made. With biting force reaching 220 and 1,250 N for the cownose ray *Rhinoptera bonasus* (Maschner, 2000) and large black drum *Pogonias cromis* (Grubich, 2005), respectively, they may easily break open the hard shells of 6–9 year-old oysters. Hence, adult oysters may suffer heavier mortality from vertebrate predators rather than crabs. Surprisingly, a large number of publications are available on the molluscan response to predation by crabs. But more publications are required for the molluscan response to predation by chelonian and piscine predators.

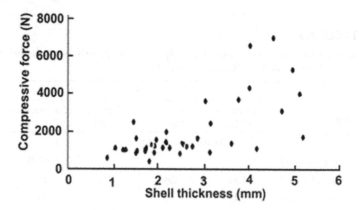

FIGURE 2.9

Effect of shell thickness on compressive force required to crack valves of 4- and 6-y old *Crassostrea virginica* and *C. ariakensis* (redrawn from Lombardi et al., 2013).

Nippers: Burying under the sediment surface provides protection for the vulnerable bivalves against potential predators and the level of protection increases with depth in the sediment (Peterson and Quammen, 1982). However, the need to protrude the inhalant (to acquire oxygen and food) and exhalant (to eliminate feces, urine and carbon dioxide) siphons just below or above the sediments surface renders the burying not as an absolute escapade from predators, especially nippers (see Fig. 2.10). Suspension feeders protrude the inhalant siphon a few millimeters below the sediments surface to acquire algae from the water column (de Goeij et al., 2001). Hence, they suffer only the nipping off the tips of the protruded siphon (e.g. *Protothaca staminea*) in clean sand (Peterson and Quammen, 1982). But the deposit-feeders have to protrude the siphons above the sediment surface; the longer the protruded siphons, larger are the feeding radius (de Goeij et al., 2001). Consequently, they suffer heavily on being nipped (e.g. *Tellina tenuis*, Trevallion, 1971). This is also true of protobranchs such as *Nucula*, whose gills are not specialized for filtration but adapted to suck up deposit, using an elongated palp proboscis.

Notably, the nipped siphons/palp proboscis are completely regenerated (see Chapter 4) fairly rapidly (@ 0.3 mg/week [w], Trevallion, 1971) but at the cost of body growth and/or reproduction. For example, regeneration

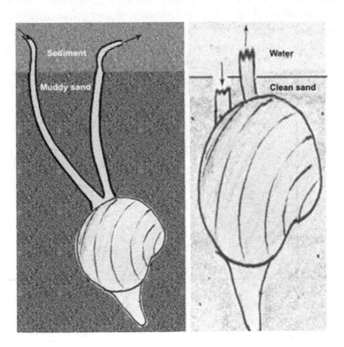

FIGURE 2.10

Free hand drawings to show the locations of inhalant and exhalant siphons of sediment-inhabiting bivalves in muddy (left panel) and clean sand (right panel) substrate.

of the siphon *Scrobicularia plana* at the rate of 6 mm/w (~ 20% of its length containing 0.2 kJ) requires 0.6 kJ (Hodgson, 1982). As suspension feeder, *P. staminea* is buried in clean sands, where it is subjected to an intense nipping. It may also occur buried in muddy sands, where it co-inhabits with *Macoma nasuta*, whose much protruded siphons are more readily vulnerable to nipping. The body growth of *P. staminea* decreased from 0.4–0.5 cm to zero level at the sizes of ~ 3.5 and ~ 4.5 cm body length in clean and muddy sands, respectively (Fig. 2.11). On exclusion of predation by nippers in clean sands, young bivalve (2.0 cm body length) grew four-times faster than when exposed to nippers. However, larger bivalves (> 3.0–4.0 cm long) grew almost equally, when either excluded from or exposed to nippers. Hence, smaller and younger bivalves seem to suffer more heavily from nippers than the larger/older ones. Apparently, the nipped siphons are unable to acquire adequate food and the bivalve has to channel its stored nutrients towards regeneration of siphons at the cost of body weight. This is also confirmed by the fact that the body growth of *P. staminea* did not differ, when it was either excluded from or exposed to nippers in the clean sands (Fig. 2.11), as the nippers preferred to nip the siphons of more vulnerable *M. nasuta* (Peterson and Quammen, 1982).

From a detailed field-cum laboratory studies, Trevallion (1971) reported that in the deposit-feeding bivalve *T. tenuis*, a large fraction of the siphon was continuously nipped and regenerated. The siphon weighed as much as one 16th of the body weight and regenerated at the rate of 0.3 mg/w;

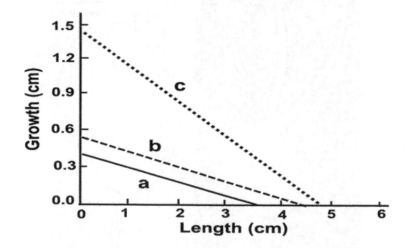

FIGURE 2.11

Growth of the bivalve *Protothaca staminea* as a function of body size in a. muddy habitat offering alternate prey to nippers, b. natural clean sandy habitat exposed to nippers and c. natural clean sandy habitat excluded from nippers (compiled and redrawn from Peterson and Quammen, 1982).

hence, regeneration of the complete siphon may require a period of 3–4 weeks, depending on the amount of siphon nipped. During the period of siphon regeneration, the bivalve may not be able to acquire as much food as it requires. Consequently, it allocates about 4.3% of the available assimilated energy for regeneration in both 2- and 4-year old animals, which otherwise could have been channeled for body growth and/or reproduction. Briefly, siphon nipping reduces body growth and reproductive output.

2.3 Shell and Resource Allocation

Major constituents of molluscan shells are conchin and calcium carbonate ($CaCO_3$). Conchin constitutes up to 7% of shell weight in gastropods (Hyman, 1967, p 161) and costs up to 26% of somatic growth energy in bivalves (e.g. *Aulacomya alter,* Griffiths and King, 1979). Hence, it is a metabolically costlier component of shell material (Palmer, 1983). Understandably, accessibility to food may affect the shell size (see Fig. 5.4A). Depending on the tidal levels, the duration of submergence and hence accessibility to food acquisition varies greatly. Consequently, the shell length of *Choromytilus meridionalis* is decreased (from 80 mm) by 9 and 50%, when the duration of submergence is reduced by 4 and 29%, respectively (Griffiths, 1977). $CaCO_3$ is the major inorganic component of the shell. Yet, its accumulation, transportation and precipitation costs ~ 5% the cost of proteinaceous conchin fraction of molluscan shell on a per gram basis (Palmer, 1992). Palmer (1981) has proposed that maximum rate of body growth is determined by maximum rate of shell production. Publications on ecological energetics of bivalves have indicated a clear shift from somatic to reproductive growth with increasing body size and advancing age (Fig. 2.12). Briefly, shell shape, size and thickness have profound effect on body growth of shelled molluscs. However, only a few publications directly address shell-body size relationship (e.g. gastropods; Hughes, 1971a, b, Geller, 1990; bivalves; Kautsky, 1982). As changes in shell thickness and volume of water retained within the mantle cavity vary widely growth of flesh biomass of shelled molluscs can be estimated only after sacrificing the animal (Griffiths and Griffiths, 1987).

In allocation of assimilated energy, a shelled mollusc may trade off between shell matrix, somatic growth and reproductive output. Allocation for conchin constituent of the shell ranges up to 7% in gastropods (see below) but up to 27–30% in bivalves (see Griffiths and Griffiths, 1987). Ovary somatic index (OSI) of *Haliotis iris* is > 60% (Poore, 1973) (Fig. 2.12A). Assimilated energy allocated for gonadal output in *Patella vulgata* is ~ 67% (Wright and Hartnoll, 1981). Data reported for the reproductive output ranges from 54% in *Tellina tenuis* (Trevallion, 1971) and 60% in *Mytilus edulis* (Kautsky, 1982) to 70% in *Ostrea edulis* (Rodhouse, 1978). Hence, allocation values of 10 and

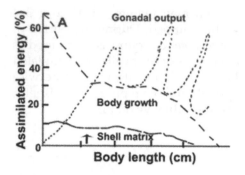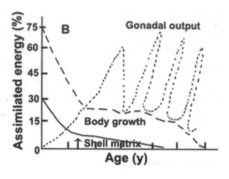

FIGURE 2.12

Allocation of assilimated energy as a function of body length in A. gastropods and age in B. bivalves. Energy allocation for gastropods is based on *Patella vulgata* (Wright and Hartnoll, 1981) and *Haliotis iris* (Poore, 1973) and that for bivalves on *Mytilus edulis* (Kautsky, 1982) and *Ostrea edulis* (Rodhouse, 1978). Arrows indicate the body/age, at which reproduction is commenced.

30% for shell matrix (inclusive of repair and regeneration) may be justified for gastropods and bivalves, respectively. The value of 60% for gonadal output for gastropods and bivalves is also justifiable. Until sexual maturity, the assimilated energy is shared between the shell and somatic growth. Following sexual maturity, a remarkable shift in allocation occurs. Notably, the allocation levels for shell and somatic growth is almost totally reduced beyond a particular age/size (Fig. 2.12B). During non-breeding period, the limited available assimilated energy in periodic spawners may, however, be shared between shell matrix and somatic growth. Consequently, a cyclic pattern of allocation occurs for reproduction.

Rare data reported for eight populations of sinistral *Busycon perversum* by Wise et al. (2004) clearly indicate a linear relationship between shell length and body weight (Fig. 2.13B), i.e. for every mm increase in shell length, the body weight increases by 4.3 g. Figure 2.13A shows a significant positive relationship between shell length and dry body mass of a few neritid gastropods. For every mm increase in shell length of the neritids, in which the shell is entirely composed of aragonite, body mass grows at ~ 0.1 mg dry weight. But the increase in body mass of *Fissurella barbadensis* is as much as 1.2 mg/mm.

In a unique contribution, Geller (1990) studied the effect of shell thickness on body mass in a thaidid snail known for polymorphism. He collected presumably sexually mature snails *Nucella emarginata* from exposed (to predatory crabs) area inside the harbor Bodega Bay, California and the protected (from crabs) area outside the harbor. He focused more on the effects of shell thickness on reproduction of snails. However, he provided rare data on the effects of shell thickness on body mass too. From Fig. 2.14A the following may be inferred: 1. The maximum size of thick shell morphs available in the field was ~ 873 mg dry biomass and ~ 8.7 g shell weight,

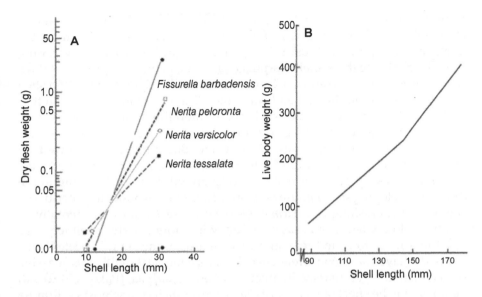

FIGURE 2.13

A. Growth (of dry body weight) as a function of shell length in selected dextral neretids (redrawn from Hughes, 1971a) and *Fissurella barbadensis* (redrawn from Hughes, 1971b). B. Growth of live body weight as a function of shell length in the sinistral whelk *Busycon perversum* (redrawn from data reported by Wise et al., 2004).

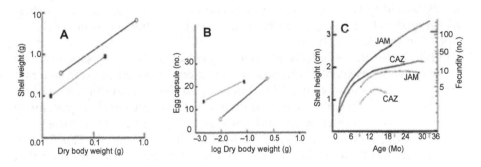

FIGURE 2.14

A. *Nucella emarginata*: Shell weight as a function of dry body weight in thick (indicated by thick line) and thin (indicated by thin line) shell morphs (redrawn from Geller, 1990). B. Effect of dry body weight (in log) on reproductive output, as measured by number of egg capsules (redrawn from Geller, 1990), C. *Viviparus georgianus*: Effect of age on shell growth (indicated by continuous dark lines) and fecundity (indicated by dotted thin lines). JAM = Jamesville Reservoir population and CAZ = Cazenovia lake population (compiled and redrawn from Browne, 1978).

respectively. But it was ~ 390 mg dry biomass and ~ 870 mg shell weight for the thin shell morph. Briefly, for growth of every mg dry biomass, the thick shell morph added just 9.6 mg shell weight but it was only 2.3 mg shell weight in the think shell morph, i.e. increase in biomass growth of thick shell morph required to add ~ 4.2-times more shell weight than the thin shell morph. With the production and carrying 4.2 times thinner shells, the thin shell morphs had to encounter such heavy mortality that not a single thin shelled morph larger than 1.1 g size could be collected in the field.

From an experimental series, Dalziel and Boulding (2005) showed an increase in shell thickness from 5.5 to 6.0 mm in the smooth, unarmed shell without a thick lip but a decrease in body growth from 53 to 45 mg in the directly developing snail *Littorina subrotundata*, when exposed to water-borne cue of a shell-crushing crab *Hemigrapsus nodus* for 60 days. Incidentally, *L. sitkana*, a close relative of *L. subrotundata*, with more massive shell required five times more force to break in a mechanical testing unit. Expectedly, *H. nodus* displayed a strong preference for *L. subrotundata* with less massive shells than that of massively thick shelled *L. sitkana* (Pakes, 2002). To demonstrate the effect of fish (*Tinca tinca*) cue on shell characteristics, Rundle et al. (2004) estimated the increases in shell mass and length in smaller (5 mm) and larger (10 mm) freshwater snail *Lymnaea stagnalis* exposed to water containing relatively poor calcium concentration (45 mg/l) for 7 days. On exposure to the fish cue, the snail increased its shell mass from 28 to 42 mg/mm and shell length from 3.0 to 6.5%. Apparently, the snail increased in shell thickness along with the shell length. On the other hand, *Thais* (*Nucella*) *lamellosa* held in water containing crab cue for 76 days, decreased aperture width or developed larger apertural teeth but not changed shell thickness Appleton and Palmer (1988).

Publications on the effects of shell thickness on body mass growth of bivalves are not available. Yet, some information can be extracted from a rare publication by Sanford et al. (2014). They reared Olympia oyster *Ostrea edulis* from hatching to juvenile stage at ambient (relatively thicker shell with mean calcite and aragonite Ω of 5.8 and correspondingly large body mass) and elevated (relatively thinner shell with mean calcite and aragonite Ω of 3.5 and 35% smaller body mass) CO_2 levels. Sanford et al. have also indicated that the number of oysters drilled by the Atlantic oyster drill, *Urosalpinx cinera* increased from 28 to 42 at ambient and elevated CO_2 respectively, i.e. the drill consumed more oysters at elevated CO_2 levels to compensate the 35% smaller sized thin-shelled oysters. The growth rate of thin shelled oysters reared at elevated CO_2 was slower than thick shelled oysters at ambient CO_2.

Increased allocation for the shell thickness, strength and/or weight is also accompanied by reduced allocation for body growth and reproductive output. Being plastic phetotypic character, the shell weight of the protobranch bivalve *Yoldiella valettai* makes to 62, 52 and 37% of the total weight in the Amunden Sea, Scotia Sea and Weddell Sea, respectively. Correspondingly, the weight of soft body mass also varies in these bivalves from 38% in the former and 63% in the last one (Reed et al., 2013). In the snail *N. emarginata*, the predation induced thick-shelled morph commenced reproduction at much smaller body size. However, the reproductive life span of thin shelled morph was much longer than that of thick shelled morph (Fig. 2.14B). Consequently, thin shelled morph produced one and a half times more number of egg capsules than thick-shell morph. Despite having a larger capsule, both morphs produced equal number of juveniles per capsule. Hence, thin-shelled morphs were more frequent, due to (i) longer reproductive life span and (ii) a consequent increased production of one and a half times more number of egg capsules. Briefly, energy allocation for shell thickening reduced body growth to 40% and reproductive output to 50%.

Intraspecific variations in shell shape, especially height impose profound changes in fecundity. The snail *Viviparus georgianus* inhabits through the length of Mississippi River from Massachusetts to the Gulf of Mexico. With life span of 36–48 months in Jamesville Reservoir (JAM) (42° 58' N), the snail grew to 3.8 cm height (Fig. 2.14C). But those inhabiting Cazenovia Lake (CAZ) (42° 55' N) grew to 2.2 cm shell height only during the 21–22 months life span. Reflecting the consequences of viviparity, fecundity of *V. georgianus* is uncommonly low for a snail. It increased to 40 offspring during the second year of its life at JAM but < 5 during the corresponding second year at CAZ. Notably, the snail at CAZ did not survive to reproduce during the third year.

The foregone account has substantiated the observations of Trussell and Nicklin (2002) that to improve fitness against predators, both gastropods and bivalves allocate more of their assimilated energy on heavily armed thicker shells at the cost of body growth and/or reproductive output. In his hypothesis, Palmer (1981) proposed that parallel trends become apparent for maximum growth rate with increasing ration/food supply and the trend for thick shell morph is always at the lower level than that for thin shell morph (Fig. 2.15). The emerging picture from bits and pieces of information summarized in this account permits a model for growth and reproduction in thin- and thick shelled morphs. Firstly, with every unit increase in body weight, the increase in shell weight is four times greater so that the difference between body weight and shell weight progressively becomes wider with increasing body weight. Secondly, reproductive output of thin shelled morph is one and a half times more, due to a relatively longer reproductive life span and larger body weight. Thirdly, thin shelled morphs commence to reproduce at much smaller body size than thin shelled morphs.

2.4 Symmetry and Sinistrals

Symmetry: From an analysis of transitions from symmetry to asymmetry and anti-symmetry that have occurred in metazoans, Palmer (1996) has

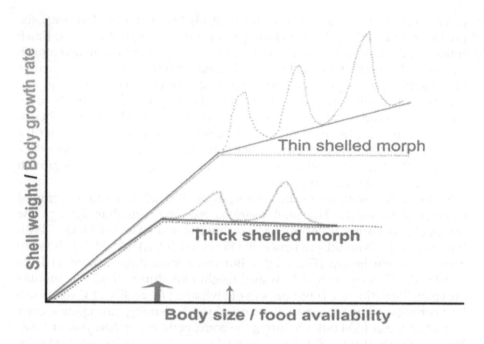

FIGURE 2.15

Proposed model for growth measured in units of shell weight and reproductive output as functions of body size/food availability. Shell weight is shown for thick (red) and thin (light green) shell morphs. In these morphs, reproductive output is shown by dotted wavy lines. Arrows indicate the body size at sexual maturity. Note nearly one and a half times greater reproductive output by thin shell morph and the widening gap in shell weight between these morphs with increasing body size. The model proposed by Palmer (1981) is superimposed showing the faster growth in thin shell morphs with increasing food availability.

recognized two patterns. The transition from symmetrical (SYM) ancestors to directional asymmetry (DS) (e.g. chirality) has occurred frequently and independently throughout the evolution of gastropods. This transition is determined irrevocably at the first cleavage. In contrast, Anti-Symmetrical (AS) transition occurs during post-larval stage; for example, the claw asymmetry occurs at or after the fourth post-larval stage of lobster (cf Feng et al., 2011). It may also be influenced more by environmental factors. Within molluscs, shell symmetry is intimately tied to early cleavage stage in conic-spirally coiled gastropods. However, it arises much later in nautiloid and ammoniod cephalopods and unequivalved bivalves, in which either the left or right valve is larger (Vermeij, 1975).

Chirality is a phenomenon, in which three dimensional (i.e. the bilateral, antero-posterior and dorso-ventral axes, see Palmer, 1996) asymmetric forms can arise in an almost mirror image forms within a single species (McManus, 2002). Besides chirality, the developments in some molluscs is further complicated by a torsion, in which visceral mass rotates by 180°, and bring the anus, genital openings and others to an anterior position. Sinistrals are not uncommon among monoplacophorans; for example, *Archaeospira arnata* is sinistral (see Palmer, 1996). A vast majority (90–99%) of gastropod species are dextrals (Robertson, 1993, Asami, 1993). Rarely, a sinistral is found in dextral gastropod species (e.g. the sacred chank *Turbinella pyrum*, Lipton et al., 2013). Similarly, sinistral lymnaeids occur very rarely in natural populations, though sinistral *Lymnaea stagnalis* predominates sporadically (Vinarski, 2007). Likewise, a dextral may also be rarely found in sinistral species (e.g. *Albinaria cretensis*, Orstan and Welter-Schultes, 2002). Sinistral species exist as lone groups within dextral genus (e.g. *Busycon*, Wise et al., 2004) or as groups related sinistral species within a dextral genus (e.g. *Diplommatina*, Peake, 1973). Some genera (e.g. *Dyakia*, Laidlaw, 1963) and families (e.g. Clausiliidae, Nordsieck, 1963) are almost entirely sinistrals. Incidentally, dextral populations of *Partula suturalis*, co-inhabit with sinistral species *P. mooreana*, *P. tohiveana* and *P. olympia* without interspecific hybridization. However, sinistral population of *P. suturalis* may hybridize in the transition zone, which corresponds with the transition from sinistral to dextral (Johnson, 1982).

Busycon, *Conus* and *Turbinella* are some examples of marine genera, in which sinistral is reported. The paucity of sinistral relative to dextral gastropod species in marine realm is an enigma (Hendricks, 2008). A reason suggested for it is that many marine gastropods are broadcast spawners and have external fertilization. Consequently, the frequency-dependent selection against rare sinistrals should be wanting in the marine molluscs (Schilthuizen and Davison, 2005). Broadcast spawning followed by planktonic stages may not afford the sustenance of small and isolated populations, an important feature obligatorily required to stabilize sinistrals (see Vermeij, 1975). However, sinistral snails do occur in freshwater. Translucent eggs and simple rearing procedure have afforded *Lymnaea* as the most popular genus for many valuable investigations on genetics and embryology of molluscs. Yet, sinisitral and dimorphic chiralities are more prevalent in terrestrial gastropods. Perhaps, the internal fertilization has stabilized the dimorphism.

Distribution of sinistrals and dimorphics: Field observations indicate that in comparison to dextrals, the number of species is lower for the sinistrals and lowest for the dimorphics in freshwater and terrestrial habitats. For example, of 13 recognized species in the genus *Partula*, 62% are monomorphic dextrals, 22% monomorphic sinistrals and only ~ 15% are dimorphics (Murray and Clarke, 1980). This observation is also confirmed by larger samples. Asami et al. (1998) determined chirality and shell shape for 889 genera in 46 families across 12 super-families. Of 422 high spiral tall stylommatophore species, (i) that mate by mounting and (ii) that mate in parallel alignment, 60 and 26% are monomorphic dextrals and sinistrals, respectively; the remaining 8% are dimorphics. The proportion for monomorphic sinistrals (1.98%) and dimorphic (0.4%) are still lower in low spired globular-flat shelled species that mate face-to-face. Evidently, shell shape and direction of coiling have profound effects on the number and proportions of monomorphic sinistrals and dimorphics, as well.

With delayed onset of helical spindle inclination and spiral blastomere deformation (Shibazaki et al., 2004), *L. stagnalis* sinistrals suffer from reduced hatching success, increased developmental aberrations and a strongly expanded body whorl (Asami, 2007). Similarly, pleiotropic effects have also been reported for *Bradybaena similaris*. In *P. suturalis*, Davison et al. (2009) found that the shell width/height ratio is determined by coiling phenotype, i.e. mother's genotype but the height is determined by its own genotype. From shell shape differences in chiral dimorphic *Amphiodromus inversus*, Schilthuizen and Hasse (2010) concluded that the differences in shape are small, as the pleiotropic effects on the chirality gene are of importance during early embryonic period only.

An important reason for the low number and proportion of dimorphics is the incompatibility of interchiral mating. A dextral individual has its gonadal opening on the right side of its body but a sinistral has it on its left. This sexual asymmetry inhibits interchiral mating, as the genitalia exposed by a sinistral on its left side cannot be joined with that exposed by a dextral on its right. This contrasting asymmetrical difference creates an insurmountable barrier for interchiral reciprocal and non-reciprocal mating (see Soldatenko and Petrov, 2012). However, with small adjustments in mating positions, the sexual asymmetry allows interchiral copulation in long spired snails. The tall-shelled species mate non-reciprocally. In many of these simultaneous hermaphrodites, the snail acting as 'male' copulates by mounting the 'female' shell mutually aligned in the same direction (Fig. 2.16). Contrastingly, the short flat shelled species mate face-to-face. Apparently, asymmetry caused by coiling direction and high spiral shell shape reduces interchiral mating success. However, it may allow interchiral mating, if one of the individuals alone rotates its body side by side with the partner. However, these interchiral matings lead to only 13% success in *P. suturalis* as against 90% mating success in intrachiral mating (Johnson, 1982). In another series of experiments, Asami et al. (1998) found 12 and 17% success in interchiral matings in *P. suturalis* and *B. similaris*, respectively, against 44 and 88% success in intrachiral matings. Despite penial insertion in some successful matings, only a few coiled 'males' successfully transfer his spermatophores to the 'females'. Notably, the failure of interchiral matings is more due to mechanical difficulty of penial insertion rather than reduced mating activity.

Reduced offspring: Even after successfully passing through the complex hazardous matings, interchiral 'hybridization' results in reduced production of progenies. In a rare contribution, Johnson (1982) reported the number of progenies produced by intrachiral and interchiral parents in *P. suturalis* as a function of age (Fig. 2.17). This non-selfing hermaphroditic ovoviviparous

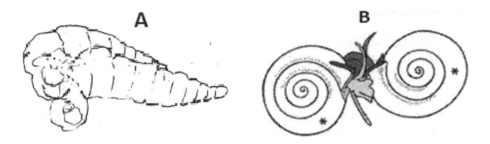

FIGURE 2.16

Contrasting mating behavior associated with sinistrals. A. Mating by shell mounting in a high spired sinistral species (redrawn from Asami et al., 1999). B. Face-to-face mating in a low spired sinistral species (redrawn from Soldatenko and Petrov, 2012).

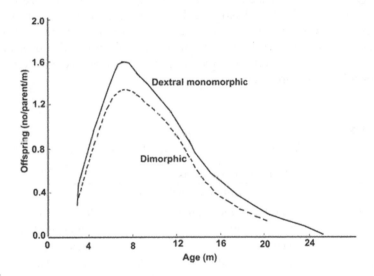

FIGURE 2.17

Birth rate in the intrachiral dextral monomorphic and interchiral dimorphic (dextral and sinistral) parents of *Partula sututralis* as a function of age (modified and redrawn from Johnson, 1982).

snail produces one young at a time from 3 months to 2 years of its life. In the laboratory, the interchiral parents produced 28% less number of offspring in their life time but those in the field produced 41% less offspring than the respective intrachiral parents.

2.5 Chemical Defense

In molluscs, the shell is a protective defense against predation but the shell-less opisthobranchs and cephalopods have to necessarily depend on chemical defense to escape from predators. For example, *Placobranchus ianthobapsus* is a poison-secreting shell-less nudibranch (Johannes, 1963). In view of their pharmacological and economic importance, defense chemicals of shell-less molluscs have attracted much attention from natural product chemists since 1980's (Derby, 2007). We seem to know more about the origin and distribution of these defensive chemicals, albeit their toxicity against predators remains untested. Defensive secretions are complex mixtures of chemicals that are distasteful, harmful and toxic or some combination of these. The newly discovered defensive chemicals/molecules are too many; achain (e.g. Ehara et al., 2002), aplysianin A (e.g. Kamiya et al., 1986, Jimbo et al., 2003), aplysianin E (e.g. Kisugi et al., 1987), dolabellanin A (e.g. Yamazaki et al., 1989) and escapin (Yang et al., 2005) are some examples. Table 2.4 is a representative brief summary of anti-microbial and anti-tumor activities detected in shell-less molluscs. Again the toxicity of these chemicals remains to be tested against predators.

TABLE 2.4

Detected anti-microbial and anti-tumor activity in shell-less aquatic molluscs

Species	Reported Observations	Reference
Anti-microbial activity		
Aplysia dactylomela	Purple gland-toxic protein	Melo et al. (2000)
A. juliana	Glycoprotein	Kamiya et al. (1988)
Dolabella auricularia	Peptide	Iijima et al. (2003)
D. auricularia	Albumen gland-Dolabellarnin A	Kamiya et al. (1986, 1992)
A. kurodai	Eggs: Aplysianin E-Glycoprotein	Kisugi et al. (1989)
Loligo pealei	Accessory nidamental and egg cases, L-amino acid activity	Barbieri et al. (1997)
A. kurodai	Eggs: Aplysianin E, fungicidal activity	Iijima et al. (1995)
D. auricularia	Dolabellarnin A-fungicidal activity	Iijima et al. (1994)
Anti-tumor activity		
A. dactylomela	Eggs: Aplysianin E	Kisugi et al. (1987)
A. kurodai	Eggs: Aplysianin E	Kisugi et al. (1987)
Aplysia sp	H_2O_2 from ink toxin kills tumor cells	Butzke et al. (2004)
Bursatella leachi	Ink-anti-retroviral activity	Rajaganapathy et al. (2000)
B. leachi	Purple fluid-anti-HIV protein	Rajaganapathy et al. (2002)
Illex argentinus	Ink-tyrosinase activity	Naraoka et al. (2003)

Passive defense: Basically these bioactive substances may be grouped into passively present and actively released defensive chemicals against predation. Deterrents may be present as distasteful *Aplysia brasiliana* (Ambrose and Givens, 1979), *A. dactylomela* (DiMatteo, 1981), *Dolabella auricularia* (Pennings et al., 1999), anti-feedant *A. brasiliana* (Kinnel et al., 1979) or in mucus (e.g. dovid nudibranch *Doriopsilla albopunctata*, Reel and Fuhrman, 1981), skin (e.g. *Aplysia* spp, Cummins et al., 2004) or egg capsules/masses (e.g. masses of 39 nudibranch species, Benkendorff et al., 2001). Apart from its mechanical protective effect, mucus can carry chemical cues, enhance their persistence and reduce their dispersal rate in water (Kicklighter et al., 2005). The skin of marine gastropods has different chemicals like terpenoids (e.g. *Stylocheilus longicauda*, Paul and Pennings, 1991). Many of these defensive chemicals are diet-derived (e.g. *Elysia halimedae*, Paul and Van Alstyne, 1988); others are synthesized *de novo* (e.g. nudibranch *Melibe leonina*, Barsby et al., 2002).

Active defense: Encountering a predator, the shell-less molluscs actively release defensive chemicals, which may be brought under the following heads: (i) alarm signals, (ii) feeding deterrents and (iii) escape-aid inks. The skin of many opisthobranchs secretes and releases acids, especially sulfuric acid with varying effects on predators (e.g. *A. californica*, Shabani et al., 2007). Attacked by predators, many animals including shell-less molluscs release chemicals that serve as an alarm signal to nearby conspecifics and sometimes even to heterospecifics to elicit predator avoidance behavior (Jacobsen and Stabell, 2004). Such an alarm signal may arise from passively released fluids from injured/killed prey or actively released secretion (e.g. ink and opaline secretions from *A. californica*). They evoke escape responses like moving away from the stimulus and escape locomotion (Nolen et al., 1995). To an alarm released from *Octopus bimaculoides* or *Loligo brevis*, *A. californica* is reported to respond and escape (Kicklighter and Derby, 2006).

Inking is a famous example for a chemical defense. It is reported from cephalopods and aplysiads. Cephalopods are known for their fast jetting escape by changing body color and inking. However, sea hares are sluggish and their purely visual smoke screening may provide only a short-term escape. The nearly contrasting features of ink secretions in cephalopods and aplysiads are listed in Table 2.5. It is not known whether the ink sac is present and used by bathypelagic cephalopods inhabiting at depths of ~ 2,400–4,800 m depths (see Table 1.6).

No comparative study has yet been made to know whether the presence of defensive shell or chemical saves more of assimilated energy and makes it available for body growth and/or reproduction. The presence of chemical defense, especially the diet-derived once released only on encountering a predator may save considerable investment on protection from predators. Hence, the shell-less opisthobranchs and cephalopods may be more successful. However, the presence of 930, ~ 75,000 and ~ 9900 species of shelled Polyplacophora, Prosobranchia and Bivalvia, respectively clearly shows the

TABLE 2.5

Nearly contrasting features of inking in sea hares and cephalopods (compiled from Derby, 2007)

Sea hares	Cephalopods
Ink is a mixture of 2 glandular secretions (Flore et al., 2004): the ink and opaline glands. It is expelled through a siphon	Ink is secreted and released from ink sac, which is a modified hypo-branchial gland
The purple ink (aplysioviolin) is diet-derived but opaline is *de novo* synthesized. Opaline becomes highly viscous upon contact with water. These glands are under the control of different ganglia and other secretions can be released independently or simultaneously	The component of ink is melanin, which is synthesized. The ink also contains dopamine secreted from different compartment of mature ink gland cells
Acts as a visual mimic distracter or smoke screen against attacking predator, while the sluggishly moving sea hare escapes, as the predator is focusing on the ink	As a deterrent, the ink smoke acts as a decoy. The cephalopod changes its color, while releasing the ink and the potential predator is kept in illusion that the cloud of ink is the prey, which has already left the site
Ink secretion acts as feed deterrent and contains aversive compounds. It can produce alarm signal (e.g. *Alysia californica, A. fasciata*)	Ink can also elicit alarm signal. It is a cytotoxin. Tyrosinase, a component of ink, has toxic properties

shelled molluscs are more abundant (see Table 1.5), speiose and successful than the shell-less Opisthobranchia (~ 2,000 species) and Cephalopoda (~ 800 species). The role of shell on life span, iteroparity and body size of molluscs is discussed in Chapter 9.

3

Sexual Reproduction

Introduction

Sexual reproduction is characterized by the presence of sex, gametogenesis, meiosis and fertilization of gametes arising from usually two conspecific individuals. It generates offspring with different mix of alleles than their parents. It also includes parthenogenics, in which females are always present and males do appear but sporadically by a not yet known cytological mechanism. In molluscs, parthenogenesis is limited to a few species and arises more due to parasitic elimination of males in populations. Self-fertilization is not uncommon among molluscan hermaphrodites but it includes gametegenesis and fertilizaton, of course, all within a single individual.

Barring a single gastropod species, in which asexual reproduction is claimed (Chapter 4), virtually all molluscs reproduce sexually (Table 3.1). Sexual reproduction is characterized by the presence of sex, meiotic gametogenesis and fertilization. In parthenogenic molluscs, females generate unreduced (diploid) eggs alone. A vast majority of hermaphroditic molluscs are Simultaneous Hermaphrodites (SH). However, selfing is enforced in them, when they are isolated in natural habitats or kept in isolation. Otherwise, they opt to mate and inseminate reciprocally or unilaterally. Sequentials and serials are not uncommon among molluscs. Some gastropods and a few freshwater mussels are characterized by protandry. Serials occur in a few mussels but rarely in gastropods. Most remarkably, protogyny, which occurs commonly in other aquatic invertebrates like crustaceans (see Pandian, 2016, p 105, 221–222), is not thus far reported from any mollusc.

Molluscan development commences with spiral cleavage. Hence the fate of each blastomere can be traced to one or other organ by a study of cell lineage. In prosobranchs, for example, the gonad proliferates from the pericardial (representing true coelom) wall (Hyman, 1967, p 313). However, no cell lineage study has been made to trace the blastomere responsible for the origin of gonad. It is now known that Primordial Germ Cells (PGCs) manifest sex in aquatic vertebrates like fishes and invertebrates like Chaetognatha and Crustacea (e.g. *Pandalus platyceros*, *Macrobrachium rosenbergii*). In a pioneering study, Feng et al. (2011) have traced the maternal origin of PGCs from two-

TABLE 3.1

Structural organization of reproductive system in molluscs (compiled from Hyman, 1967, Beauchamp, 1986, Runham, 1988, Lutzen et al., 2015)

Sexuality	Structural Organization
	Aplacophora
Hermaphrodites	A pair of undelimited ovotestes, each leading to gonoduct opening into cloaca
	Chaetodermids are gonochores with a single or paired ovaries/testes
	Polyplacophora
Gonochores	Dorsally located median anteriorly paired but posteriorily fused ovary or testis opening into gonopores. No intromittent organ
	Trachydermon raymondi is a hermaphrodite
	Monoplacophora
Gonochores	Two pairs of lobulated ovaries/testes. Each lobule discharges by a short gonoduct into nephridium
	Porsobranchia
Gonochores	Single ovary or testis. Lower prosobranchs have a short gonoduct, opening into right nephridium. But, the gonoduct opens directly in monotocardian prosobranchs. Intromittent organ present. Parthenogenesis occurs in 4 genera (e.g. *Melanoides*)
	Protandric hermaphrodites are not uncommon. E.g. *Patella vulgata, Crepidula fornicata*
	Opisobranchia
Hermaphrodites	A single large ovotestis with fairly complicated reproductive system. Simultaneous Hermaphrodites (SH) (e.g. *Aplysia*) or protandric (e.g. *Caliphyla*) hermaphrodites.
	Strubella paradoxa and *Microhedyle* are gonochores
	Pulmonata
Hermaphrodites	SH selfers, reciprocals or unilaterals. A single undelimited ovotestis with fairly complicated reproductive system is present. Median testis and peripheral ovaries
	Scaphopoda
Gonochores	Single median gonad is connected to kidney by a short gonoduct
	Bivalvia
Gonochores/ Hermaphrodites	9% of molluscs are hermaphroditic bivalves. Hermaphroditism ranges from SH to serial and Marian. Gonochoric: *Nutricola tantilla* median ovary surrounded by visceral mass. Two gonoducts. Pericardial voluminous testis with two ducts.
	Gynogenic: *Lasaea subviridis* median ovotestis. Oogenesis and spermatogenesis occur in anterior and posterior parts of ovotestis with no visible boundary
	Cephalopoda
Gonochores	Relatively simple reproductive system

celled stage, migration and establishment of PGCs in gonad of *Fenneropenaeus chinensis*. A similar study has also been made in oyster *Crassostrea gigas* (Fabioux et al., 2004). Hence, the PGCs manifest sex in molluscs too, as in other aquatic invertebrates.

3.1 Reproductive Systems

In animal kingdom, molluscs display an array of most complicated reproductive systems. Some representative features of their reproductive system is briefly described in Table 3.1 and Fig. 3.1. Reasons for the presence of these complicated reproductive systems are: 1. Unlike ~ 96% of crustaceans brooding their eggs on one or other part of their body (see Pandian, 2016,

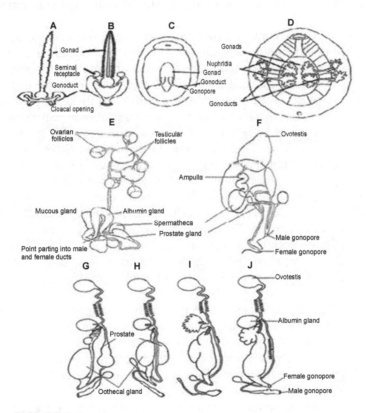

FIGURE 3.1

Scheme of reproductive systems of aplacophorans, A. gonochoric *Chaetoderma nitidulum*, B. Hermaphroditic *Neomenia carinata*, C. Polyplacophora and D. Monoplacophora. Reproductive system of E. monaulic *Okadaia*, F. diaulic *Lobiger* of opisthobranchs; pulmonates G. *Lymnaea peregra* H. *Planorbarius corneus* I. *Ancylus fluviatilis* and J. *Physa fontnalis* (free hand drawings from Hyman, 1967).

section 2.7), a vast majority of molluscs literally leave the oviposited eggs almost totally uncared. However, the eggs are enveloped either in a thick chorionic capsule or embedded in gelatinous mass prior to oviposition in an attempt to provide some sort of protection. This enveloping process requires the construction and maintenance of many special organs like the mucous gland, pallial gland and so on. 2. Molluscs are relatively less motile. Many of them are either sedentary or sessile. In them, the reproductive assurance demands the construction and maintenance of storage organs like the bursa copulatrix, seminal receptacle (some authors also use the term seminal vesicle) and so on for long term storage of viable sperm at the cost of ovarian reproductive system. 3. An alternative is the manifestation of hermaphrotism, which demands the establishment and maintenance cost of dual genital systems in an individual mostly at the cost of somatic growth—a reason for relatively smaller body size of many opisthobranchs and pulmonates (see Table 1.13).

With the need for (i) enveloping process of eggs, (ii) sperm storage organs and/or (iii) maintenance of dual sexes in an individual, the molluscan reproductive system has to accommodate more and more component organs within the relatively smaller body. In fact, these demands have driven the molluscs to explore several means to accommodate these extra component organs in one way or other. For example, prosobranchs accommodate them in two types of reproductive systems. The primitive ones discharge the gametes through the nephridium. The advanced type is often more complicated with provisions for receiving and storing sperm and internal fertilization (Hyman, 1967, p 289). In opisthobranchs, the simplest monaulics have a single gonopore, as in many cephalapsids, anapsids, notapsids and pteropods (Fig. 3.1E). In diaulics, the distal part of the gonoduct is separated into the sperm duct and oviduct, as in *Lobiger* (Fig. 3.1F), some pleurobranchids and onchidiaceans. In the most complicated triaulics, the oviduct is subdivided into an egg carrying part and a copulatory part as well as the sperm duct and all of them open separately but closely together to the exterior, as in Sacoglossa (Hyman, 1967, p 487). As pulmonates, the freshwater snails are characterized by the presence of a complicated hermaphroditic reproductive system (Fig. 3.1G–J). Depending on the presence (absence of oothecal gland in *Ancylus fluviatilis*), location (e.g. spermatheca) and enlargement (e.g. prostate gland in *Lymnaea peregra*, *Planorbarius corneus*) of the component organs, the reproductive systems in these snails are grouped into four types (Hyman, 1967, p 593).

In bivalves, the reproductive system is characterized by the presence of a single median ovary or testis in gonochorics and ovotestis in gynogenics and hermaphrodites. The single gonoduct arising from the ovary/testis or ovotestis is paired and opens directly. Female reproductive system of cephalopods is perhaps is the simplest in molluscs. The system includes an ovary, oviductal meander, ovarian gland and oviduct. The male reproductive system is simple and consists of a testicular complex, Needham's sac (e.g.

Loligo vulgaris). It may also include an accessory gland and seminal vesicle (Runham, 1988).

3.2 Gonochorism

A vast majority of molluscs are gonochorics. Gonochorism is distributed across the major taxons of molluscs. It occurs in chaetodermid aplacophorans, polyplacophorans (1,930 species), monoplacophorans (30 species), scaphopods (572 species), cephalopods (800 species) and majority of prosobranchs (> 60,000 species). Among bivalves, mussels and clams are gonochorics. In all, ~ 75% of molluscs are gonochorics. In these taxons, sex ratio may be altered by parasites, which is explained in Chapter 7. Otherwise, the ratio does not depart too far from the expected 1 ♀ : 1 ♂ at population level (however, see Chapter 6). In some riverine mussels, the ratio is rarely skewed in favor of males. It is 0.36 ♀ : 0.64 ♂ in *Quadrula asperata* and *Q. pustulosa* (Haag and Staton, 2003). In these mussels, fertilization of eggs depends on chance encounter of water-borne sperm. Hence, it may be adaptive to skew the ratio in favor of males. Expectedly, skewed sex ratio also occurs in protandric Pacific oyster *Crassostrea gigas*, in which the ratio is 0.37 ♀ : 0.62 ♂ (Allen and Downing, 1990). In some gonochoric bivalves, a small proportion of hermaphrodites do occur; for example, the ratio in *Mytilus galloprovincialis* is 0.495 ♀ : 0.495 ♂ : 0.1 ⚥ (Kiyomoto et al., 1996). In gastropods (e.g. *Pomacea canaliculata*, Yusa and Suzuki, 2003, Yusa, 2004b) and bivalves (e.g. *M. galloprovincialis*, Saavedra et al., 1997), experimental observations have shown that sex ratio remains 1 ♀ : 1 ♂ at population level, but the ratio varied greatly from nearly all males to all females among different parents. Yusa (2007) suggests that the mechanism responsible for this pattern is oligogenic sex determination, i.e. sex is determined by the small number of genes.

Maintenance of males is a luxury. Not surprisingly, molluscan males are smaller in size and weigh less than females. Perhaps, it is for this reason that protogynics, which have evolved from gonochoric lineages, and are not uncommon among other aquatic invertebrates (e.g. crustaceans, Pandian, 2016, p 105, 221–222), do not occur in any molluscs. Yet, one exception is the riverine mussel *Obliquaria reflexa*, in which female and male measure 45 and 63 mm, respectively (Haag and Staton, 2003). In prosobranch *Lacuna pallidula*, a male is one third the length and one 10th the weight of a female (Hyman, 1967, p 172). Dwarf males are reported from many bivalves (e.g. *Ostrea puelchana*, see also Table 3.8). In cephalopods, the dwarf males are known from three families: Tremotopodidae, Ocythoidae and Argonautidae (Laptikhovsky and Salman, 2003). In some argonaut species, the minute dwarf male (~ 6 cm length) is one eighth the length and one 600th the weight a female (see Finn and Norman, 2010). Apart from body size, sexual dimorphism in molluscs may include one or other morphological structure (Dillon, 2000) or shell color. For example, the males are brown in color and females red in prosobranch *Cypraea gracilis* (Hyman, 1967, p 172). In the apple snail *P. canaliculata*, male and female can be distinguished by the presence of convex and concave opercula, respectively (Gamarra-Luques et al., 2013). By the presence of heterocotylus arm, males are readily identified in all cephalopods, except in cirrate octopods (Mangold, 1987). In some of these cases, sex may be identified by external morphology. However, visceral morphology (e.g. Heard, 1975) may not be reliable for sex identification. Inspection of the gonad and related organ is preferable. Off from these, parasitism has not only 'sacculinized' the male into a 'testis' but also accommodated it in simplified body cavity of female entoconchid prosobranchs, which are parasitic on echinoderms. A pygmy male larva of these entoconchids is implanted on the wall of the pseudopallial cavity of the females (Fig. 3.2). Following the loss of ciliation, the larva enters the female through a ciliated tube, the siphon, which is connected to the female's oesophagus or body cavity. Subsequently, the non-ciliated male larva is 'sacculinized' into 'testis' (Lutzen, 1968).

In gonochoric molluscs, fertilizations can be external or internal. External fertilization is associated with broadcast spawning, which costs huge investment on gamete production. To increase fertilization success, sedentary/sessile molluscs like abalones, mussels and oysters increase population density, and synchronize spawning and milting. Other motile molluscs swarm to a site, where they spawn synchronously. Broadcast spawning molluscs

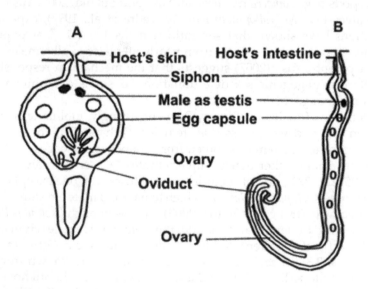

FIGURE 3.2

Anatomy of entoconchid prosobranch gastropods. A. *Entocolax* and B. *Thyomicola* (redrawn from Lutzen, 1968).

adopt one of the following strategies: 1. Huge investment on gamete/larvae. 2. Morphological, physiological and behavioral adaptations to bring the mating partners close together by aggregation/stacking or guiding them through visible and/or chemical cues. With an Ovarian Somatic Index (OSI) of > 65, the broadcast spawning abalone *Haliotis iris* spawns > 11.5 million eggs (Poore, 1973). These broadcasters may ensure fertilization success at sperm density of 100,000–190,000 sperm/ml (see Shepherd, 1986), although 90–95% monospermic fertilization can experimentally be achieved in the giant scallop *Placopecten magellanicus* with 150,000 spermatozoa + 5,000 ova/ ml (see also Desrosiers et al., 1996). High values of 33–40 are reported for both OIS and Testicular Somatic Index (TSI) in another broadcast spawner *Patella vulgata* (Wright and Hartnoll, 1981). In the Australian abalone *H. laevigata*, sexually mature individuals display aggregative behavior just before and during the spawning season (Shephard, 1986). The Antarctic limpet *Nucella concinna* deliberately stacks and each stack may consist of up to eight adults, one resting over the shell of the one beneath (Picken and Allan, 1983). Hickman and Porter (2007) vividly describe a spectacular swarming of swimming vestigastropod *Scissurella spinosa* accompanied by epidemic nocturnal synchronized spawning and milting (Fig. 3.3A) during November in a shallow back-roof in Moorea, French Polynesia.

Other molluscs characterized by internal spawning come in contact with conspecific mating partners attracted by photo cues. For example, light organs, as in bathybenthic cephalopods like *Eledonella pygmaea* and *Japetella diuphana* attract mating partners from distances and bring them together

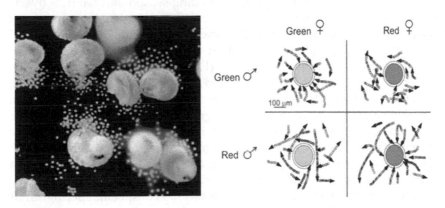

FIGURE 3.3

A. Spawning and milting in vetigastropoda *Scissurella spinosa*. Note the reddish orange eggs in water spawned by a female and a cloud of sperm from an adjacent male at the lower right corner (from Hickman and Porter, 2007). B. Swimming behavior of red abalone *Haliotis rufescens* and green abalone *H. fulgens* near an isolated conspecific or heterospecific egg. Note the attraction of sperm toward conspecific egg and detraction against heterospecific egg (from Riffell et al., 2004).

(Mangold, 1987). In other benthic molluscs, chemical cues like pheromones attract and bring conspecific mating partners together. To detect the presence of sex attractive chemical cues, two types of approaches have been made namely the use of T-maze (e.g. littorinid limpet, *Littorina littorea*, Seuront and Spilmont, 2015) and mucus trail (e.g. mangrove littorinid snail, Ng et al., 2011). From a series of T-maze experiments in *L. littorea*, Seuront and Spilmont (2015) have inferred the following: 1. During breeding season, the odor and mucus arising from males and females contain sex pheromones. 2. The pheromones arising from females are soluble and volatile but those arising from males soluble only. 3. These pheromones are sex specific and limited to breeding season alone.

From their investigation on mud snail *Ilyanassa* (= *Nassarius*) *obsoleta*, Moomjian et al. (2002) postulated that the breeding aggregation in the snail is organized by at least three sources of water-borne pheromones and one source of kairomone. The pheromones arise from sexually active females and males, copulating pairs and capsule depositing females. Painter et al. (1998) made an elaborate study to identify and characterize the first water-borne peptide pheromone called 'attractin' in a mollusc. To establish and maintain aggregation, egg cordons of *Aplysia californica* secretes 'attractin'. The full length peptide 'attractin' has been isolated from the albumin gland, in which package of the eggs into cordon is made. The peptide has 58 amino acids in length. It contains N-linked glycosylation and six cysteine residues, which contribute to the hydrophility and stability of the peptide. Temptin is another water borne peptide pheromone that acts in concert with 'attractin' to stimulate male attraction (Cummins et al., 2004).

In many of the 60 abalone species, breeding season and habitats overlap. Yet hybrids are rare (Leighton and Lewis, 1982). Chemical communication between gametes is a critical factor mediating sexual reproduction. The discovery of L-tryptophan as a potent attractant to red abalone *Haliotis rufescens* sperm is a landmark discovery (Riffell et al., 2004). Riffell et al. (2002) have established that the free amino acid L-tryptophan is the natural sperm 'attractin' for the red abalone. Behavioral responses of sperm to manipulated natural tryptophan gradient around eggs revealed that both chemotaxis and chemokinesis play a significant role in bringing contact between sperm and egg. The experimental observations on endogenous pathway and chemical communication between egg and sperm of these abalones are briefly summarized: 1. The freshly spawned eggs of these abalones release L-tryptophan, which evokes positive response by conspecific sperm (Riffell et al., 2002). 2. Following the release, L-tryptophan evokes maximum response up to 30 minutes and subsequently diminishes to zero by ~ 50 minutes (Krug et al., 2009). 3. The red abalone egg mimic, with a microcrystalline cellulose core surrounded by a layer of L tryptophan, also evokes within 10 µm distance correct orientation and doubly speeded positive response, as those by natural eggs (Himes et al., 2011). 4. Fertilization Success (FS) is dependent on the ratio of sperm to egg, which ranges from 0% with 1 sperm: 1 egg to

100% in 10,000 sperm: 1 egg. At intermediate ratios (10–1,000 sperm per egg), FS is increased significantly as a function of L-tryptophan. 5. The optimal L tryptophan concentration that evokes maximum response is 10^{-8} M (Riffell et al., 2004). 6. On exposure to heterospecific sperm, the naturally spawned eggs of red and green abalone evoke no positive response by these sperm (Fig. 3.3B). Clearly, sperm activation and chemotaxis are species specific (Riffell et al., 2004). 7. Thus reproductive isolation seems to reside at the level of membrane recognition-proteins (Vacquier, 1998). 8. Surface proteins involved in sperm-egg interaction have been earlier characterized for abalone displaying strong selection for species specific gamete recognition (Vacquier, 1998). The findings of Riffell et al. (2004) indicate that reproductive isolation resides at the level of membrane recognition protein. Species specific attractants like L-tryptophan may locate the right target within mixed gamete suspensions of closely related species. Briefly, the benthic sedentary molluscs use pheromones as chemical cues to navigate and come in contact with each other. Conversely, the broadcast spawning molluscs use free amino acids like L-tryptophan as a potent attractant in the navigation of sperm to conspecific egg. Briefly, the amino acid L-tryptophan and peptides have been identified as pheromones employed by sexually active molluscs in pelagic spawning abalones and peptides in benthic spawning sea hares, respectively.

3.3 Parthenogenesis

It is characterized by the following features: (i) Females are always present and produce unreduced (diploid) eggs, which develop without the need for a trigger by an auto- or -allosperm and (ii) Following sporadic appearance of males, sexual reproduction occurs. Typically, two types of parthenogenesis are recognized. Ameiotic or apomictic parthenogenesis, in which synapsis of homologous chromosomes does not occur. Hence there is no segregation or crossing over. This ameiotic oogenesis generates diploid eggs, requiring no fertilization. Conversely, normal meiosis occurs in automictic parthenogenesis. However, diploidy is restored in the egg by fusion of two haploid homologous nuclei of the first mitotic division or by fusion of the haploid egg nucleus with non-homologous nucleus of the polar body. In animal kingdom, parthenogenic reproduction is limited to 0.1% of all species (Bell, 1982). Not surprisingly, it is reported to occur only in four genera *Melanoides, Campeloma, Potamopyrgus* and *Hydrobia* belonging to three prosobranch families among the known 133,000 molluscan species. These parthenogenics and their broad geographical distribution are summarized in Table 3.2. The marine bivalve *Lasaea* (see Beauchamp, 1986, Crisp and Standen, 1988) was suspected to be a Simultaneous Hermaphrodite (SH) and parthenogenic. However, it is now shown as a gynogenic (O'Foighil and

TABLE 3.2

Parthenogenic prosobranch gastropods, their characteristics and geographical distribution

Species, Characteristics & References	Geographic Distribution	Reported Observations
Melanoides tuberculata, 2n = 32, 6n = 90–94 (Jacob, 1954)		In India, out of 10,000 specimens, no male was present
Viviparous. Matures at 7 mm shell height. Males readily recognized by red gonad Heller and Farstey (1990)	From Far East to Africa	Of 44 Isreali populations 7 had no male, had < 10, < 36, 46 and > 66% males were present in 15, 10, 1 and 1 populations, respectively
M. lineatus 71–73 allopolyploids (Jacob, 1954)		When present, males were sterile
M. scabra 76–78 allopolyploids (Jacob, 1954)		
Campeloma decisum, Ovoviviparous see Johnson et al. (1995)	In North America from Indiana to Lousiana	Geographic parthenogenesis in northern glacial habitats, sexual in southern non-glacial habitats
C. rufrum see Heller and Farstey (1990)	As above	As above
Potamopyrgus antipodarum, Co-existing diploid sexual and triploid parthenogenics Lively and Jokela (2002)	New Zealand	Males constitute 0 to 40% of populations with higher male frequency in lakes than streams
P. jenkensi (see Heller and Farstey, 1990)	Introduced from New Zealand into Europe	Only one male out of 8,000 specimens in Poland. 1% males in 8,500 snails in Britain and Netherland

Thiriot-Quievreux, 1991). Hence parthenogenesis in molluscs is limited to prosobranchs alone.

For providing the first cytological evidence for the existence of parthenogenesis in molluscs, the credit goes to Annamalai University, Tamil Nadu, India. Through a series of publications, Jacob (1954, 1958a, b) showed the presence of allodiploidy, allotetraploidy (?) and allohexaploidy in *M. tuberculata*. In *M. lineatus*, males were totally absent in Tamil Nadu and Andhra Pradesh. When occurred at the frequency of 0.1%, they were sterile due to degeneration of spermatid. No sperm was found in testis, oviduct and brood pouch. Verifying chromosome number before and after the first cleavage division in the oocyte and polar body, apomictic parthenogenesis with two equational maturation divisions was confirmed in *M. tuberculata*. In *P. antipodarum*, both diploid sexual and triploid parthenogenic clones are characterized by similar heterozygosities and the clones do not display any new allele (Dybdahl and Lively, 1995). Hence, origin of the clones is likely to be the result of fertilization of diploid egg by a haploid sperm or haploid egg by a diploid sperm (see Chapter 6). But the clones have not arisen from inter-specific hybridization (see Jokela et al., 1997). It is likely that diploids and polyploids of *Melanoides* have originated by inter-specific hybridization but the triploids of *Potamopyrgus* and *Campeloma* by fertilization of one of the unreduced gametes. Regarding triploid parthenogenesis, a distinction must also be made. Both the obligately diploid sexual and triploid apomictic parthenogen of *P. antipodarum* coexist in New Zealand (Lively, 1992, Lively and Jokela, 2002, Jokela and Lively, 1995). But *C. decisum* represents a classical geographic parthenogenesis. The distribution of triploids is limited to the northern glacial habitats of the USA, whereas that of diploid sexual to the southern non-glacial habitats.

Reduction in male frequency in molluscan parthenogenics is traced to (i) conductivity in freshwater and (ii) trematode parasitic load. For reasons not yet known, increasing concentrations of magnesium, sodium potassium and chloride progressively reduce the male frequency in *M. tuberculata* (Fig. 3.4A). Sexual reproduction is favored in geographical locations, where the risk of parasitic infection is high, as it may provide some progeny, which may evade the co-evolving parasites. However, parthenogenesis is favored in populations with low risk of parasitism. The gonochoric snail *P. antipodarum* serves as the first intermediate host to at least 14 species of digenetic trematodes and all of them sterilize both sexes (see Jokela and Lively, 1995). From his investigation on 66 populations of the snail from natural lakes and streams of New Zealand, Lively (1992) showed that with increasing parasitic load, the male frequency progressively increases (Fig. 3.4B). In a subsequent publication, Jokela and Lively (1995) found that the parasitic infection induced precocious maturity and brooding. A significant correlations between the prevalence of trematode parasites in the past and present set of data in *A. antipodarum* (Lively and Jokela, 2002) reveals that

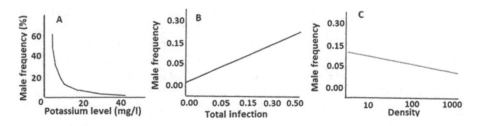

FIGURE 3.4

Male frequency as functions of A. potassium level in *Melanoides tuberculata* (redrawn from Heller and Farstey, 1990) and B. total infection, and C. density of *Potamopyrgus antipodarum* infected by trematode parasites (redrawn from Lively, 1992).

selective pressure imposed by the parasitic infection has remained persistent. Hence, the trematode parasite may be an important factor to selection of sexual reproduction.

With no need to invest on males, the parthenogenic clone may grow faster than sexual males and females. Consequently, the clone may be able to displace sexual males and females. An estimate suggests that the clone beginning with a single female will replace a sexual population of 10^5 individuals in < 50 generations and 10^6 individuals in < 60 generations (Jokela et al., 1997). From an elegant competitive experiment to assess the cost of sex on growth and potential displacement of sexual *P. antipodarum*, Jokela et al. (1997) showed the clones and sexuals attained maturity at equal 4.25 mm shell size. Being autotriploids, the clones did not suffer from any developmental incompatibility and poor hatching success. All the tested 14 clones grew faster at variable rates and might displace sexual males and females under certain conditions.

In contrast to *P. antipodarum*, parasitic prevalence in *C. desicum* is specifically higher in parthenogenic population than in sexual population. For example, all 26 examined parthenogenic populations were infected by digenic trematode *Leucochlorimorpha constantiae* but all 14 examined sexual populations remained uninfected. Notably, parthenogenesis is not maintained in locations, where the parasites are rare or absent. Accessibility of sperm for ingestion by parasites and severe sperm limitation seem to have driven *C. desicum* to employ parthenogenesis and to assure reproduction. Analysis of enzymatic loci indicates that the northern parthenogenics reproduce apomictically, while the southern ones by automictically. Being genetically identical at 19 enzyme loci, the homozygous clones of North Carolina reproduce by apomictic parthenogenesis. All the parthenogenics from Indiana and Michigan are heterozygous at 6–7 enzyme loci. These heterozygous are automictic clones.

Data accumulated in these parthenogenics provide an opportunity to test the validity of many hypotheses proposed in recent years (see Johnson et al., 1995). For example, the Red Queen Hypothesis predicts that sexual should be more common in populations with high risk of parasitic load than those without it. The findings of Lively (1987, 1992) in *P. antipodarum* from New Zealand support the hypothesis. Conversely, from data collected for both past and present *M. tuberculata* in Isreal, Ben-Ami and Heller (2005) have found no correlation between the frequencies of male and trematode infection. Hence, the findings in *M. tuberculata* do not support the Red Queen hypothesis. It must be indicated that the origin of parthenogenesis in these two species group differs. *Melanoides* have arisen from inter-specific hybridization but *Potamopyrgus* by elevated ploidy. Reproductive assurance hypothesis predicts the evolution of mechanism to assure reproductive success in populations, where access to male is severely limited or prohibited from fertilizing eggs. The correlation between male frequency and population density of

P. antipodarum in New Zealand (Fig. 3.4C) was negative, instead of the expected positive one. Hence the findings in the New Zealand snail do not support the reproductive assurance hypothesis. Tangle bank hypothesis predicts selective advantage of sexual reproduction in heterogenous environments, where the inter-specific competition for resource is intense. As geographic distribution of selfers and parthenogens is more common in homogenous but less stable habitats like high altitudes or latitudes (e.g. *C. desicum*), the hypothesis is partially supported.

3.4 Hermaphroditism

The expression of both female and male reproductive functions in a single individual either simultaneously or sequentially results in hermaphroditism. In molluscs, hermaphroditism occurs in 99–100% of aplacophorans (263 species), opisthobranchs (~ 2,000 species) and pulmonates (~ 24,000 species) as well as 3% (1,560 species out of ~ 52,000 species) and 9% (810 species out of ~ 9,000 species) of prosobranchs and bivalves, respectively. It occurs in ~ 40% of the 5,600 molluscan genera (Heller, 1993). Of about 115,000 molluscan species, ~ 25% (~ 28,370 species) are hermaphrodites. This approximate value may be questionable until otherwise proved. Incidentally, this 25% hermaphroditism occurs mostly in sedentary/sessile molluscs. This value may be compared with > 6% and > 2% hermaphroditism reported for the relatively more motile crustaceans (Pandian, 2016, p 222) and highly motile teleostean fishes (Pandian, 2012, p 222), respectively.

Simultaneous hermaphroditism (SH): As in fishes and crustaceans, hermaphroditism in molluscs also ranges from simultaneous to sequential and to serial. Figure 3.5 describes the known seven ontogenetic pathways, through which simultaneous, sequential, serial or Marian hermaphroditism are differentiated in bivalves. Most remarkably, SH, in which the two opposing sexual tendencies are mutually tolerated, is functional and successful in aplacophorans, opisthobranchs, pulmonates and some bivalves. Surprisingly, > 92% of molluscan hermaphrodites are simultaneous. Contrastingly, SH is limited to a dozen species in fishes (see Pandian, 2010, p 76–83). In crustaceans, it is limited to androdioecious taxons Notostraca, Spinicaudata, Thecostraca and some 18 protandric simultaneous hermaphroditic caridean species (Pandian, 2016, p 105–109, 111–112). In these crustaceans too, it is diluted by the presence of males and androdioecious mating system. In the absence of the clasping organ, the cincinnuli on endopodite of the first pleopod, the caridean hermaphrodite has to necessarily depend on a mating partner for insemination. Thanks to (i) the high level of polygamy, (ii) presence of special pouches (e.g. bursa copulatrix, seminal receptacle) for long term sperm preservation (for > 1 year) and (iii) their ability to selectively

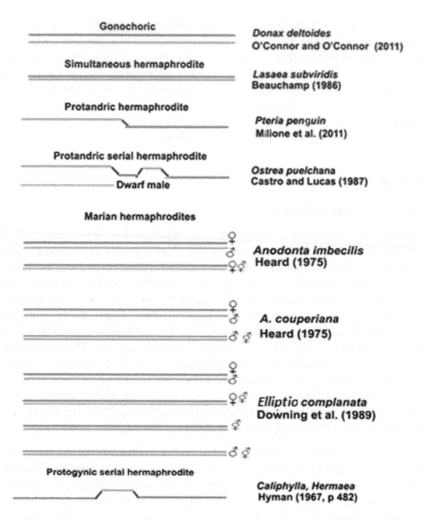

FIGURE 3.5

Ontogenetic pathways of sex differentiation in bivalves. The suggested pathway for a couple of opisthobranchs is also included. Red line indicates female, green males and violet transdifferentiating hermaphrodite. Note the presence of additional dwarf male in *O. puelchana* and male like hermaphrodite and female like hermaphrodite in Marian hermaphroditism.

mix sperm received from more than one suitors, these SH molluscs generate much genetic diversity in their offspring.

Simultaneous hermaphroditic molluscs can either be an obligate (e.g. *Lasaea subviridis*, Beauchamp, 1986) or facultative selfers (Fig. 3.6). Selfing is widespread among freshwater prosobranchs; of 48 genera, which self, 60% are inhabitants of freshwater (Heller, 1993). The freshwater snails

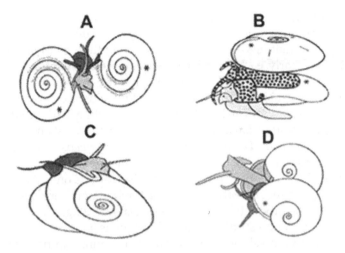

FIGURE 3.6

Mating positions of A. reciprocal *Bathyomphalus contortus*, B. unilateral *Segmentina oelandica*, C. reciprocal position of *Planorbis planorbis*, D. mounting position of unilateral *P. planorbis* (Free hand drawing from Soldatenko and Petrov, 2012).

Biomphalaria glabrata (Vianey-Liaud, 1995) and *Potamopyrgus antipodarum* (Jokela et al., 1997) are capable of self- and cross fertilization. Within a single mating, the reciprocal suitors inseminate each other. But one of the unilateral suitors acting as male alone inseminates the other, which acts as a female. For example, the planorbid pulmonates *Anisus vortex* and *Bathyomphalus contortus* (Fig. 3.6A), in which the suitors stop feeding and crawling, and mate face to face by positioning themselves away from the substratum, are reciprocals. But other planorbids *Chromphalus riparius* and *Segmentina oelandica* (Fig. 3.6B), in which the female-acting suitor continues to feed and the male-acting suitor mount upside down, are unilaterals. In still others *Planorbis planorbis* (Fig. 3.6C, D) and *P. duryi*, the suitors stop feeding and crawling, can be either reciprocals or unilaterals, depending up on the suitor's choice to orient mating position from shell mounting to face to face mating (Soldatenko and Petrov, 2012). For detailed descriptions on shell shape and mating behavior, and terminology of mating behavior in pulmonate gastropods, Jordeans et al. (2009) and Derbali et al. (2009) may be consulted. The sea hares copulate in chains, in which suitors act as male to the preceding one, as female to the succeeding one (Hyman, 1967, p 492, e.g. *Aplysia californica*, Angeloni et al., 2003). This type of sequential turners is more occasional (1–3%) than regular. Though many opisthobranchs and pulmonates are reciprocals, the sea hares are usually obligate unilaterals. In cephalapsid opisthobranch *Chelidonura tsurugensis*, yellow, black and white morphs occur. Of 16 matings between yellow and black morphs, the yellow acts as donor in five matings, black in four and reverse reciprocal in five and no insemination in two. In yellow versus white matings, yellow donates five times, white six times, and reverse reciprocal in two matings and no insemination in three matings (Turner and Wilson, 2012). Clearly, each morph can act as either unilateral or reciprocal.

In general, the freshwater pulmonates belonging to the families Acroloxidae, Ancylidae, Lymnaeaidae, Planorbidae and Physidae are capable of selfing, when reared in isolation. The isolation-enforced selfing is reported in a dozen freshwater pulmonates: *Biomphalaria boissyi, B. glabrata* (Vianey-Liaud et al., 1987), *Bulinus contortus, Ferrissia shimeki, Helisoma trivolis, Lymnaea columella, L. corneus, L. ovata, L. palustris, L. stagnalis, Physa fontinalis, P. gyrina* and *P. sayii.* Hyman (1967, p 599) narrates the amazing reports on isolated rearing of *L. stagnalis* for four generations, *L. columella* for 47 generations over a period of 9 years and yet the production of viable egg capsules (e.g. *L. corneous*). Firstly, corresponding isolation studies on opisthobranchs are wanted. Secondly, the demarcation between selfing and unilateral insemination behavior appears a flexible one. For example, of 108 pairs of sea hare *A. californica* observed, three pairs were mating both simultaneously and unilaterally (Angeloni et al., 2003). However, the observations on copulations behavior have to be complemented with anatomical studies on reproductive system. These studies alone can explain why the isolated *Ancylus fluviatalis* is sterile (Hyman, 1967, p 599) and *C. riparius* and *S. oelandica* remain unilateral (Soldatenko and Petrov, 2012). Incidentally, recent studies on enforced isolation of terrestrial snails show that the isolation reduced life span and reproductive life span but extended generation time and senescent post-reproductive life span of *Bulimulus tenuissimus* (Table 3.3). Consequent with the reduced reproductive life span, the number of broods and fecundity of *B. tenuissimus* are significantly reduced. These observations have to be extended urgently to freshwater pulmonates too. Incidentally, long day photoperiod is reported to reduce LS and thereby also fecundity in *L. stagnalis* (see Table 3.25).

TABLE 3.3

Comparative accounts on life span and reproduction of *Bulimulus tenuissimus* reared in isolation or group (compiled from Silva et al., 2013, approximate values)

Parameter	Group	Isolation
Life span (d)	880	725
Generation time (d)	250	475
Reproductive life span (d)	230	30
Post-reproductive life span (d)	400	220
Spawning season (no.)	3	1
Egg production (no./ ♀ /season)	16	9
Cumulative fecundity (no.)	48	9

The pulmonate SH snail species belonging to the genus *Bulinus* serve as intermediary host of the dreadful trematode *Schistostoma*. These tetraploid snails are distributed over a wide range of freshwater systems in northern and western Africa and the Middle East. The snails *B. contortus* and *B. truncatus* display dimorphic phallism namely euphally and aphally. The euphallics develop an ovotestis and fully functional male and female tracts. In contrast, aphallics do not develop the distal portions, i.e. the copulatory organ, prostate gland and vas deferns of the male tract but functional sperm are produced in the testicular tissues of the ovotestis. Hence, aphallics cannot donate sperm, whereas euphallics can do it. Phally is developed prior to sexual maturity but cannot change from one to other, once they are developed. Phallic dimorphism has arisen 14 independent times (see Johnson et al., 1995). The available information on euphally may be summarized hereunder: 1. As in *Potamopyrgus antipodarum* (see p 81) the overall prevalence of the trematode infection is positively correlated with the frequency of euphally. 2. There is no positive correlation between euphally and snail density (see also Lively, 1992). 3. As in *Melanoides tuberculata* (see p 81), the frequency of euphally is decreased with increasing concentrations of magnesium, potassium and other ions. 4. Phally is sensitive to temperature, i.e. colder temperature favors the development of euphally. With no need to invest and maintain distal male organs, selfing aphallics are able to invest more in reproduction. Table 3.4 provides a comparative account on the fitness parameters, as assessed by fecundity and survival of phallic morphs of *B. contortus*. Aphallics are able to allocate more resources for reproduction.

Marian hermaphroditism: Through a series of publications, St Mary (1993, 2000) has brought to light a new dimension to the sex allocation theory. She described a mixed or preferably called Marian hermaphrodites in sedentary gobiid fishes. For example, there are pure females, female-like hermaphrodite with 5% allocation to testicular tissue, hermaphrodites, male-like hermaphrodites with 5% allocation to ovarian tissues and pure males in

TABLE 3.4

Fecundity and survival of euphallic and aphallic
Bulinus contortus (compiled from Delay, 1992)

Parameter	Euphallic	Aphallic
F_2 offspring		
Capsule (no./ ♀ /d)	4498	5609
Egg (no. capsule)	0.83	0.87
Fecundity (no./♀/d)	5438	6296
F_3 offspring		
Hatchability (%)	95.4	97.5
Size at hatching (mm)	1.21	1.07

the goby *Lythrypnus dalli*. Similarly, some freshwater bivalves also consist of pure females, hermaphrodites and pure males as well as female-like- and male-like-hermaphrodites and display Marian hermaphroditism. The term continues to be used, as it was already named earlier (Pandian, 2010, p 117–120). Typically, Marian hermaphrodites commence to differentiate into hermaphrodites initially and female, or male subsequently (Pandian, 2010, p 121). Simultaneous hermaphroditic molluscs possess an ovotestis, in which the ovarian and testicular tissues are accommodated in (i) completely (delimited type) or incompletely separated (undelimited type, e.g. *Chaetoderma nitidulum*) anlage. However, the Marian hermaphrodites have a single globular ovotestis containing both ovarian and testicular tissues in different proportions. They are frequently found in bivalves that (i) occur at low densities, brood their embryos, (ii) have small genetically isolated populations and (iii) are sedentary as adults (cf Ghiselin, 1969, Charnov, 1982). Not surprisingly, microscopic examination of 97 species of the American freshwater bivalves revealed the presence of female and male tissues in the same gonad of almost all populations in four species and in a few populations within a species of 22 species, while the remaining 71 species were gonochorics consisting of pure females and pure males. In nine species belonging to the genus *Elliptio*, four were pure gonochorics. In the remaining five species, the frequency of Marian hermaphrodites averaged to 2% (see Downing et al., 1989). Hence, populations of these freshwater bivalves surviving at low densities in isolated aquatic systems have explored almost all possible combinations of sexuality. As a consequence, sexuality in them is a population specific rather than a species specific trait. Incidentally, male-like and female-like hermaphrodites are also reported from induced triploid *Crassostrea gigas* (Allen and Downing, 1990).

The sex of most unionids can be determined by morphology of the outer marsupial demibranchs, where embryos are incubated. For description of these marsupial demibranchs, Heard (1979) may be consulted. However, the phenotypic demibranch-based sex identification may not coincide with that of the visceral sex. In eight anodontids examined, *Anodonta wahlamartensis*, *A. californiensis*, *A. gibbosa* and *A. couperiana* collected from Myakka River were pure gonochorics (Table 3.5). From both Grant and Taiquin Lakes, *A. imbecilis* was unique and consisted of pure females and female-like hermaphrodites. *A. hallenbeckii*, *A. peggyae* and *A. couperiana* from Appalachicola River consisted of pure females, pure males and male-like hermaphrodites. Further, the following may also be inferred, when data reported by Heard (1975) in his Tables 3 and 5 are combined: 1. It is likely that *A. gibbosa*, *A. hallenbeckii* and *A. peggyae* are protandric sex changers. 2. *A. peggyae* is differentiated as male-like hermaphrodite initially and as male subsequently. 3. On the other hand, *A. imbecilis* is differentiated as female-like hermaphrodite initially and as female subsequently. For insemination, both gonochoric females and female-like hermaphrodites of *A. imbecilis* have to depend on the male-like

TABLE 3.5

Sexuality of individuals in American unionids bivalves of the genus *Anodonta*. Values in brackets indicate age of individual, as estimated by the number of annuli on the shell. Thick arrow lines suggest protandric sex change from male to female, thin line from male-like hermaphrodite to male and broken line from female-like hermaphrodite to female (compiled from Heard, 1975).

Species, Area	Examined (no.)	Female (no.)	Male (no.)	Hermaphrodite-like	
				Female (no.)	Male (no.)
Gonochorics					
A. californiensis	14	7 (4)	7 (4)	0	0
A. gibbosa	53	18 (15)	35 (13)	0	0
A. wahlamartensis	7	3 (10)	4 (9)	0	0
A. couperiana					
Myakka River	60	38 (7)	22 (7)	0	0
A. corpulenta	14	7	7	0	0
Gonochorics with female-like hermaphrodite					
A. imbecilis					
Grand Lake	41	17 (8)	0	24 (8)	0
Taiquin Lake	70	36 (13)	0	34 (9)	0
Gonochorics with male-like hermaphrodite					
A. couperiana					
Apalachicola River	16	10	0	5	1
A. hallenbeckii	31	20 (15)	10 (13)	0	1
A. peggyae					
Holmes Creek	102	47 (10)	45 (10)	0	10 (8)
Taiquin Lake	123	42 (14)	80 (10)	0	1 (6)

hermaphrodites alone. Hence, there may be intense competition among these gonochoric females and female-like hermaphrodites to secure insemination.

Examining resource allocation for sexuality of a dense population (28/m²) of *E. complanata* in a Canadian freshwater Lake, Downing et al. (1989) found that only about 20% of this bivalve was either pure females (7.3%) or pure males (11.7%). However, considering individuals with 90% ovarian and testicular tissues as females and males, it was found that 58% were males within 15–50 mm size range and 36% were females within 50–90 mm size range and the remaining 6% were hermaphrodites (Table 3.6). These values suggest that *E. complanata* is a protandric hermaphrodite. However, *E. complanata* population consists of pure females, pure males and hermaphrodites with different proportions of ovarian and testicular tissues at any given point of

TABLE 3.6

Frequency variations and combinations of sexuality in Lac de l'Aehigan population of *Elliptio complanata* (compiled from Downing et al., 1989)

Sexuality	Size Class (mm)	♀ Tissue (%)	Frequency (% of Population)
Male	15–50	0–10	58–80
♂-like hermaphrodite	–	10–40	1–3
Hermaphrodite	50–90	40–60	2–5
♀-like hermaphrodite	–	60–90	3–5
Female	50–90	90–100	36–60

time. Notably, many males and male-like hermaphrodites continued to be present in the largest size class and may not change sex during the life time (Downing et al., 1989).

Sequential hermaphrodites are further divided into (i) male to female sex changing protandrics and (ii) female to male sex changing protogynics. In fishes, monogynic (secondary females only) and digynic (with primary and secondary females) protandrics as well as monandric (secondary males only) and diandric (with primary and secondary males) are known (Pandian, 2010, p 98–103). However, digynics and diandrics are not reported to occur in Mollusca and Crustacea (Pandian, 2016, p 110) as well. The dwarf males of *O. puelchana* do not change sex to females (Castro and Lucas, 1987). Interestingly, protogyny is limited to 16% in crustaceans (Pandian, 2016, p 105) but is not thus far reported to occur in any mollusc. Conversely, its frequency is as high as 74% of hermaphroditic fishes (Pandian, 2012, p 105). The preponderance of protogynics in fishes but their limited presence in crustaceans and total absence in molluscs may be due to: 1. Ketotestosterone (KT) is shown to be responsible for the expression of aggressive behavior in large/dominant male fishes (Pandian, 2013, p 199–121). In its absence, males of these invertebrates may not be as aggressive as fishes are and 2. Sperm competition in these sedentary and relatively slow motile aquatic invertebrates may not also be as intense as in highly motile fishes.

The occurrence of protandrics is limited to marine gastropods and bivalves; however, 'protandrics' occur rarely in Marian hermaphroditic freshwater bivalves. The reason(s) for the rarity of protandrics in freshwater is not yet known. Protandrics are reported from prosobranchs (e.g. Archaegastropoda: *Patella vulgata*, Baxter, 1983, Mesogastropoda: *Crepidula fornicata*, Hyman, 1967, p 293), opisthobranchs (e.g. Pteropoda: *Limacina helicina*, Gannefors et al., 2005) and parasitans (e.g. Aglossa: *Balcis shaplandi*, Morton, 1979) as well as bivalves (e.g. oysters: *Ostrea edulis*, Walne, 1964, pearl oyster: *Pteria penguin*, Wassnig and Southgate, 2011, clams: tridacnids, Hardy and Hardy, 1969, deep sea: Bathymodiolinae, *Idas modiolaeformis*, Gaudron et al., 2012,

Pholadoidea: *Xylophaga supplicata*, Haga and Kase, 2013). Among 53 known wood-boring pholadid species belonging to the genus *Xylophaga*, 19 species were considered to brood 'juveniles'. Many deep sea pholadids, which were earlier considered to brood their 'juveniles', are now shown as protandrics and the 'juveniles' as protrandric males. From their study on the deep sea *X. supplicata*, Haga and Kase (2013) found that the pholadids are indeed protandrics. The 'brooded juveniles' are dwarf males (3.7 mm) that settled on the dorsal margin of the posterior slope of the shell. The mature male (5.7 mm) aligns its siphon turned toward the dorsal of the female host to facilitate fertilization adjacent to the siphon. As it grows to > 5 mm, it changes sex to female (8 mm).

Considering the economic importance of black pearl oyster *Pinctada margaritifera* as a representative, a brief account on our understanding of protandry is provided. In the French Polynesian Islands spread over from 14° 38′ S to 23° 0.6′ S, the pearl oyster is cultured in 800 farms covering an area of 10,000 ha at 30 atolls, providing gainful employment to 7,000 people; from these farms, an export of 10 tonnes of pearls fetches euros 87 million (Chavez-Villalba et al., 2011) and thereby has brought an economic revolution in these Islands. The oyster grows to > 150 mm (shell height) lives for > 12 years and commences sex change from ~ 80 mm size at the age of 3+ (Fig. 3.7A) and female ratio progressively increases to 41% at the age of 10+. In all protandrics including the pearl oyster, the first sex differentiation is fixed and determinate, i.e. the young ones can differentiate strictly into males only and the process is not affected by environmental factors. For example, no biotic and abiotic factor is known to influence sex differentiation into male

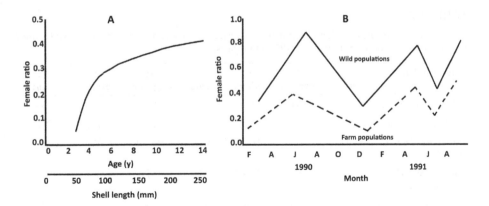

FIGURE 3.7

A. Female ratio as functions of age and size in natural and farm populations of black pearl oyster *Pinctada margaritifera* in French Polynesian Islands. B. Female ratio in wild and farm populations of the pearl oyster in French Polynesian Islands (redrawn from Chavez-Villalba et al., 2011).

during the first two years of life in *P. margaritifera* (Chavez-Villalba et al., 2011). The only exception to this are the slipper limpets *Crepidula fornicata, C. onyx* and *C. plana* (see Table 3.7). The second differentiation process namely sex change to female is labile, and protracted over a period, and is influenced by factors like temperature, food availability and so on. Grafting involves cleaning and implantation of a spherical (4 mm²) bead of shell and a piece of mantle, the nacre. Since *P. margaritifera* female is more sensitive to stress like cleaning and nacre implantation, males predominate in farms; in fact, some stocks contain as few as 5% females. Figure 3.7B shows that in comparison to natural populations, female ratio is lower in farm population throughout the years 1990–1991. Rearing at 8°C for 1 year resulted in the production of more *Crassostrea gigas* females. In tropical Cortez oyster *C. cortenziensis* too, there were more males at 18°C than for *C. gigas* at 9°C; apparently, low temperatures promote the production of more females. Not surprisingly, the female ratio of *P. margaritifera* increases from 0.10 at 14° 38′ S to 0.30 at 23°0.6′ S within the Polynesian Islands (Chavez-Villalba et al., 2011). Similarly, availability of food is reported to alter sex ratio. The ratio of *Mytella charruana*, when starved for 1 month, shifted toward male biased ratio (Stenyakina et al., 2010), suggesting that the females need to acquire more food energy to develop.

Understandably, male ratio is greater in smaller and younger protandrics and decreases with increasing size and age (Fig. 3.8A). The following are interesting protandrics. For reasons not known, females are more abundant than males in all size classes of *M. charruana* (Stenyakina et al., 2010). In reef-forming protandric *Vermicularia spirata*, males are motile but on settling, the

TABLE 3.7

Reproductive patterns of sedentary gastropod *Crepidula* and wood-boring bivalves *Bankia* and *Psiloteredo* (compiled from Hoagland, 1978 and others)

Characteristics	*C. adunca*	*C. convexa*	*C. plana*	*B. gouldi*	*P. megotara*	*C. fornicata*	*C. onyx*
Planktonic larvae	No	No	Yes	Yes	Yes	Yes	Yes
Substrate limited	Yes	Yes	Yes	Yes	Yes	No	No
Juveniles attracted by adults	No	No	Weak	Wood	Wood	Yes	Yes
Sex of isolated young ones	♂	♂	♂ or ♀	♂	♂	♂ or ♀	♂ or ♀
♀ dwarfed at high density	Yes	Yes	Yes	Yes	Yes	No	No
♀ delayed sex change of ♂	No	No	Yes	–	–	Yes	Yes
Density-dependent sex ratio	No	No	–	–	–	Yes	Yes

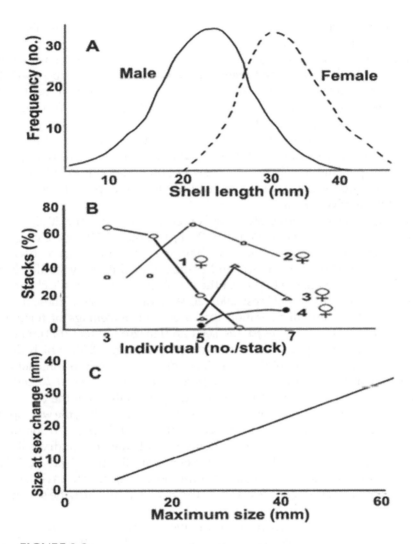

FIGURE 3.8

A. Frequency distribution of males and females as a function of shell length in protandric *Crepidula fornicata* B. Proportions of stacks harboring 1 to 4 females/ stack as a function of stacking density (redrawn from Collin, 1995), C. Calculated size at sex change as a function of maximum attainable shell size of 27 populations belonging to19 species of *Crepidula* (redrawn from Collin, 2006).

females become sessile (Bieler and Hadfield, 1990). In *C. fornicata*, younger slipper limpets that settle on larger female differentiate into males but solitary limpets into females (Hoagland, 1978). This observation clearly indicates that (i) Primordial Germ Cells (PGCs) of the limpet and all other protandrics retain bisexual potency and (ii) the young limpets are sensitive to an unidentified chemical arising from female limpet that inhibits female differentiation in newly settling limpets.

Details of anatomical changes in sex changing *Crepidula* have been described. The relatively large penis and regenerative male follicles are absorbed. The small oocytes (Oogonial Stem Cells, OSCs) that have been present in the gonad, begin to enlarge into oocyte (Hyman, 1967, p 295). With regard to source of OSCs, there are differences. In the parasitic aglossan gastropod *Mucronatia fulvescens*, all other male-related organs are digested excepting the seminal groove, from which the pallial oviduct emerges (Morton, 1979). In another parasite *Stylifer linckiae,* Lutzen (1972) has suggested the origin of female reproductive system anew and not from any part of degenerating male reproductive system.

Among aquatic invertebrates (e.g. crustaceans, Pandian, 2016, section 3.5.2) and teleostean fishes (Pandian, 2010, Chapter 7–10), sex change is a widespread phenomenon. The sex changing sequentials have been the subject of a number of theoretical and empirical evolutionary studies. Many attempts have been made to predict the size/age, at which sex is changed at the sequentials (Allsop and West, 2003, 2004). Notably, some of these theories have not taken into consideration that a certain percentage of females in protogynic fishes (e.g. *Pagrus pagrus,* see Pandian, 2011, p 166–167) and males in protandric gastropods like *Patella vulgata* (Baxter, 1983), which do not change sex until the end of their life, has not been taken into consideration. As teleostean fishes undergo reproductive senescence between 70 and 80% of their life span, Pandian (2012, section 2.13) estimated that the protogynic fishes change sex, when they attain ~ 75% of life span. In fishes, the widespread occurrence of confirmed protogyny occurs in about 140 species belonging to 50 genera and 15 families but the incidence of confirmed protandry is limited to 25 species belonging to 17 genera and eight families (see Pandian, 2011, p 167). However, protandry being more common among aquatic invertebrates, molluscs offer a better scope to estimate the size/age, at which the sex change occurs in protrandrics.

Sex change occurs, when reproductive output of the changed sex increases faster with increasing size/age than it does with the same sex. To maximize life time reproductive output, an individual begins life as the sex characterized by least increase in reproductive output with size/age and subsequently changes to the other sex. Protandry occurs, when female reproductive output increases with size/age, while the male's ability to fertilize egg is not dependent on size. Because males invest a little on their offspring relative to females, the parental investment theory predicts that males grow small (Trivers, 1972). The standard explanation for protandry resulting in macrogyny is that large females produce large broods/eggs. In microandric male, gamete production is not constrained by size. The size advantage model has successfully explained this adaptive significance of large female size (Ghiselin, 1969). In species, in which males compete for mates, territories and other ecological resources to attract females, the contest and/or competition favors large males (Anderson, 1994). In many benthic fishes, males build nest and guard the embryos/young ones (Pandian, 2011, p 13–15, 2013, p 121). In majority of crustaceans, the females carry their embryos on one or other parts of their body. No male mollusc is yet known to guard/brood eggs/embryos sired by him. There is no report from molluscs on the presence of large, dominant males with nest and ecological resources to attract females. Not surprisingly, the need for protogynic sequentials in these aquatic crustaceans and molluscs is minimal or nil.

In gregarious (stacking) protandrics like *C. fornicata*, 50% of males changed sex anytime, when they attained a size from 20 to ~ 28 mm shell length (Fig. 3.8A). Hence, there were considerable variations in the size at sex change. Notably, the number of females increased from one female per stack with three-five individuals to four females per stack with five–six individuals (Fig. 3.8B). Besides, there were wide variations in sex ratio within a stack. In stacks with five individuals, some held one male and four females but others four males and one female. Surprisingly, the mean sex ratio in these stacks remained constant for stacks of different densities. Interestingly, a similar stacking situation occurs in wood-boring protandric bivalves. Considering selected reproductive characteristics of *Crepidula* species and wood-boring bivalves, Hoagland (1978) noted that the responses ranged from one to other end of a spectrum. Briefly, the responses of gregarious *C. fornicata* and *C. onyx* were on one end but *C. adunca* and *C. convexa* were on the opposite end (Table 3.7).

Wisely using data on size, sex and gregariousness (stacking) of 27 populations belonging to 19 species of protandric calyptraeid gastropods, Collin (2006) has chosen to predict the size at sex change in these protandrics. Examination of these data to assess variations in size at sex change were related to (i) skewed male ratio during the male phase, (ii) ratio of size at sex change in relation to maximum attainable size and (iii) variable levels of stacking and its relation to size at sex change. He found that across 19 species, the sex ratio was not significantly skewed towards male ratio. Secondly, the species that form largest stacks or mating groups displayed more variations in size at sex change within a population rather than species that were generally solitary. Variations in population density seem to induce different responses in sex changing solitary individuals of both *Crepidula* species and wood-boring bivalves (Table 3.7). Hence, size/age at sex change is more a population trait rather than a species trait. With this limitation, Collin found that the size at sex change occurs in selected populations of calypteraeid gastropods around 50% of its maximum attainable size (Fig. 3.8C). Remarkably, the protogynic teleostean fishes change sex anytime between 67% (Allosop and West, 2003, 2004) and 75% (Pandian, 2012, p 85) of their life span but the protandric molluscs do it around 50% of the maximum attainable body size.

Serial hermaphrodites: In molluscs, sex changing hermaphrodites include sequentials and serials. Sequentials undergo natural sex change only once in one direction but serials do it more than once in either direction. The presence of rhythmic but preferably called serial hermaphroditism is reported from a few bivalves. The simple serial occurs in wood-boring *Teredo navalis* (Coe, 1941). Four to six weeks following settlement, the teredo matures and functions as male. It changes sex to female at the age of 6 months. A second sequence of male and female phases may continue during favorable conditions (Fig. 3.5). *Ostrea edulis* breeds as male during the first part of the season and as female during the second part of the season (see Walne, 1964). Hence, *O. edulis* may also be a serial hermaphrodite. Oyster groups *Crassostrea angulata*, *C. gigas*, *C. virginica* and *O. cuculatta* are gonochores but they retain bisexual potency holding Primordial Germ Cells of both sexes in the gonad. In a 5-year study, Galtsoff (1964) found an individual, which changed sex four times. In *O. puelchana*, the serial is complicated with the presence of additional dwarf males, which do not change sex (Fig. 3.5). At the age 1+, the oyster functions as a typical male. Subsequently, it changes sex to female at the age of 2+. Thereafter, the sex change is repeated from female to male and to female with advancing age. In this process of rhythmic sex change, the female phase is progressively prolonged with increasing age (Castro and Lucas, 1987). Field experiments suggest that the bearer female inhibits the somatic growth of epibiont dwarf males (Pascual et al., 1989).

The dwarf male of the oyster *O. puelchana* settles on the bearer oyster, grows to 20–26 mm shell height, functions as male. On detachment from the bearer, it dies at age 1+. In a few sedentary bivalves, the dwarf male phenomenon has been described (Table 3.8). Depending on the bearer's location, on which the dwarf male larvae settle, their size is reduced and one or more organs and systems are lost with simultaneous enlargement of testis. For example, *Zachsia zenkewitschi* settled in the pallial cavity have lost maximal somatic organs but with the huge enlargement of testis.

3.5 Mates and Mating Systems

Molluscs display all the three recognized polygamic mating systems. Monogamy is defined one female exclusively for one male. A computer search with 'monogamy in molluscs' as keywords revealed its total absence. With the presence of > 25% hermaphrodites, the terms of monogamy, polygyny and polyandry may have to be redefined to suit Simultaneous Hermaphroditic (SH) molluscs. Accordingly, monogamy occurs not between male and female individuals but in a single individual of an obligatorily selfing SH clam *Lasaea subviridis*. In this clam, only a few gametes are simultaneously produced. Fully matured sperm and oocytes have never been observed outside the gonad (Beauchamp, 1986). In parasitic prosobranchs, a male is miniaturized to a 'sacculina' male (Lutzen, 1968). In *Thyomicola*, a single dwarf male is

TABLE 3.8

Examples for loss of organs and enlargement of testis in dwarf males of some serial hermaphroditic bivalves

Species, Reference	Reported Observations
Males attached to left valve	
Ostrea puelchana Calvo and Morriconi (1978)	Juvenile sized to males; completely developed testis
Males housed in mantle cavity	
Montacuta floridana, Deroux (1960)	Shelled; large foot
Entovalva sp, Jenner and McCrary (1968)	Shelled; large foot
Pseudopythina subsinuta, Morton (1972)	Shelled; large foot
P. rugifera, O'Foighil (1985)	Large foot; no demibranchs
M. percompressa, Jenner and McCrary (1968)	No shell; massive gonad
Males housed in pallial cavity	
Ephippondonta oedipus, Morton (1976)	Small shell; foot with byssus gland but large testis
Zachsia zenkewitschi, Yakorlev and Malakhov (1985)	Incomplete shell; reduced circulatory system, adductor muscle, labial palps, foot, pedal ganglia; larval-like digestive system; simple but incompletely developed gills and inhalant siphon; absence of excretory organs; extensively large gonad
Males housed in branchial cavity	
Chlamydoconcha orcutti, Morton (1981)	Well developed visceral mass contains large foot with byssus gland and testis. Shell partly enveloped by reduced mantle; absence of labial palps inhalant and exhalant siphons, ctenidia and kidney

harbored within this pseudopallial cavity of a female (Fig. 3.3). Hence, *Thyomicola* may be considered monogamous. However, polygamy, as per the definition as well as modified definition to suit hermaphrodites, occurs more commonly in both gonochoric and hermaphroditic molluscs. For example, with multiple copulations, i.e. polygyny is reported in many gonochoric gastropods. The neogastropod whelk *Buccinum undatum* displays a high degree of polygamy. Of a dozen whelks observed, seven females copulated more than once, three at least three times and two four times. In this whelk, the males as well as females copulated many times. In an extreme case, simultaneous copulation by a male with two females of *Neptunea antiqua* and *Eupleura caudata* is reported (see Martel and Larrivee, 1986).

In *Bulinus contortus*, the euphallic (possessing male organs) SH can self or get inseminated by more than one male-acting SH (polyandry) and also inseminate more than one female-acting (polygyny) SH. Hence, the euphallic

SH are polygynic at a point of time and polyandric at another point of time. They may be considered as polygynandrics. However, aphallics, lacking male organs, can self or get inseminated but cannot inseminate female-acting SH. These aphallics may be considered as polyandrics. Another factor, which may also alter sex ratio, is the androdioecious mating system consisting of hermaphrodite and male. Androdioecy is reported from a few bivalves; but data on the sex ratio is not yet available (e.g. Castro and Lucas, 1987). In molluscs, the most prevalent mating system is polygamy, especially polygynandry. SH being unique to molluscs, their polygynandric mating system is highlighted here.

Through a series of simple but meaningful experiments, Vianey-Liaud (Vianey-Liaud, 1976, 1989, 1995, Jarne et al., 1993, Vianey-Liaud and Dussart, 2002, Vianey-Liaud et al., 1987, 1989, 1991, 1996) made significant contribution to our understanding of mates as selfers and unilaterals, and the consequences on cumulative fecundity and Mendelian inheritance in simultaneous hermaphroditic (SH) planorbid *Biomphalaria glabrata*—an intermediate host of the blood fluke *Schistosoma*. Incidentally, the freshwater *B. glabtrata* may also serve as a representative of marine SH opisthobranchs and terrestrial pulmonates. It is a unilateral SH and is capable of self and cross fertilization. In *Bulinus*, a single insemination provides adequate sperm to fertilize 1,000–2,000 eggs (Wethington and Dillon, 1991). When mated more than once, an individual may transmit allosperm received from an earlier male-acting SH (Vianey-Liaud et al., 1987). With increasing autosperm storage, *Physa fontinalis* SH initiates male-acting behavior but with high allosperm reserve, it rejects a male-acting SH. In *P. fontinalis,* reciprocal and multiple inseminations occur and sperm from many can be stored for 50 days. The shell color of *B. glabtrata* may be either homogenous recessive cc albino (Alb) or heterozygous (Cc) dominant pigmented (Pig) or homozygous (CC) dominant Pig. The pigmentation is a genetic marker and transmitted as a single Mendelian trait. Both Alb and Pig snails self within themselves or unilaterally outcross inseminating multiple number of times. Successive selfings may lead to reduced genetic diversity and inbreeding depression. Following outcrossing Alb preferentially use (donor) allosperm from Pig partner. When allosperm are exhausted, selfing is resumed in Alb. However, multiple inseminations usually occur in most SH. For example, the same male-acting SH inseminates 94% of the female-acting *B. globosa* daily. When inseminated by two male-acting SH, *P. heterostropha* uses the two batches of allosperm. Inseminated by two 'fathers', an isolated *B. africanus* faithfully sires offspring of the two 'fathers' (see Vianey-Liaud et al., 1996).

In *B. glabrata*, the Pig snail grew faster but both Pig and Alb attained sexual maturity at the same age. Cumulative fecundity, as measured in units of clutch, egg, egg/clutch, viable embryos and hatchability, was significantly higher for Alb (Table 3.9). When virgin dominant CC homozygous Pig and recessive cc homozygous were paired for different durations for 1–20 days, Alb sired Pig offspring. The use of (donor) allosperm lasted for 50 days after isolation of the pair (Vianey-Liaud et al., 1989). From a series of 20 combinations, in which the number of potential Alb + Pig increased from 19 Pig + 1 Alb to 1 Pig + 19 Alb, Vianey-Liaud (1989) found that the increase of potential Alb from 1 to 19 reduced to the number of female using allosperm, as indicated by increased number of Alb progeny from 20 to 46% and decreased Pig offspring from 80 to 44% (Fig. 3.9A).

To characterize the preferential use of allosperm, Vianey-Liaud (1995) carried a series of five experiments by introducing (i) different durations of delay between the first and second pairing as well as (ii) different male-acting SH to pair with Alb in successive matings. An Alb was first paired

TABLE 3.9

Fecundity and hatchability of offspring sired by albino and pigmented 'fathers' of *Biomphalaria glabrata* (compiled from Vianey-Liaud, 1989)

Parameter	Pigmented	Albino
Clutch (no./ ♀ /d)	0.38	0.52
Cumulative fecundity (no.)	25,603	32,200
Eggs (no./clutch)	20.20	21.41
Sterile eggs (no./ ♀ /d)	5.95	9.03
Hatchability (%)	77.9	81.6

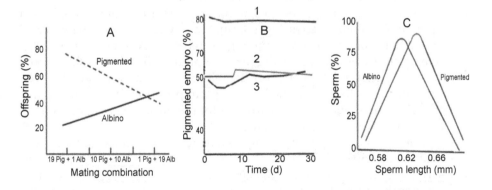

FIGURE 3.9

A. Proportions of albino and pigmented offspring with increasing number of potential albinos and decreasing number of pigmented strains of *Biomphalaria glabrata* (redrawn from Vianey-Liaud et al., 1987). B1. Excess production of pigmented embryos in selfing heterozygous *B. glabrata*. B2. Excess production of pigmented embryos in heterozygous *B. glabrata* after pairing with homozygous albino. B3. Excess production of pigmented embryos after pairing with pigmented heterozygous *B. glabrata*. C. Proportions of sperm as function of sperm length in homozygous albino and pigmented *B. glabrata* (redrawn from Vianey-Liaud et al., 1996).

with Pig on the first day and then by Alb on the second day. The second experiment was the reverse of it. Alb preferentially used Pig allosperm, as indicated by the presence of 85 and 98% of egg masses containing Pig embryos. Interestingly, the delay of 1–10 days between the first and second mating did not alter the 98% Alb preferentially and selectively using Pig allosperm and siring. In the fourth and fifth series, Alb was sequentially inseminated by a Pig partner on the first day (first group) and then by a series of Alb_1 (second group), Alb_2 and Alb_3 (fourth group) on the second, third and fourth day, respectively. Virtually, Vianey-Liaud created an increasing sperm competition between allosperm of Pig and Alb at the ratio of 0.5 Pig : 0.5 Alb in the second group and 0.25 Pig : 0.75 Alb in the fourth group. The sired Pig offspring progressively decreased from 100% in the first group to 0% in the fourth group. Further, when given an option between autosperm and allosperm within selfing albino or pigmented snails, allosperm were always selected by these snails. Clearly, the allosperm were preferentially selected to fertilize the eggs, irrespective whether they are from Pig or Alb 'father'.

To verify whether or not the Pig trait is transmitted strictly following Mendelian inheritance, Vianey-Liaud et al. (1996) accomplished another series of experiments. Exclusive selfing in Cc heterozygous snail is expected to produce 75% Pig and 25% homozygous cc recessive Alb progeny. However, selfing always produced 80–83% Pig offspring (Fig. 3.9B), i.e. significantly (10–13%) more Pig progeny. Similarly, mating between Cc Pig x cc Alb in alternative combinations also sired 2–4% but significantly more Pig progeny, where the expected ratio was 50% Pig and 50% Alb offspring. The cause for the excess Pig offspring production in these matings was studied by Vianey-Liaud et al., Classical DAPI staining of spermatozoa revealed the rare presence eupyrene and apyrene spermatozoa. More interestingly, the length of eupyrene CC homozygous Pig spermatozoa was always longer (0.625 mm) than that (0.617 mm) of cc homozygous recessive Alb sperm (Fig. 3.9C). Expectedly, the Cc heterozygous Pig SH may produce 50% longer C and 50% short c sperm. A correlation between flagellum length and its speed is well known. Hence it is likely that the longer sperm produced by homozygous/ heterozygous Pig snail may be responsible for the presence of Pig progeny in more than the expected Mendelian ratio.

Unaware of the importance of Oogonial Stem Cells (OSCs) and Spermatogonial Stem Cells (SSCs) in gametogenesis, Vianey-Liaud and Dussart (2002) gonodectamized SH *B. glabrata* by gently breaking open central whorls of the shell on the left hand side and thereby exposing the gonad. After ligating the hermaphroditic duct, the ovotestis was surgically removed and the shell opening was sealed with dental cement. The snail easily withstood the surgery. The pairing of the gonodectamized Pig with an intact Alb produced Alb progeny alone. Had there been unilateral insemination either from the intact Alb to the gonodectamized Pig or from the gonodectamized Pig to the intact Alb, Pig progeny could have been generated. The fact that no Pig progeny was generated revealed that the gonodectamized Pig did neither produce oocytes nor sperm. Briefly, the gonodectamized Pig became sterile, as it has lost the entire basket of OSCs and SSCs along with the surgically removed ovotestis. Hence, Vianey-Liaud and Dussart have brought for the first time an evidence for the presence of OSCs and SSCs in ovotestis alone and not in any other organ of the snail. As the snail is amenable to surgical removal of the entire ovotestis, it must also be possible to surgically separately ablate ovarian or testicular tissues alone. Alternatively, administration of busulfan-supplemented algal diet may chemically eliminate SSCs alone (see Pandian, 2011, p 150–151). When ovariectomized or castrated Pig SH is paired with Alb SH, the proposed assumption shall also be confirmed.

3.6 Alternative Mating System (AMS)

In polygynous mating system, competition for male becomes more and more intense. Males of lower competitive ability may adopt AMS to make use of bad situation (Taborsky, 2001) and pave ways for the evolution of two or more mating morphs (Gadgil, 1972). "However, the struggle between the male morphs for possession of females....results not in death of unsuccessful competitors but in a few or no offspring" (Darwin, 1859). Fishes (Oliveira, 2006) and crustaceans (Pandian, 2016, p 118–120) displaying AMS are classified into three groups:

1. *Plastic reversibles*, which switch reversibly in either direction back and forth from one morph to another during the life time. Among cichlids, *Astatotilapia burtoni* male, for example, reversibly acts as 'submissive' or 'dominant' morph. Similarly, *Neolamprologus pulcher* male reversibly assumes 'territorial' or 'helper' morph, when he holds a territory or does not hold it. Among molluscs, the nearest seems to be the unilateral and facultative Simultaneous Hermophrodites (SH), which change from unilateral to reciprocal. Yet they may not qualify plastic reversible, as the change crosses the boundary of sex.

2. *Plastic transformants* irreversibly switch from one morph to another in a single direction during their life time. For example, *Macrobrachium rosenbergii* changes from pink to orange and then to blue morph in a single direction during its life time (Pandian, 2016, p 118–120). The nearest to the plastic transformants are a couple of squids: *Dorytheuthis plei* (Marian, 2012) and *Loligo bleekeri* (Iwata et al., 2015). In *L. bleekeri*, males occur in two morphs namely consort and sneaker. Large sized consort guards the selected female partner, mate in parallel position and produce larger spermatophores containing more but smaller sperm (Fig. 3.10, left panel), which encounter the eggs first at the oviduct (Table 3.10). Conversely, the smaller sneakers produce smaller spermatophores containing less but larger sperm, which

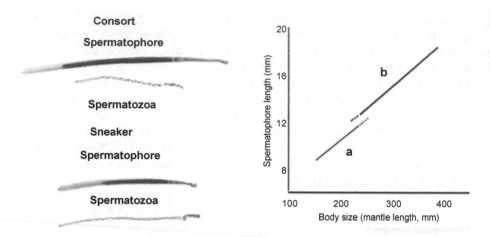

FIGURE 3.10

Loligo bleekeri: left panel shows the spermatophores (~ 25 mm) and spermatozoa (~ 35 μm) of consort male and sneaker male (18 mm, 45 μm) (Redrawn from Iwata et al., 2011). Right panel shows the relation between spermatophores length and body size of *L. bleekeri*. a and b represent sneaker and consort males, respectively (Redrawn from Iwata et al., 2015).

TABLE 3.10

Characteristics of spermatophore and sperm of consort and sneaker males of *Loligo bleekeri* (compiled from Iwata et al., 2011, 2015)

Characteristics	Consort	Sneaker
Spermatophores (mm)	~ 10	~ 6
Sperm (no./spermataphore)	14.6×10^{11}	3.1×10^{10}
Spermatozoa size		
a) Head length (μm)	7.50	8.47
b) Flagellum length (μm)	64.3	90.5
c) Total length (μm)	71.8	99.0
Velocity (μm/s)	167.7	167.1
Fertilization site	Internal	External

encounter the eggs at the buccal membrane around the seminal receptacle. The sneaker mates in head to head position. Following ejection, the clustered sperm of sneaker (by CO_2 chemotaxis, Hirohashi et al., 2013, Hirohashi and Iwata, 2013) move *en masse* to achieve a greater share in fertilizing eggs. However, the egg–sperm encounter occurs at different times and different sites, providing more opportunities to the consort to achieve a greater share in fertilization. For example, of 266 females sampled, 159 spermatangia were attached inside the oviduct but 56 (21%) had them attached on the buccal membrane around the seminal receptacle (Iwata et al., 2015). Consequently, the share of sneaker in fertilization success is 13% only, in comparison to 87% for the consort (Iwata et al., 2005). Through a series of their publications, Iwata et al. have not traced the plastic transformance in *L. bleekeri*. However, it is possible to infer from two of their observations that *L. bleekeri* is a plastic transformant: 1. The allometric relationship between body size and spermatophore length of *L. bleekeri* displayed a clear switch from sneaker to consort at 236 mm size (Fig. 3.10, right panel). 2. Iwata et al. (2005) reported that a C-M2 male (217 mm size) mated in parallel position and subsequently in head to head position. Obviously, this male is a transient from sneaker to consort morph, as it could mate in either position. Incidentally, a deviation is to be made to narrate the transfer of spermatophores in some squids. There are two heterocotyli to transfer spermatophores in *Idiosepius paradoxus* (Sato et al., 2013). Another squid *Todarodes pacificus* uses only one mating behavior to transfer spermatophores into the buccal membrane of females (Takahama et al., 1991). In the deep sea squids (e.g. *Meroteuthis ingens*, Hoving and Laptikhovsky, 2007), spermatophores are either transferred by the male's hypocotlylus or by a long penis, which is present in species that lack hypocotlylus (Nesis, 1983). These spermatophores have an autonomous mechanism that enables to penetrate the skin and subsequently into the body tissue. For more details, Equardo and Marian (2002) may be consulted.

3. *The fixed alternative sex-linked morph* is a specific irreversible strategy for life time. For example, the morphotypes such as hooknose and jack among salmonids have genetic basics. The jack grows faster, precociously matures and parasitizes hooknose by sneak mating. Hybridization between jack and hooknose produces both jacks and hooknoses but the higher proportions are determined by the male morph (see Pandian, 2013, p 122). No mollusc is yet reported to display alternative sex linked morph. More research is required in this area.

3.7 Multiple Paternity

In molluscs, one or a few matings are adequate for a female to fertilize all the eggs and maximize reproductive output during a reproductive season. For example, a single mating provides enough sperm to fertilize 1,000–4,400 eggs of the snail *Bulinus globosus* (Rudoph, 1983). Hence females are expected to be 'monogamous'. A classic paradigm in evolutionary biology is that females benefits little from mating with more than one male. Contrary to this expectation, polyandry is a ubiquitous phenomenon in animals, especially molluscs with > 25% Simultaneous Hermaphrodites (SH). Sexual selection can also occur in hermaphrodites. Unlike gonochorics, a hermaphrodite can adjust resource allocation to mating in female and male roles (Angeloni et al., 2003). Though mating can be costly in terms of time (Vianey-Liaud et al., 1989, Hanlon and Messenger, 1996) and energy (Franklin et al., 2012), multiple mating leading to multiple paternity and polyandry seem to be a significant adaptive reproductive strategy. In fact, multiple mating with more than one male facilitate 'trade ups' to a more desirable partner and/or 'bet-hedge' to avoid committing the entire resource with one partner (Makinen et al., 2007). The evaluation of female promiscuity and consequent multiple paternity is promoted by the direct benefits of nuptial gifts, fertilization assurance as well as indirect benefits like potential sperm competition to select high quality sperm and genetic diversity among siblings (Xue et al., 2014).

In this context, it is relevant to note an experimental study on the effects of single and multiple mating by the same or two males, each only once on the reproductive characteristics of the dumpling squid *Euprymna tasmanica* (Squires et al., 2012). A female was mated by the same male once and twice as well as another female was mated by two males, each only once (Table 3.11). As an evidence for reproductive assurance, the female mated twice by the same male produced the highest number of eggs and hatchling. Despite its smallest egg size of 0.28 mg, its hatching success was the highest (65%) and size at hatching (0.14 mg) was equal to others. Incidentally, the interpretation of Squires et al. does not comply with their data.

During recent years, parentage studies and family construction have become increasingly important to have a better insight into mating behavior and reproductive dynamics. Due to obvious difficulties in observing mating behavior in the wild, parentage determination in natural population is difficult. With advent of molecular techniques and availability of reliable softwares for statistical analysis, it has become possible to determine paternity in a brood, often consisting of eggs (embedded in capsules, which are interwoven into a long string) spawned in the field and/or laboratory. Indeed, paternity analysis has greatly facilitated genetic investigation of mating systems and evaluation of each male's ability to achieve fertilization success (Iwata et al., 2005).

During the last 10 years, publications on mating system, paternity analyses and paternity of many representative molluscs have become available (Tables 3.12, 3.13, 3.14). Fortunately, these publications provide adequate representation to (i) major taxonomic groups: seven for prosobranchs (*Littorina saxatilis, Li. obtusata, Rapana venosa, Busycon carica, Neptunea arthritica, Crepidula fornicata, C. coquimbensis*), one each for opisthobranch (*Aplysia californica*) and pulmonate (*Bulinus globosus*), two for bivalves (*Ostrea edulis, Hyriopsis cumingii*) and three for cephalopods (*Octopus vulgaris, Loligo pealei, Lo. bleekeri*), (ii) sexuality ranging from gonochores (all the tested prosobranchs, cephalopods and *H. cumingii*) to protandrics (*C. fornicata, C. coquimbensis, O. edulis*), unilateral SH (*A. californica*) and to facultative SH (*B. globosus*), (iii) from freshwater (*H. cumingii, B. globosus*) to marine habitats and (iv) from no parental care (e.g. *R. venosa*) to initial brooding and subsequent

TABLE 3.11

Reproductive success of females *Euprymna tasmanica* mated by a male once or twice as well as by two males once each (compiled from Squires et al., 2012, some calculated* data are also added)

Parameter	Single Male Mated		Two Males Mated Once Each
	once	twice	
Clutch (no.)	5.2	4.7	4.8
Egg (no./clutch)*	38.4	45.5	37.9
Cumulative fecundity (no.)	200	214	182
Cumulative fecundity (mg)	80	60	60
Inter-clutch interval (d)	12	10	8.4
Incubation period (d)	39	40	36
Hatching success (%)	63	65	62
Hatchlings (no.)*	113	139	113
Egg weight (mg/egg)*	0.4	0.28	0.33
Hatching size (mg)	0.14	0.14	0.15

release of free living planktonic larvae (e.g. *C. fornicata*) but subsequent parasitic glochidia larvae (e.g. *H. cumingii*) as well as brooding up to juvenile stage (e.g. *C. coquimbensis*) and to ovoviviparity. Interestingly, the males include (a) mate guarders (e.g. *Li. obtusata*) (b) males that displace/remove the sperm of preceding males (c) sedentary (e.g. *C. fornicata*) and (d) alternative mating morphs (consort and sneaker males). Notably, fertilization is internal in all these molluscs, except perhaps in *H. cumingii*. Publications on paternity of molluscs characterized external fertilization and planktotrophic larvae are wanting.

To analyze and determine paternity, the low exclusion powered enzymatic loci were initially used (Gaffney and McGee, 1992). With availability of molecular markers, the microsatellites afforded considerable improvement over allozyme markers. Notably, new microsatellite markers (Emery et al., 2000) and improved accuracy over the presently used markers (Charrier et al., 2013) are being continuously developed. In molluscs like *Lo. pealei*, the presence of > 4 alleles, i.e. two maternal + two paternal revealed that the progeny were sired by > two males (Buresch et al., 2001). However, with extensive sharing of alleles from sires and dams, the confirmation of paternity with 80% confidence level required statistical analyses by one or other software system. GENEPOP v 3.4, CERVUS 2.0, GERUD 1.0 and 2.0 and COLONY 1.0 and 2.0 developed Raymond and Rousset (1995), Marshall et al. (1998). Jones (2005), Jones et al. (2002) and Jones and Wang (2010), respectively were used to determine the paternity of some molluscs (Tables 3.12, 3.13, 3.14). With advancing time, the confidence level of paternity determination increased from ~ 80% with the use of CERVUS 2.0 in

TABLE 3.12

Comparative account on multiple paternity in gonochoric prosobranchs

Species, Characteristics & Reference	Experimental Procedures	Reported Observations
Littorina saxatilis, Large thick (E) & small thin (S) shelled morphs from 2 locations, Ovoviviparous, Makinen et al. (2007)	Field collected 4 S + 4 E females, 214 S and 168 E juveniles genotyped. 5 microsatellite loci (L sub 62, 32, 8, 6 & 23) each with 5 alleles detected paternity using allelic count, GERUD 1.0 & COLONY software systems	Frequency of multiple paternity was 100%. Number of fathers were 3–5, 3–8 and 4–10, as determined by allelic count, GERUD 1.0 & COLONY 1.0. Mean paternity was 7.6 sires/female
L. obtusata, sperm stored in bursa initially and subsequently in seminal receptacle. Sperm alive for > 3 mo, Mate guarding males, Paterson et al. (2001)	Females isolated from 4–5 ♂ in lab. Paternity analysed by 3 microsatellites at loci L sub 8, 32 & 62 each with alleles 8 to 14. It was assessed using CERVUS at 80% confidence level.	Non-sampled males contributed 22.5% progeny/brood suggesting competitive ability of long stored sperm of preceding males. Absolute minimum paternity was 4–6 sires/brood
Rapana venosa. Females mate and spawn many times to produce 184–410 strings with 975 eggs/female. No parental care, Xue et al. (2014)	Egg strings (19) were raised in lab. 2–6 capsules were randomly selected. Variable polymorphic alleles displaying 11–32 alleles/locus and COLONY 2.0 software were used	Frequency of multiple paternity was 89.5%. Paternity ranged from 1 to 7 and averaged to 4.3 sires/brood
Busycon carica, no parental care, 36 mm string laid for several days. Each string consists of 89 capsules (50 embryos/capsule). Scope for female to selectively fertilize eggs Walker et al. (2007)	Females copulating with 2–9 males, each inseminating equal amount of sperm. Field collected 12 strings were analyzed by 3 microsatellites (2 autosomal + 1 sex-linked marker) using GERUD software system	Each sire was fathered by constant proportion of embryos. No observed mating males sired the analyzed embryos in 2 strings. Multiple paternity ranged from 2 to 7 and averaged to 3.5 sires/capsule
Avise et al. (2004)	Embryos displayed alleles 285, 300, 309 and 312 at bc 2.2	At least 4 fathers sired a brood
Pomacea canaliculata (Yusa, 2004a)	Female mated twice/oviposition	> 2 sires/brood
Neptunea arthritica, Female with 'conduit-type' bursa. Post-copulatory guarding. Males selectively inseminate the already copulated females up to 3 times, Lombardo et al. (2012)	Mating trails in lab. A male selectively copulated same female thrice. On insemination, sperm of preceding male oozed out. With five microsatellites, paternity confirmed by COLONY 3.7 software	In 13 analyzed clutches, single, dual and triple paternity occurred in 7, 3 and 3 clutches. Due to removal of sperm of preceding male, paternity averaged to 1.7 sire/clutch

TABLE 3.13

Comparative account on multiple paternity in gonochoric bivalve and cephalopods

Species, Characteristics & Reference	Experimental Procedures	Reported Observations
	Gonochoric bivalve	
Hyriopsis cumingii, Freshwater pearl mussel. First larval stage is brooded. Subsequent glochidia larvae are parasites on gills of fish, Bai et al. (2011)	Gravid mussels (20) brought to lab. Each of the 5 catfish *Pelteobagrus fulvidraco* were used to host brown glochidia from a single mussel. Five microsatellites each with 16.8 alleles & COLONY were used	Ten of 15 (66.7%) broods were sired by > one male. In 10 broods, the sire number ranged from 2 to 4 and averaged to 2.7 sires/brood
	Gonochoric cephalopods	
Octopus vulgaris, Spoon-like ligula on heterocotylized arm removes the sperm (from seminal vesicle) of preceding male, Quinteiro et al. (2011)	Field collected 124 adults in lab. Four clutches of egg strings analyzed by 2 microsatellites Oct 3 with 5 alleles + Ov 12 with varying number of alleles	Alleles of two males were present in all the assayed egg strings
Loligo pealei, Consort & sneaker males fertilizing internally at oviduct and seminal vesicle, respectively, Buresch et al. (2001)	Five capsules, each containing ~ 150 eggs extruded by 5 females were field collected. Presence of > 4 alleles, i.e. 2 maternal + 2 paternal revealed that the progeny are sired by > 2 fathers	Mean paternity of 495 assayed embryos was 3.2 sires/brood. Two capsules were sired by 2 fathers, 1 by 3 fathers and 2 by 4 fathers
L. bleekeri. Mating positions of consort male is parallel but head to head in sneaker males. Extra-pair mating behavior also occurs. Iwata et al. (2005)	In lab, 3 dams were mated with 1 or 2 males in 3 trails. Ten progeny from each of 74 capsules were genotyped. 5 microsatellite loci (Lb 1, Lb 3, Lb 4, Lb 5 & Lb 6) each with 13–30 alleles & GENEPOP v 3.4 & GERUD 1.0 software's were used (exclusion of probabilities of 0.989 to 0.996) to estimate paternity	Fertilization success > 99%. Up to 13% embryos sired by sneakers. The same male acted as consort at one time but as sneaker at another time

Li. obtusata by Paterson et al. (2001) to 99.99% using COLONY 2.0 in many molluscs. Iwata et al. (2005) used GENEPOP v 3.4 and GERUD 1.0 software systems to achieve the highest confidence level of paternity determination in *Lo. bleekeri*. Makinen et al. (2007) found that the paternity estimates increased from 3–5 to 5–8 and 4–10 sires with the use of Allele Lx23, GERUD 1.0 and COLONY software systems. Notably, the estimated paternity is limited to 76, 77 and 92% of the assayed embryos of *A. californica*, *N. arthritica* and *B. carica*, respectively. However, it covers 100% of the genotyped embryos *Li. saxatilis*.

Considering motility, sexuality and other features, the hitherto estimated paternity in some molluscs are brought under selected groups (Table 3.15). In gonochoric cephalopods, motility affords a wider scope for selection of

TABLE 3.14

Comparative account on multiple paternity in protandrics and simultaneous hermaphrodites

Species, Characteristics & Reference	Experimental Procedures	Reported Observations
Protandric prosobranchs		
Crepidula fornicata, Gregarious, Broods early larvae & releases planktonic veligers, Multiple copulations by different males. Dupont et al. (2006)	From brooding females, larvae collected by filters. Paternity determined by 4 microsatellite loci with 10–32 alleles & CERVUS 2.0 software	8.5% larvae sired by external males. In 14 of 18 broods, larvae sired by 2–5 fathers, with a maximum 5 males in 11–16 larvae/brood. Largest males were the most successful fathers
C. coquimbensis, Gregarious, Juveniles are released from brood. Brante et al. (2011)	Females (5) + 1 extra male kept in lab for 6 mo. Polymorphic 5 loci were employed to assign paternity of 528 embryos using GERUD 2.0 with 92% exclusion at 95% confidence level	In only one female, extra male sired maximum progeny. Paternity ranged from 2.0 to 4.7 and averaged to 4 sires/brood. About 5–14% embryos were not assigned to any known father
Protandric bivalve		
Ostrea edulis, Gregarious with subsequent planktonic larvae Lallias et al. (2010)	Complementary studies on wild & lab populations. 4 microsatellites from 13 brooders were used	Paternity ranged from 2–40 sires/brood
Unilateral SH		
Aplysia californica, Motile, Selection prior to copulation. Mates repeatedly as male and female. Males displace sperm of others. Large ones with more resource for reproduction. Angeloni et al. (2003)	Field collected 20 sperm recipient kept in lab. Samples of large egg masses alone assayed. *Apl* microsatellites locus with 23 alleles was adequate & used to assign paternity	In 33% egg masses, 76 and 27% embryos were sired by 2 and 3 fathers with paternity averaging to 2.1 sires/egg mass. In 15 of 20 egg masses, the last mated male had precedence. No effect of body size on paternity
Facultative SH		
Bulinus globosus, Freshwater snail. Jarne et al. (1992)	DNA finger printing was used	Selfed & outcrossed snails were unambiguously distinguished. 3 paternal bands were detected
Bulinus africanus Vianey-Liaud et al. (1996)	"Inseminated by 2 'Fathers'; an isolated SH faithfully sired offspring of 2 fathers"	Multiple paternity = 2 sires/brood
Biomphalaria obstructa Mulvey and Vrijenhoek (1981)	Isozyme markers 6–phosphogluconate dehydragenase	Multiple paternity = 2 sires/brood
Physa fontinalis Wethington and Dillon (1991)	LAP locus marker	Multiple insemination and possible paternity from 2 male-acting SH

TABLE 3.15

Paternity of selected groups in Mollusca

Species	Paternity	
	Sires/Brood	Mean
Gonochorics with sperm removing males		
Neptunea arthritica	1.7	
Octopus vulgaris	2.0	1.9
Slow motile unilateral/facultative SH		
Aplysia californica	2.1	
Bulinus globosus	2.0	2.0
Motile gonochorics with consort and sneaker males		
Loligo bleekeri	2.0	
Lo. pealeii	3.2	2.6
Sedentary gonochorics		
Hypriosis cumingii	2.7	2.7
Sedentary protandrics		
Crepidula fornicala	4.0	
C. coquimbensis	4.0	4.0
Ostrea edulis	> 4.0	
Slow motile gonochorics		
Littorina saxatilis	7.6	
Li. obtusata	5.0	5.1
Busycon carica	4.3	
Rapuna venosa	3.5	

superior mating partner and to generate offspring of high fitness. On the other hand, the obligate need is greater to produce progeny with as much genetic diversity as possible. Not surprisingly, the option of sedentary and sessile molluscs is to 'bet-hedge', multiple mating and paternity. The presence of multiple storage organs and ability of stored (for > 3 months–1 year, Baur, 1998) sperm to compete for fertilization (e.g. *Li. obtusata*) considerably increase multiple paternities to > 5.1 sires/brood. Surprisingly, the option of 25% molluscs to SH has not limited multiple paternity; they too have at least two sires/brood. Removal of the sperm of preceding males by the last male effectively reduces multiple paternities to ~ two sires/brood in *N. arthritica*, *Oc. vulgaris* and *A. californica*. Regarding paternity, population density can be an important feature, especially in sedentary *C. fornicata* with internal fertilization and gamete broadcasting bivalve *Hyriopsis cumingii*. In *C. fornicata*, the level of paternity increased from 2.2 to 3.1 sires/brood with increasing population density from 100 to 600 g/m^2 (Table 3.16). In the context of other aquatic invertebrates, molluscan paternity is discussed elsewhere.

TABLE 3.16

Effect of population density on paternity in *Crepidula fornicata* (compiled from Dupont et al., 2006)

Parameter	Keraliou	Rosegat	Roscanvel
Density (g/m²)	100	600	2000
Sex ratio ♀ : ♂	0.43 : 0.57	0.40 : 0.60	0.49 : 0.51
Adult (no./stack)	8.64	7.31	6.90
Adults analyzed (no.)	87	112	86
Brooding females (%)	81	57	77
Unassigned (%)	15.4	23.3	5.2
External paternity (%)	8.6	1.4	15.9
Multiple paternity (%)	80	100	50
Paternity (sires/brood)	2.2	2.2	3.1

3.8 Encapsulation and Nurse Eggs

Encapsulation: With the need to oviposit benthic eggs left with almost no parental care, majority of gastropods and cephalopods envelope their eggs in enclosing structures. These structures range from multi-layered capsule to fragile gelatinous mass, strand and string. In cephalopods, for example, each egg of *Loligo opalescens* is enveloped in a thick chorion (Hixon, 1983). In *Octopus tetricus*, each egg is enveloped by a chorion with a long stalk and the stalks are twined into a central stem to form a string (Joll, 1983). Each egg of *Oc. briareus* has a stalk, which is intertwined on to a central strand that binds the eggs together as a cluster (Hanlon, 1983a). The droplet-shaped opaque orange egg capsules of *Sepiola atlantica* are laid in a single layer or piled up into clusters of seven layers (Rodriques et al., 2011). *Illex illecebrosus* oviposits its benthic spherical egg mass (up to 1 m diameter) containing 100,000 eggs in a homogenous jelly. With a change in water content of the jelly as small as 0.004 g/cm², the egg mass begins to float (O'Dor, 1983). An egg hull of *Ischnochiton streamineus* is formed by polygonal, perforated plates (~ 30 μm width); micropores (0.2–0.5 μm) are randomly distributed over the entire surface (Liuzzi and Zelaya, 2013). In gastropods too, morphology of the capsules varies greatly. For example, the nassarid capsules may be shaped in the form of a bottle, vase, bell or blister, or it may also be bulliform, vasiform or triangular. These capsules may also be ornamented with ridges and spines (see Borysko and Ross, 2014). Amazingly, a tubular string of the West Indian *Strombus* reaches a length of 50 m and contains upto 460,000 eggs (Hyman, 1967, p 301). Cypraeids lay gelatinous mass of 500–1,500 capsules each containing 200 to 500 eggs (Hyman, 1967, p 303). Opisthobranchs spawn gelatinous egg masses (Przeslawski, 2004). For example, the cephalapsid *Haminoea japonica*

(= *callidegenita*) produces cylindrical egg mass (15 x 5 x 5 mm), in which the individually encapsulated 200–700 eggs are arranged in a spiral string embedded in jelly (Fig. 3.11F). In the sacoglossan *Alderia willowi* (= *modesta*, see Krug, 1998, 2001), eggs are (densely in planktotrophics and less densely in lecithotrophics, see later) packed in sausage shaped gelatinous egg mass (Fig. 3.11A). In *Elysia* spp, the brightly orange colored eggs are individually packed in capsules tightly and spirally embedded inside a jelly string within a gelatinous mass surrounded by a tough membrane (Krug, 2009). Figure 3.11 shows egg masses of some opisthobranchs. The nudibranch *Chromodori* spp lay planar transparent egg mass with center spiraling outwards (Trickey et al., 2013). Apparently, the formation of egg capsules or masses is structurally and chemically complex and costs energy. Not surprisingly, it has required highly modified/specialized female reproductive anatomy, function and behavior (Pechenik, 1986). For example, the capsular gland is replaced by jelly gland in littorinids that lay their eggs in gelatinous masses (Hyman, 1967, p 291).

The capsules may protect the embryos against desiccation, predation, osmotic stress (see Ojeda and Chaparro, 2004) and UV-irradiation (Rawlings,

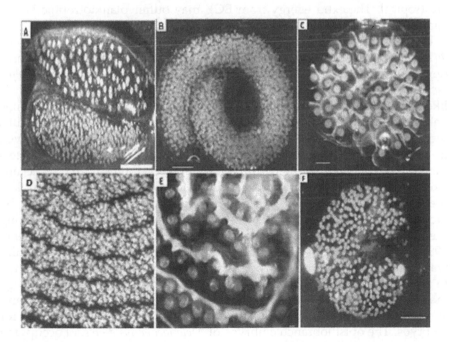

FIGURE 3.11

A. Planktotrophic (below) and lecithotrophic (above) egg masses of *Alderia willowi* (from Krug et al., 2007), B. planktotrophic and C. lecithotrophic egg masses of *Elysia pusilla* (Vendetti et al., 2012), D. planktotrophic and E. lecithotrophic egg masses of *E. zuleicae* (from Krug, 2009) and F. egg mass of *Haminoea japonica* (from Gibson and Chia, 1989).

1996). They may also provide nutrients, especially the critically required calcium to build shell in freshwater snails (Pechenik, 1986). In cephalopods and many gastropods, the eggs within the capsule solely rely on yolk energy. However, provision of extra-embryonic nutrients is a characteristic of gastropods. For example, many caenogastropods are provided with extra embryonic nutrients drawn from capsular fluids, nurse eggs and in some cases sibling embryos. *In lieu* of nurse eggs and others in prosobranchs, the strategy of opisthobranchs is to provide Extra Capsular Yolk (ECY). In all species of *Elysia*, egg mass contains extra capsular yolk (ECY), contacting every egg capsule, within which the veliger develops. The ECY granules enter the capsule through minute tears at the point of contact. The ECY is provided in the form of thick orange ribbons in *E. papillosa, E. pratensis, E. subornata* and *E. tuca*, or pale yellow blobs/ribbons in *'B'. mercusi* and *E. cornigera*, or thin white zig-zags and thick white ribbons in planktotrophic and lecithotrophic *E. zuleicae*, respectively (Krug, 2009). In *Halimeda discoides*, the ECY gradually disappears, as larva completes intracapsular development (Vendetti et al., 2012). The ECY may be akin to nurse eggs. Both nurse eggs and ECY are adaptive mechanisms to accelerate cleavage by reducing egg size, allowing the embryo to grow faster (Clark and Jensen, 1981). Most species with ECY are tropical. The extra energy from ECY may buffer planktotrophic larvae against oligotrophic waters.

The capsular wall is permeable to water and salts. For example, *Crepidula fornicata* egg is reported to absorb 0.52 µg salt during its embryonic development (Pandian, 1969). An individual egg of the cuttlefish *Sepia esculenta* absorbs 0.4 ml water just prior to hatching (Lei et al., 2014). It is likely that more water is absorbed by embryos to be hatched as pelagic veligers and less by those emerging as benthic juveniles. Understandably, the inflowing water may irrigate and supply the required oxygen. However, no research has yet been undertaken to know how the inflowing water meets the increased oxygen required with advancing development. In crustaceans like *Macrobrachium nobilii*, for example, a fanning frequency to ventilate the pleopod-borne eggs is accelerated with advancing development by increasing the frequency from 2,676 times/hour to 8,760 times/hour during the initial and terminal periods of incubation (Pandian and Balasundaram, 1980). In fact, the need to provide inflowing water through the capsular wall at different velocities may be a reason for affording parental protection to the capsular embryos in many gastropods (see later). As jelly in the egg mass represents a non-dissolvable barrier between developing embryos and surrounding water, the gelatinous egg mass may pose diffusion problems for embryos. The opisthobranch embryos of *Melanochlamys diomedea* occupying the periphery of their globular egg mass develop earlier than those more deeply embedded within the mass. Conversely, the embryos of another opisthobranch *Haminoea vesicular* embedded in a ribbon-shaped strand provides relatively more surface area and hence the embryos are developed synchronously (Chaffee and Strathmann, 1984, see also Krug, 2009).

Capsular nutrition: Irrespective of the presence or absence of nurse eggs, the fluid bathing the embryos within a capsule contains amino acids, proteins and polysaccharides. That the embryos of *Concholepas concholepas* (Constella and Henley, 1971), *Conus pennaceus* (Perron, 1981) and *Nucella lapillus* (Lord, 1986) can successfully be reared after artificial removal from capsules suggests no nutritive role for the fluid in these gastropods. The fluid does not also suppress bacterial infection (e.g. *N. lapillus*, Pechenik et al., 1984). However, *C. fornicata* embryos, following the removal from capsules, died within 2 days (Hoagland, 1986). The nutritive value of the fluid has also been reported for *Urosalpinx cinerea* (Rivest, 1983). It appears that the nutrients arising from the capsular wall during the course of embryonic development may be nutritive. Using gravimetric, histochemical and biochemical analyses and TEM/SEM pictures, Ojeda and Chaparro (2004) have made a very detailed study on the role of capsular wall in providing nutrients to developing embryos/ trochophore of *C. fornicata*. Hence capsular wall of *C. fornicata* is considered as one of the nutrient sources for gastropod embryos. The capsular wall contains an external fibrous layer, whose thickness does not change during development. But the internal spongy wall is greatly reduced from 12 µm at the zygote stage to 3 µm, when the embryo attains the veliger stage (Table 3.17). That the wall weight loss of 0.50 and 0.46 µg at trochophore and veliger stages, respectively suggests the transfer of considerable materials including proteins from the inner spongy layer is transferred to the capsular fluid. The transfer accounts for 15 µg/capsule equivalent to 0.015 J/capsule during the period from the veliger stage to pediveliger stage alone. Flouresent microscopic study has revealed the endocytic uptake of proteins like albumin from the capsular layer. Incidentally, protein uptake by encapsulated larvae of gastropod (e.g. *Littorina*, Moran, 1999) has also been reported. In fact, lecithotrophic larvae of gastropods are known to acquire amino acids and sugars directly from water (e.g. *Haliotis rufescens*, Welborn and Manahan, 1990).

Nurse eggs: In aquatic invertebrates, nurse eggs are reported from polychaetes (e.g. Rasmussen, 1973), nemerteans (e.g. Schmidt, 1932) and crustaceans

TABLE 3.17

Decreases in capsular wall thickness and contents during development from zygote to veliger (shell length of 400 µm) of *Crepidula fornicata* (compiled from Ojeda and Chaparro, 2004)

Parameter	Zygote	Trochophore	Veliger
Capsular wall thickness (µm)	12.0	7.0	3.0
Dry weight of capsule wall (µg)	0.12	0.11	0.04
Organic matter (µg)	0.075	0.065	0.035
Protein (µg)	0.42	0.43	0.20

(e.g. *Streptocephalus dichotomus*, Munuswamy and Subramonian, 1985). The molluscan nurse eggs are potentially normal ova, whose development is aborted or prevented by an unknown mechanism. Hence they remain as one of the mysterious aspects of gastropod development. Within the genus *Crepidula* alone, the uncleaved eggs are ingested as a whole in some species (e.g. *C. dilatata*, Gallardo and Garrido, 1987, see also *Searlesia dira*, Rivest, 1983). In others like *C. capensis*, *C. coquimbensis* (see Collin et al., 2005), the embryo sucks the yolk out from the nurse eggs, which remain as ciliated yolk-filled balls, following the arrested development after gastrulation. In still others like the vermetid gastropods *Vermetus triquertrus* (Calvo and Tempelado, 2004) and *Thylaeodus regulosus* (?) (Strathmann and Strathmann, 2006), the large larvae feed not only on non-developing nurse eggs but also the smaller siblings, in which the development was arrested after the larvae passed through the initial protoconch stage.

The expected bimodal egg size distribution was not apparent in many gastropods. Rarely, *Crepipatella dilatata* displayed a bimodal egg size distribution with smaller (154–200 µm) and larger (200–300 µm) eggs but the smaller eggs were not nurse eggs (Zelaya et al., 2012). Further, the incubation period, during which nurse eggs were consumed, varied widely from 2–5% of total incubation period (~ 135 days) in *Buccinum undatum* (Smith and Thatje, 2013) to as much as 100% of the period (~ 22 days) in *Crepidula dilatata* (Chaparro and Paschke, 1990). Within 3–7 days out of the 135 day incubation period, *B. undatum* consumed all the nurse eggs, whereas it took all the 22 days of incubation for *C. dilatata* to consume the required nurse eggs. Hence the mechanism that determines the nurse eggs/siblings, is not known in which development is arrested after gastrulation and protoconch stage, respectively. Rivest (1983) provided circumstantial evidence for genetic predetermination of nurse eggs. However, no direct cytological and/or cytogenetic evidence is thus far available for genetic determination of nurse eggs/nurse siblings.

During incubation period, there are wide intra-specific and inter-specific variations in the number of nurse eggs consumed by an embryo. In general, large maternal size results in more eggs and larger capsuls as well as more eggs per capsule (see Nasution et al., 2010). The number of embryos emerging from a capsule ranges from one in *Petaloconchus montereyensis* (Hadfield, 1989) to ~ 100 in *Nucella crassilabrum* (Gallardo, 1979). Within a species, for example, it varies from 11 to 46 embryos across capsules of *S. dira* (Rivest, 1983). The number of eggs consumed by an embryo ranges from 1.7 nurse eggs in the Pacific shallow water muricid *Acanthinucella spirata* (Spight, 1976) to 50,000–100,000 nurse eggs in the north Atlantic deep sea buccinid *Volutopsius norwegicus* (see Smith and Thatje, 2013). From a detailed study, Rivest (1983) showed that the number of nurse eggs available for an embryo within a capsule of *S. dira* decreased from 172 in a capsule with one embryo to seven in a capsule with 72 embryos. Consequently, the hatchling size also decreased from 2.14 to 1.15 mm with increasing number from 1 to 15 of hatchlings/capsule.

Table 3.18 summarizes some data on the level of correlations between selected parameters of reproductive characteristics like those between the number of eggs or embryos on one hand and shell length of females or capsular size on the other. As capsule size is a critically limiting factor in holding more eggs/embryos or less, the capsule has been measured in terms of its length, volume and inner surface area (cf providing capsular fluid). The correlations between cumulative fecundity on one hand and capsular surface area range from $r^2 = 0.70$ in *Crepipatella dilatata* to volume $r^2 = 0.76$ in *B. undatum* and capsule size $r^2 = 0.82$ in *Nucella crassilabrum*. The correlation levels of egg number vs capsule in *N. crassilabrum* ($r^2 = 0.82$) and vs capsular volume in *B. undatum* ($r^2 = 0.76$) decreased to $r^2 = 0.68$ and $r^2 = 0.57$, respectively, when the correlations are related to embryos/veligers. This decrease may be due to wide variations in embryos losses during incubation period (e.g.

TABLE 3.18

Summary correlations reported for different reproductive characteristics of gastropods

Relationship Between	Correlation	Level ($r^2 =$)
Crepidula dilatata, Indirect, Lecithotrophic, Chaparro et al. (1999)		
Capsule wall weight vs Shell length	Positive significant	0.15
Egg no./brood vs Shell length	Positive significant	0.44
Egg no./capsule vs Shell length	Positive significant	0.45
Egg consumed/embryo vs Hatching size	Positive significant	0.36
Hatching size vs Shell length	Positive significant	0.32
Crepipatella dilatata, Direct developer, Zelaya et al. (2012)		
Dry weight vs Shell length	Positive correlation	0.37
Capsule no. vs Shell length	Not significant	0.06
Capsule no. vs dry weight	Not significant	0.05
Capsule surface area vs Shell length	Positive significant	0.42
Cumulative fecundity vs capsule area	Positive significant	0.70
Nucella crassilabrium Direct developer Gallardo (1979)		
Egg no. vs capsule size	Positive significant	0.82
Embryo no. vs capsule size	Positive significant	0.68
Buccicum undatum, Direct developer, Smith and Thatje (2013)		
Egg no. vs capsular volume	Positive significant	0.76
Veliger no. vs capsular volume	Positive significant	0.57
Nerita melanotragus, Direct developer, Przeslawski (2011)		
Larval stage vs incubation days	Positive significant	0.83

24–36% embryo loss in *Nerita melanotragus*, Przeslaswki, 2011). Otherwise, the egg number per capsule and per brood held positive relations (r^2 = 0.45) in *Crepidula dilatata*. Notably, the correlations between capsule number and shell length or dry weight of female *Crepipatella dilatata* are not significant (r^2 = < 0.06). Clearly, large females opt to invest more eggs within a capsule rather than to increase the number of capsules, as construction cost of capsules is costlier. In *B. undatum*, larger females produced more eggs/capsules and larger capsules. In hermaphroditic opisthobranch *Aplysia kurodai* too, this relationship is apparent (Fig. 3.12B). When expressed in number of eggs produced by a hermaphrodite per day, fecundity is increased linearly as a function of body weight but the number of string per hermaphrodite per day decreases after 80 g size, clearly indicating that large hermaphrodites opt not to invest more on string (Fig. 3.12A).

Hatchling size is an important factor in offspring fitness. After consuming 5,000 nurse eggs during incubation period of 5.5 months, *Neptunea antiqua* juveniles emerge at the size of 12.7 mm length. This hatchling size is > 42 times the egg size (300 µm). For 74 species that do not produce nurse eggs, hatchling size is twice the egg diameter (see Rivest, 1983). Many gastropods like *Crepipatella dilatata* develop directly into juveniles. In others like *Crepidula dilatata*, development is indirect, i.e. the earlier trochophore larvae are brood protected and subsequent lecithotrophic veliger is planktonic for a short period. In still others, development is indirect involving planktotrophic trochophore, veliger and pediveliger stages (Table 3.18). Hence modes of development may have important consequences for dispersal, gene flow, geographical range of distribution as well as rates of speciation and extinction (see p also 30). In an excellent review, Collin (2003) has analyzed data on developmental characteristics of 78 calyptraeid species (53 *Crepidula*. 11 crucibulum and 9 Calyptraeae and five others) and arrived at the following conclusions: 1. Egg size is positively correlated with incubation period and hatchling size in species without nurse eggs. 2. There is a significant difference in egg size between direct developers with and without nurse eggs. 3. Hatchling size

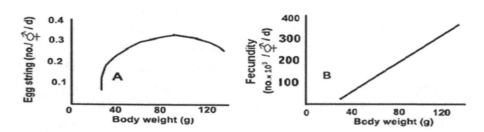

FIGURE 3.12

A. String number and B. Fecundity as function of body weight in simultaneous hermaphrodite *Aplysia kurodai* (redrawn from Yusa, 1994).

does not differ between directly developing lecithotrophics with or without nurse eggs. 4. Egg- and hatchling-size distributions are positively correlated. The correlation becomes strong, when data for species with nurse eggs are not considered. 5. The frequency of planktotrophic species decreases with increasing latitude, while that of direct developer's increases. 6. Strikingly, almost all species with nurse eggs occur in the southern hemisphere.

Prior to Collin (2003), similar analyses have been made covering more taxons of prosobranchs and opisthobranchs. Table 3.19 represents a simplified version of these analyses. Briefly, (i) the minimum egg size of planktotrophic, lecithotrophic and directly developing caenogastropods are two-three times larger than the corresponding eggs of opisthobranchs, (ii) hatchling size of directly developing muricids, calyptraeids and conids are almost four-times larger than the respective planktotrophic caenogastropods, (iii) the directly developing opisthobranchs are, however, are only about two-times larger than planktotrophic nudibranchs. It is not clear whether the differences between prosobranchs and opisthobranchs are related to sexuality. Protandric calyptraeids act as male or female at a given point of time. Hence, almost all prosobranchs are 'gonochoric', whereas almost all opisthobranchs are simultaneous hermaphrodites. Having to invest two times on gonadal tissues and gonadal tracts, opisthobranchs produce smaller eggs and hatchlings than prosobranchs, who invest more to produce larger eggs and hatchlings.

TABLE 3.19

Developmental characteristics of marine gastropods

Taxon, Species (no.)	Egg Size (µm)			Hatching Size		Reference
	Plankto-trophic	Lecitho-trophic	Direct Developer	Plankto-trophic	Direct Developer	
Caenogastropoda						
Muricids 53 species	179	–	497	312	1280	Spight (1976)
Calyptraeids 78 species	189	321	336	343	1295	Collin (2003)
Conids 25 species	~ 225	390	~ 470	391	1130	Kohn and Perron (1994)
Opistrobranchia						
Cephalapsids 38 species	~ 80	200	~ 280	–	–	Schaefer (1996)
Nurdibranchs 37 species	~ 88	~ 170	~ 113	162	239	Hadfield and Switzer-Dunlap (1994)
Opisthobranchs 53 species	~ 88	~ 165	205	147	268	Hadfield and Switzer-Dunlap (1994)

3.9 Poecilogony and Dispersal Dimorphism

In molluscs, the mode of development is usually divided into Planktotrophics (P) and Lecithotrophics (L). The former oviposit a large number of smaller eggs that develop into pelagic, free swimming larva prior to metamorphosis. The latter may bypass larval stages and develop directly into benthic juveniles. However, hatching occurs prior to metamorphosis in some L, which passes through a short pelagic facultative feeding stage prior to metamorphosis, whereas in others hatching of juveniles occurs after metamorphosis (Thorson, 1950). In either case, the L embryos/larvae often have lost many of their characteristic P features of their close relatives (Collin, 2012). Planktotrophic development facilitates dispersal of offspring resulting in gene flow and colonization of new habitats. However, dispersal incurs biotic risks like larval motility due to predation and planktonic food scarcity as well as abiotic risks like physical transportation away from appropriate settlement sites by ocean current competition for food and space, and extinction by local hazards. The costs, benefits and consequences of different dispersal strategies (see p 31–33) are fundamental factors in structuring the life history patterns (Krug, 2001). It is in this context, the 'intermediates' sharing the benefits of both the contrasting 'dimorphic modes' of development namely poecilogony and dispersal dimorphism are interesting, despite their rarity (Collin, 2012).

Poecilogony is a rare reproductive mode, in which the dichotomic planktotrophic and lecithotrophic egg-larval morphs are generated from the same egg mass of an individual or individuals of the same species, whose populations are seasonally or geographically separated. Dispersal dimorphism, often misinterpreted as poecilogony, occurs in opisthobranchs, in which both intra-capsular and post-hatching metamorphoses occur in the same clutch, i.e. dispersal dimorphism is limited to terminal larval stage alone but their egg size may not be dimorphic (Krug, 2001). Ever since the existence of poecilogony was proposed by Giard (1905), it has been claimed to occur in many gastropod species. Most of these claims were due to erroneous identification of species (for 64 species Hoaglant and Robertson, 1988, for 15 species Bouchet, 1989) or laboratory disturbance (see below).

Poecilogony has been hitherto confirmed to occur in six species only, of which five species are sacoglossan opisthobranchs *Alderia willowi*, *Costasiella ocellifera*, *Elysia chlorotica*, *E. pusilla*, *E. zuleicae* (Vendetti et al., 2012) and one prosobranch *Calyptraea lichen* (McDonald et al., 2014). The slugs *A. willowi* (= *modesta*, see Krug, 1998, 2001, Ellington and Krug, 2006), collected from the islands of the Washington state, USA, produced lecithotrophic egg masses during the summer and planktotrophics in winter (Table 3.20). Uniquely, about 1% of the parental individuals produced mixed clutches containing both planktotrophics and lecithotrophics. Pairing between P and L parents produced planktotrophic F_1 progeny alone. However, three lineages, arising from sibling matings produced > 90% L egg masses, while four other lineages produced about equal proportions of L and P egg masses in both F_2 and F_3 generations. Rearing *A. willowi*, Smolenksy et al. (2009) reported that about 67% of isolated slugs switched from planktotrophic to lecithotrophic clutch and vice versa; the rapidity, at which the switching occurs, poses a challenge to endocrinologists as to how the vitellogenic process is so quickly switched (Table 3.21). Only two out of 16 slugs simultaneously produced mixed planktotrophic and lecithotrophic clutches. Notably, selfers alone switched

TABLE 3.20

Reproductive characteristics of poecilogonous *Alderia willowi* (compiled from Krug, 1998)

Parameter	Planktotrophic	Lecithotrophic	Mixed
Adult producing (%)	43	56	1
Encapsulated period (d)	3.0	5.4	–
Egg no./clutch	31.1	32	44
Egg diameter (µm)	68	105	–
Egg capsule (µm)	121	247	–
Larval size (µm)	116	186	152

TABLE 3.21

Observations reported from long term rearing of *Alderia willowi* in laboratory by Smolenksy et al. (2009)

Reported Observations
➢ Most slugs produced either planktotrophic or lecithotrophic clutch but 67% of isolated slugs switched between lecithotrophic and planktotrophic clutches at least once
➢ Four out of 16 isolated slugs produced exclusively lecithotrophic clutches
➢ Seven isolated slugs produced solely planktotrophic clutches
➢ Five out of 16 isolated slugs initially produced 2–4 planktotrophic clutches but subsequently switched to lecithotrophic clutch production
➢ Two slugs produced one planktotrophic clutch but subsequently lecithotrophic clutch only
➢ One out of 16 isolated slug produced planktotrophic clutches for 22 days, then produced no clutch for 11 day but finally a single lecithotrophic clutch
➢ Two slugs initially produced lecithotrophic clutch but switched to 1 to 5 planktotrophic clutches and finally reverted to produce lecithotrophic clutches
➢ Four slugs switched from lecithotrophic to planktotrophic clutches a few days prior to death
➢ Only two out of 16 isolated slugs simultaneously produced planktotrophic and lecithotrophic clutches
➢ No paired slugs switched from planktotrophic to lecithotrophic clutch or vice versa

from lecithotrophy to planktotrophy or vice versa, depending on the resource available to sustain vitellogenesis. Regarding diversity, individuals of one reproductive mode were more divergent from each other than those from different reproductive modes.

Of two different populations of *E. chlorotica* from the east coast of USA, smaller (7.6 mm length) morph from Ipsvich produced the capsular-metamorphosing lecithotrophic egg masses only but the larger morph (20.0 mm) from Martha-Vineyard (Table 3.22) oviposited planktotrophic egg masses alone. Pairing between adults of L and P produced P egg masses alone in F_1 progeny, as in *A. willowi*. When siblings of L and P were paired, the fertile F_2 offspring produced larval sizes of both extremes (West et al., 1984, see also Krug, 1998). *E. pusilla* from the Indo-Pacific region can spawn lecithotrophic clutch containing a few larger eggs or planktotrophic clutch consisting of a large number of smaller eggs (Table 3.22). One clutch from Sobe, Japan consisted of planktotrophics eggs alone but the second clutch hatched into lecithotrophic larvae alone. Of 13 deposited clutches from Guam, 11 were planktotrophics and the other two were lecithotrophics. Clearly, there was no mixed clutch but both lecithotrophics and planktotrophics are produced by an individual in the same population. Molecular phylogenetic analyses of the Japanese and Guam individuals revealed that despite high dispersal potential of the planktotrophic larvae, very little dispersal has occurred in these populations during the past two million years of their existence. Hence selection for dispersal per se is not a plausible explanation for the persistent of poecilogony in *E. pusilla* (Vendetti et al., 2012).

TABLE 3.22

Reproductive characteristics of poecilogonus gastropods (compiled from Krug, 2009, Vendetti et al., 2012, McDonald et al., 2014)

Species	Egg Size (µm)		Fecundity (no./ clutch)		Incubation (d)		Larvae Shell Size (µm)	
	Plank	Lecith	Plank	Lecith	Plank	Lecith	Plank	Lecith
Sacoglossid opisthobranchs								
A. willowi	68	106	311	32	3	5	116	186
Co. ocellifera	77	106	219	24	5	15	121	268
E. chlorotica	79	96	8092	178	6	11	146	217
E. pusilla	70	97	–	59	–	15	150	218
E. zuleicae	66	–	–	104	6	19	114	254
Caenogastropod								
Ca. lichen†	200	–	–	–	–	–	407	1000

A = Alderia, Co = Costasiella, E = Elysia, Ca = Calyptraea, Plank = Planktotrophic, Lecith = Lecithotrophic, † with nurse eggs

C. lichen is a protandrous calyptraeid prosobranch that brood thin walled transparent egg capsules. It is the very first representative for poecilogony from prosobranch and demonstrates a novel combination of planktotrophics and lecithotrophics involving nurse eggs and adelophagy. Understandably, it does not display egg size dimorphism (hence poecilogony in them is questionable), as lecithotrophics acquire extra-embryonic nutrients by ingesting nurse eggs/sibling embryos. Of five females observed by McDonald et al. (2014), three brooded planktotrophics alone. But the other two females produced (i) broods of small eggs that hatched as planktotrophic larvae (50%), (ii) broods of large eggs that hatched as juveniles or short-lived lecithotrophic larvae (13%) and (iii) broods of adelophagic embryos that hatched as crawling juveniles (20%). All three COI sequences from female were identical to GenBank sequence for the planktotrophic *C. lichen*. The two females with mixed broods shared a sequence that differed from the others by one silent substitution.

Larval dimorphism: In many gastropods, lecithotrophic development is followed either a pelagic larva, which metamorphoses after hatching or benthic juvenile, which undertakes intra-capsular metamorphosis prior to hatching. In a few opisthobranchs, both planktonic non-feeding veligers and benthic juveniles are produced from the same egg mass (e.g. *Haminoea*) or egg masses oviposited by different individuals of the same species but from seasonally separated populations, For example, *E. subornata* oviposits planktotrophic egg masses in spring, lecithotrophic pelagic larvae in summer and lecithotrophic (intra-capsular) benthic juveniles during the fall and winter (Clark et al., 1979, Clark and Jensen, 1981). The sacoglossans *E. cornigera, E. crispata, E. everlinae, E. timida, E. tuca, E* (= *'B'*) *marcusi*, the nudibranchs *Tenellia adspersa, T. pallida* and cephalapsid *Haminoea japonica* (= *callidegenita*) (Gibson and Chia, 1989) and *H. zealandica* (Clemens-Seely and Phillips, 2011) are all reported to display dispersal dimorphism. More information for a representative *H. japonica* is hereunder summarized.

An unusual larval dimorphism in *H. japonica* results in the production of half the siblings from each mass as free-swimming veligers and the other as encapsulated embryos. *H. japonica*, collected from silty substratum of islands of the Washington state, USA, simultaneously produced, after 32-days incubation, both non-feeding veliger and encapsulated embryos in approximately 1 : 1 ratio. Hatching lasted for 3–11 days. The veligers gradually became competent and metamorphosed after a brief pelagic stage (for 30 days). Following the artificial separation from the jelly, the encapsulated embryos could be reared successfully (Gibson and Chia, 1989). Capsules containing gastrulae were separated and reared; 70% of the ~ 10 day-old encapsulated embryos, on exposure to the methanol extract of the jelly, were induced to metamorphose by a compound present in gelatinous matrix of the egg mass. The inducer is a small (< 1,000 DA), polar, non-proteinaceous and stable (against changes in pH and temperature) compound. Methanol extract of the posterior reproductive gland (61% intra-capsular metamorphosis), digestive gland (49%) and parapodial lobe (56%) of *H. japonica* also induced variable proportions of the embryos to metamorphose, indicating the presence of an inducer in many adult tissues. However, the inducer failed to induce metamorphosis in many other tested opisthobranchs *H. vesicula*, *Melanochlamys*, *Alderia* and *Onchidoris*. Clearly, the inducer is a species specific compound. Incidentally, the veligers of *Phestilla sibogae* (Ruiz-Jones and Hadfield, 2011) and *Eubranchus doriae* also metamorphosed in response to inducers. Small (< 500 DA) water-borne stable compound released from its prey, the hard coral *Porties* induced *P. sibogae* veligers to metamorphose. Likewise, a small polar water soluble compound with glucosidic residue originating from its host cum prey, the hydroid *Kirchenpaueria pinnata* induced *E. doriae* veligers to metamorphose (Gibson and Chia, 1994, Chia et al., 1996).

Starvation is reported to induce a shift from lecithotrophic to planktotrophic in both poecilogenous sacoglossan *A. willowi* and dispersal dimorphic nudibranchs *Spurilla neopolitiana* and *Tenellia adspersa* (see Krug, 2009). In *A. willowi*, a period of 5 days starvation increased planktotrophics from 50 to 68 and 90% in the first and fourth clutches, respectively (Krug, 2001). Adults of *A. willowi* lost their green body color (due to algal chloroplasts) within a day and dramatically the body weight within a few days of starvation. Expectedly, these starved individuals were unable to deposit adequate extra capsular yolk in lecithotrophic egg masses and let the planktotrophic larvae to acquire the required food to support them until metamorphosis.

To bolster the 'bet-hedging' strategy, not only hatching duration is prolonged up to ~ 10 days but also lecithotrophic larval duration is extended temporally spreading larval duration. The lecithotrophics larvae of poecilogenous *A. willowi* and dispersal dimorphic nudibranchs are induced to metamorphose and settle by appropriate host plant/animal or substrate. For example, 95% of the veligers are induced to metamorphose in the presence of the yellow green alga *Vaucheria longicaulis*, which serves as both substratum and food for the adults. In the absence of the inducer, some 'desperate larvae' may undertake metamorphosis or die (Krug, 2001).

The presence of jelly within the egg mass introduces diffusion problems to the opisthobranch embryos and results in asynchronous hatching with peripheral embryos hatching earlier and others hatching gradually during the ensuing 3–10 days. The asynchronous hatching is another temporal 'bet-hedging' strategy for successful recruitment of veligers within a short duration of pelagic dispersal. However, hatching of veligers or juveniles in lecithotrophics is a highly plastic phenotypic tarit. For example, when the egg masses of the dispersal dimorphic aeolid nudibranch *Berghia veruciocornis* were reared in laboratory without aeration, both lecithotrophic veligers and juveniles were hatched from the same egg mass. However, lecithotrophic veligers were alone hatched, when the egg mass was aerated (Carroll and Kemff, 1990).

3.10 Brood Protection and Viviparity

Brood protection means affording protection and irrigation to supply dissolved oxygen to short-term brooded embryos (released as veligers) within capsules/gelatinous mass. Besides protection and oxygen delivery, female (in gonochorics and protandrics) or hermaphoroditic viviparous parents also provide nutrients to developing embryos until young ones are born. Incidentally, the supply of nutrients from capsular fluid, nurse eggs and/or sibling embryos in prosobranchs and extra embryonic yolk in opisthobranchs is distinctly different from viviparity, as these nutrients are not supplied directly by the gastropod female/parent. Brood production includes a range of behavior from embryo guarding over a period of 50% of its life time in a cephalopod (*Bathypolypus arcticus*) to carriage of egg mass until young ones are hatched (*Strombus indica*), brooding in a cloacal pouch (*Halomenia gravida*) in Aplacophora, incubatory pouch (*Planaxis*), mantle cavity (epipelagic pteropod *Spiratella inflata*) in gastropods and in marsupial gills (in freshwater unionids), gill chamber or pallial cavity in larviparous protandric oysters and wood boring bivalves (Table 3.23). Notably, the brooding sites are located on the gills and in the mantle cavity to ensure adequate oxygen supply. Remarkably, brood protection in gastropods and bivalves is limited to short-term brooding followed by release of veligers. Presumably, the limitations of brooding space (to accommodate growing larvae) and oxygen supply triggers the emergence of veligers to become pelagic for dispersal. Secondly, males play no role in egg guarding and brood protection. Viviparity occurs in prosobranchs like the vermitids, viviparids (e.g. *Hydrobia* spp), cerithioids (e.g. *Lavigera*) and bathy pelagic pteropod opisthobranch like *S. helicoides* as well as marine clams *Transennella tantilla* and *Lasaea subviridis*. Heller (1993) examined the theoretical possibility of origin of viviparity among hermaphrodites. He found that among the predominantly gonochoric prosobranchs, brooding is recorded in 85 genera (4%) but only two of them are hermaphrodites. In the predominantly hermaphroditic opisthobranchs too, only 4% (95 genera) are brooders. Hence the possibility of hermaphrodites evolving from brooding and viviparous gastropods may not be relevant. However, viviparity has evolved in cerithioids through several independent avenues (Glaubrecht, 1999).

3.11 Fecundity

The brood number or Fecundity (F) is shown to increase with body size (L) in many molluscs including viviparids like *Transennella tantilla* (Fig. 3.13A) and *Lasea subviridis* (Fig. 5.4B). Interestingly, the linear relationship also

TABLE 3.23

Egg guarding, brood protection and viviparity in molluscs

Species & Reference	Reported Observations
Aplacophora	
Halomenia gravida, Pruvoltina providens Hyman (1967, p 5)	Brooded in cloacal pouch
Polyplacophora	
Chiton barnesi, Ischnochiton imitator Hyman (1967, p 114)	Brooded in pallial grooves
Prosobranchia	
Strombus indica	Female carries 30 cm long gelatinous cylinder of capsules until young ones are hatched
Littorinid *Planaxis*, Vermetid *Pyxipoma*	Incubatory pouch
Marine *Acmaea rubella*, Freshwater *Hydrobia* spp	Mantle cavity
Viviparid *Campeloma* (Hyman, 1967, p 306)	Uterine (pallial oviduct) viviparity
Freshwater Cerithioidea (Glaubrecht, 1999)	Uterine pouch brooding/pallial pouch
Melanoides tuberculata	Pseudoplacental nourishing
Opisthobranchia	
Epipelagic *Spiratella inflata*	Mantle cavity
Bathypelagic *S. helicoides* (Lalli and Wells, 1973)	Mucous gland of reproductive tract
Bivalvia: Oysters & wood borers	
Ostrea edulis (Walne, 1964)	Larviparous in mantle cavity
O. puelchana (Castro and Lucas, 1987)	Larviparous in pallial cavity
Teredo novalis (Coe, 1941)	Larviparous in gill chamber
Bivalvia: Clams	
Transennella tantilla (Kabat, 1985)	Viviparous brooded between gills & visceral mass
Lasaea subviridis (Beauchamp, 1986)	Viviparous in sub-branchial chamber
Freshwater Unionidae *Anodonta* (Heard, 1975)	Marsupial demibranchs
Cephalopoda	
Octopus joubini (Hanlon, 1983b)	Eggs guarded till hatching, artificial incubation possible with aeration
O. maya (Van Heukelem, 1983)	Eggs are guarded
Bathypolypus arcticus (O'Dor and Macalaster, 1983)	Meticulous egg guarding for 12 months representing ~ 50% of mothers life span

holds good for the release of male and female gametes as a function of body weight (measured by adductor muscle weight) of *Choromytilus meridionalis* (Fig. 3.13B). Because F is related to volume of the body cavity available to accommodate the growing embryos, geometry suggests that the Length

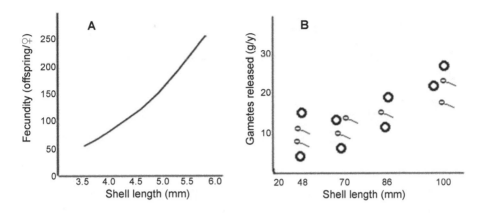

FIGURE 3.13

A. Fecundity (offspring number) as a function of shell size in viviparid clam *Transennella tantilla* (redrawn from Kabat, 1985). B. Gamete released as a function of shell length in *Choromytilus merdionalis* (redrawn from Griffths, 1981a).

(L) exponent would be 3.0. Expectedly, the exponent of *T. tantilla* is 3.02, indicating that brood bearing gill surface does not constrain the brooding capacity (Kabat, 1985). But, the exponent is 4.7 in the short-term brooding with veliger releasing *Mysella bidentata* The high exponent indicates that reproductive output is increased at a faster rate than that of body growth (volume) (Ockelmann and Muss, 1978). In fact, this is what happens in gamete broadcasting mytilids and larviparous oysters. In many iteroparous oysters, gonadal allocation after maturation and advancing age exceeds that for stomatic growth (see Fig. 2.12).

Fecundity is decisively an important factor in recruitment. For familiarization, a few terms related to molluscan fecundity are explained in the context of molluscs. 1. Batch Fecundity (BF) or clutch is the number of eggs produced per spawning. In a single spawning pulse, broadcast spawning gastropods *Patella*, *Helcion*, *Gibbula* and *Littorina neritoides* shed all the eggs directly into water (Hyman, 1967, p 299). During the 14-day period of observation, *Aplysia kurodai* oviposited 0.35–0.52 egg string/d, i.e. the inter-batch interval was 2–3 days only (Yusa, 1994). With synchronized single spawning pulse (see Table 1.9 Rocha et al., 2001), the semelparous bobtail squid *Sepiola atlantica* has been found dead after spawning her eggs capsules overnight (Rodriques et al., 2011). Within a spawning period, BF increases to a peak and subsequently decreases gradually in many iteroparous gastropods (e.g. oviparous: *Lymnaea palustris*, Fig. 3.14A viviparous: *Viviparus georgianus*, see Fig. 2.14C) or rather sharply in iteroparous *Ostrea edulis*, (Fig. 3.14C, D) but progressively in semelparous cephalopods (e.g. *Euprymna tasmanica*, Fig. 3.14B). 2. Seasonal Fecundity (SF) is the BF multiplied by the number

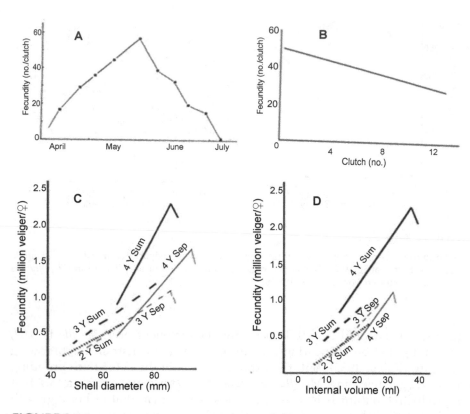

FIGURE 3.14

A. Fecundity as a function of successive clutches in snail *Lymnaea palustris* with advancing calendar month (redrawn from Hunter, 1975). B. Fecundity as a function of successive clutches in squid *Euprymna tasmanica* (redrawn from Squires et al., 2013). C. Fecundity (in number of veliger) as a function of shell diameter and D. internal volume of the oyster *Ostrea edulis* during different spawning seasons. Note the changes in the levels of direct relationships between fecundity and shell size/internal volume with age and season (Sum = summer, Sep = September) (redrawn from Walne, 1964).

of spawning during a reproductive season. In temperate waters, spawning occurs during the favorable season. For example, *O. edulis* spawns first time during June–July and the second time in September (Fig. 3.14D). Similarly, some cephalopods are characterized by synchronized intermittent multiple terminal spawning (e.g. *Dosidicus gigas*, see Table 1.9. 3. Life time fecundity (LF) is SF multiplied by the number of years. Following sexual maturity at the age of 3+ years, *Patella vulgata* spawns once a year for the rest of its life of > 11 years (Wright and Hartnoll, 1981). *O. edulis* spawns twice a year until its life time of 7+ years, after sexually maturing at the age 2+ years (Walne, 1964). Similarly, following sexually maturity at the age 3+ years, the freshwater mussel *Quadrula putulosa* continues to spawn once a year during the remaining 47 years of its life (Haag and Staton, 2003). Conversely, majority of cephalopods and opisthobranchs are semelparous. Hence, the reported BF values represent. In the squid *S. atlantica*, it amounts to 1.6 times its body weight (Rodriques et al., 2011). 4. Relative Fecundity (RF) is the number of eggs spawned per unit weight of molluscs. It provides considerable scope for comparative analyses of reproductive performance in different populations and species, in which RF varies temporally and/or spatially. Facing problems of molluscs with shell and without shell (slugs, most cephalopods) as well as the differences in weight of capsular wall/gelatinous mass, molluscan biologists have not reported RF values. Rarely, a single RF value of 97.5 oocytes/g body weight of the coleiod semelparous cephalopod *Octopus insularis* is reported (de Lima et al., 2014). 5. The total number of oocytes contributing to fecundity is assured by waves of oogonial proliferation and subsequent recruitment. 6. Potential Fecundity (PF) is the maximum number of oocytes commencing to differentiate and develop. However, due to one or other environmental factors like food supply (see Table 3.24), a fraction of these oocytes undergo atresia. The atretic oocytes form one of three sources of reserve for oocyte development (e.g. *Pecten maximus*, Le Pennec et al., 1991). 7. Realized fecundity (RlF) is a fraction of PF actually developed as eggs. Despite brooding embryos through the entire length of all the four gills, many freshwater mussels are left with undeveloped oocytes ranging from 1.3% in *Amblema plicata* to as much as 48.0% in *Fusconaia cerina* (Haag and Staton, 2003). In many bivalves, there are wide differences between PF and RlF. In cephalopod *Ocythoe tuberculata* too, PF and RlF of a small mature female (36 g with 3.6 Ovarian Somatic Index [OSI]) are 180,000 oocytes and 21,000 eggs. The corresponding values for a larger female (126 g with 8.6 OSI) are 220,000 oocytes and 25,000 eggs (Tutman et al., 2008). Irrespective of the differences in body size and OSI, the cephalopod realizes ~ 11.5% of PF only. In both freshwater mussel and cephalopods, the Oogonial Stem Cells (OSCs) seem to produce ~ 9-times more oogonia than their capacity to realize them into mature eggs. Unlike fishes and crustaceans, molluscs may not undergo reproductive senescence due to progressively reduced oogonial production by OSCs with increasing age/size.

8. Determinate Fecundity (DF) means the presence of a definite number of oocytes undergoing synchronized maturation. Consequently, the eggs are spawned in a single continuous pulse within a short time (e.g. *S. atlantica*, Rodriques et al., 2011) or intermittent pulses during a short period (e.g. 3.5 days in *Alderia willowi*, Krug, 1998). 9. Conversely, Indeterminate Fecundity (IF) means the presence of a large number of oocytes undergoing asynchronized maturation. It results in continuous spawning (*Idiosepius pygmaeus*, see Table 1.9) or polycyclic spawning pulses interspersed (with periods of feeding and growth) by prolonged intermittent periods (e.g. *Nautilus* spp, see Table 1.9). Planktotrophics may represent determinate fecund capital (see Kjesbu et al., 2010) brooders. On the other hand, lecithotrophics may be indeterminate

TABLE 3.24

Effect of selected factors affecting fecundity of molluscs (compiled from Runham, 1993)

Species	Reported Observation
	Sexuality
Lymnaea stagnalis	Rearing in groups facilitated cross breeding and advanced egg laying but reduced fecundity by 50%. Self fertilization delayed egg laying but doubled fecundity
Eubranchus doriae	Isolated aeolids died without laying eggs. When paired, fecundity depended on volume of water. The more crowded, the aelolid was less fecund
Bulinus contortus	Aphallics were more fecund (5609 eggs) than euphallics (4498 eggs)
	Food availability
Thais lamellosa	Food abundance increased fecundity from 930 to 1428 eggs
T. emarginata	Food abundance increased clutch number and fecundity
Kathurina tunicata *Patella vulgata*	Despite 5 months starvation, gametes were produced reducing somatic growth
Mytilus edulis	Equal amount of gametes produced at the expense of somatic growth
Lymnaea stagnalis	Starvation delayed vitellogenesis
Biomphalaria glabrata	Prolonged starvation ceased oviposition and resorption of oocytes
L. peregra	Vitamin E profoundly stimulated egg production
Geukensia demissa	Spinach produced more eggs that algal food
Choromytilus meridionalis Griffiths (1977)	Gametogenesis delayed in populations occupying high tides with less accessibility to water and food
Euprymna tasmanica (Steer et al., 2004)	High mortality of eggs spawned by mothers fed on low ration
	Temperature
Lymnaea stagnalis	Increase in temperature stimulated oviposition
Macoma balthica	Increase in temperature stimulated spawning
Cardium edule	Temperature ~ 14°C was required to stimulate gametogenesis
	Photoperiod
Melampus bidentatus	Photoperiod of 14 h was required for gonad maturation
Lymnaea stagnalis	Long day (16 L : 8 D) snails were 9-times less fecund due to earlier death than short day (8 L : 16 D) snails due to their longevity
Pseudosuccinea columella	Egg mass production was inversely proportional to day length
Lymnaea stagnalis	Yellow light decreased fecundity by 13% but red light increased it by 77%
Sepia officinalis (Mangold, 1987)	Short photoperiod induced precocious sexual maturity. High light intensity inhibited maturation
	Density
Bulinus tropicus	High densities reduced fecundity due to accumulated waste
	pH
Amnicola limosa	Fecundity decreased by 66% in acidic waters (pH 5.8) containing < ~ 1 mg calcium/l, a critical limit for egg laying, in comparison to alkaline waters with pH 7.6
	Shell thickness
Nucella emarginata	Thin shelled morphs were more fecund than thick shelled morphs

fecund incoming brooders (Pandian, 2013, p 47–53, 2014, 36–40, 2016, p 74–76).

Table 3.24 represents a condensed compilation from Runham (1993), which adequately explains the effects of biotic and abiotic factors on gametogenesis and fecundity of molluscs. By affording nutrients and space to accommodate ripening eggs, body size is a more decisive factor that significantly affects the relationship between it and fecundity. Uniquely, the presence of nurse eggs/siblings considerably complicates the quantitative relationship between body size/age and fecundity. In molluscs body size is measured as a function of shell length (e.g. Prosobranchs: *Lymnaea palustris*, Hunter, 1975, *Patella peroni*, Parry, 1982), Opisthobranchs: *Limacina retroversa*, Dadon and de Citre 1992, Bivalves, *T. tantilla*, Kabat, 1985), shell height (e.g. *V. geogianus*, Browne, 1978), shell diameter (e.g. *O. edulis*, Walne, 1964), mantle length (e.g. Cephalopods: *S. atlantica*, Rodriques et al., 2011), internal volume (e.g. *O. edulis*, Walne, 1964), age (e.g. freshwater bivalves, *Elliptio arca*, *Quadrula asperata* or number of mating partners (e.g. *Biomphalaria glabrata*, Vianey-Liaud et al., 1987), or capsule size (e.g. *Nucella crassilabrum*, Gallardo, 1979), capsular volume (e.g. *Buccinum undatum*, Smith and Thatje, 2013), capsular wall area (e.g. *Crepipatella dilata*, Zelaya et al., 2012) or string (e.g. *Aplysia kurodai*, Yusa, 1994).

Firstly, the level of complication introduced by the inclusion of nurse eggs within a capsule was assessed in some gastropods, considered representative species, for which adequate data are reported. In muricid *N. crassilabrum* (Gallardo, 1979), both fecundity including nurse eggs and number of embryos in the capsules increased with increasing capsular length, although at different levels (Fig. 3.15A, B). These linear and positive relationships hold true also, when capsular volume was considered in another representative species *B. undatum* (Fig. 3.15C, D). Despite varying proportions of nurse eggs are ingested by the embryos within a capsule, the linear relationship between fecundity and capsule size remains unaltered.

In majority of molluscs, a linear and positive relationship was apparent between fecundity and body size, for example, in different populations of freshwater mussels *A. plicata*, and *F. cerina* (Fig. 3.16A, C). In cephalopod *S. atlantica*, the relationship between egg number and mantle length was more strongly correlated ($r^2 = 0.865$) than that between egg volume and mantle length ($r^2 = 0.65$) (Rodriquez et al., 2011). In *Aplysia californica*, in which eggs are enveloped in string embedded in gelatinous mass, the linear relationship between fecundity (either in mg string or whole string weight) and body weight was apparent (Fig. 3.12B). The relationships holds good for both gametes release by *C. meredionalis* (Fig. 3.13B). Almost all the available evidence goes to confirm that with the positive relationship between fecundity and body size, molluscs do not undergo reproductive senescence. This would also imply that the Oogonial Stem Cells (OSCs) continue asymmetric divisions to generate OSCs and oogonial cells until the life time of molluscs (see Pandian, 2012, p 86–87). As aquatic invertebrates, the

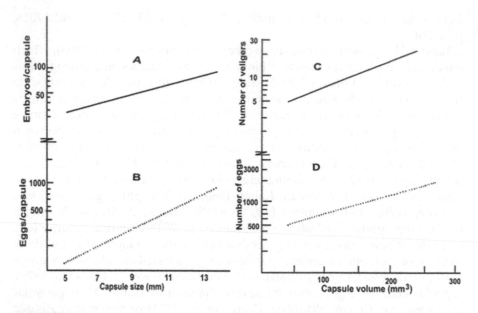

FIGURE 3.15

Fecundity [expressed in number of B. eggs/capsule and A. embryos/capsule] as a function of capsule size in muricacean gastropod *Nucella crassilabrum* (redrawn from Gallardo, 1979). Fecundity [expressed in D. number of eggs and C. veligers] as a function of capsule volume in the common whelk *Buccinum undatum* (redrawn from Smith and Thatje, 2013).

molluscs are known to enjoy indeterminate growth. Expectedly, they do not suffer from reproductive senescence. However, a few temperate freshwater mussels *Lampsilis arnata* and *Quadrula asperata*, with a relatively long life span of 20 years show a mixed, i.e. positive and negative relationships, when their fecundity is related to body length and age, respectively (Fig. 3.16D). More confusingly, the relationship is negative beyond a particular size and age of *Q. pustulosa* (~ 65 mm body length and after an age of 25+, Fig. 3.16B). Clearly, these long living mussels do undergo reproductive senescence, as they may also be characterized by determinate growth. More research is required to resolve whether the molluscs undergo reproductive senescence or not.

Besides body size, many other factors also alter fecundity. Shell colors can be one such factor. For example, the shell color of the apple snail *Pomacea canaliculata* can be brown or yellow. Crossing experiments covering three generations have revealed that color inheritance follows Mendelian inheritance and the yellow color is recessive to brown. Fecundity, egg productivity, egg size and hatchability vary widely indicating no advantage to yellow or brown color morph (Table 3.25). At the best pairing of brown female to yellow male seems to have an optimal advantage. Pairing a yellow female first by brown male and subsequently by yellow male, and vice versa

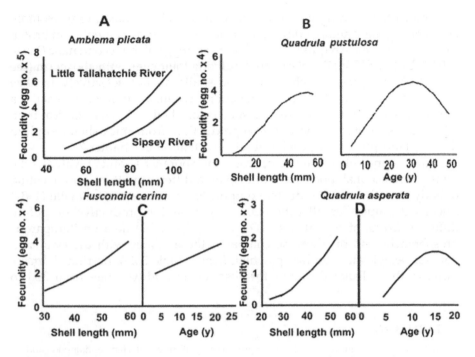

FIGURE 3.16

Fecundity as functions of shell length and age in freshwater mussels A. *Amblema plicata*, B. *Quadrula pustulosa*, C. *Fusconaia cerina* and D. *Quadrula asperata*. Note the differences in the trends of fecundity-shell length in different populations of *A. plicata*. Note the reproductive senescence displayed by *Q. pustulosa* and *Q. asperata* (redrawn from Haag and Staton, 2003).

TABLE 3.25

Fecundity, egg productivity and hatchability in dimorphically colored apple snail *Pomacea canaliculata* (0.6 g) observed for 60 days. B = Dominant brown shell color, Y = Recessive yellow shell color (compiled from Yusa, 2004a)

Pairs/First Male	Fecundity (Egg no.)	Egg Weight (mg)	Egg Productivity (g/d)	Hatchability (%)
b ♀ X b ♂	121	9.7	1.05	71
y ♀ X b ♂	110	13.5	0.36	61
b ♀ X y ♂	190	10.9	0.50	79
y ♀ X y ♂	86	10.9	0.23	64
b ♂ X y ♀	133	10.4	0.26	72
y ♂ X b ♀	182	9.4	0.37	66

has shown that pairing brown female first by yellow male produces more eggs per egg mass and per day. In sequence of male replacement from brown to yellow, the female requires 10–28 days to sire eggs of the second male (Yusa, 2004a). Unlike SH snails selectively fertilizing their eggs with allosperm (see p 99–100); the prosobranch gonochoric snails do not selectively fertilize their eggs with sperm from the dominant brown male. In molluscs, shell is a flexible phenotypic trait. More than shell color, shell weight is a critical trait, which alters resource allocation to reproduction, from thick shell morph to thin shell morph. This aspect has received attention in Chapter 2.

Unlike some of these iteroparous prosobranchs and bivalves, the semelparous cephalopods seem to display a different pattern for allocation of body growth and reproductive output. From limited data, Saville (1987) calculated values for allocation for growth and reproductive output in different taxons of cephalopods. For want of suitable data on iteroparous prosobranchs and bivalves, he compared these values with the estimated data for pelagic and benthic spawning fishes (Table 3.26). Barring *Todarodes pacificus*, the calculated values for absolute fecundity range from 300 to

TABLE 3.26

Comparative account on growth and reproductive output in selected cephalopods and representative pelagic and benthic spawning fishes (compiled from Saville, 1987)

Taxon/Species	Ratio of growth to maximum body weight	Absolute fecundity (egg no.)	Approximate relative fecundity (no./g)
Cephalopods			
Squids			
Loligo opalescens		4,250	60
L. pealei	0.66	40,000	80
Illex illecebrosus	–	–	–
Todarodes pacificus	0.77	400,000	1,000
	0.99		
Octopods			
Octopus vulgaris	0.80	250,000	200
Eledene moschata	0.86	300	0.60
Cuttlefish			
Sepia offcinalis	0.80	350	0.45
Fishes			
Gadus morhua	0.10	2,000,000	400
Melanogrammus aeglefinus	0.11	400,000	600
Pleuronectes platessa	0.07	–	–
Clupea herangus	0.12	70,000	385

250,000 eggs and approximate relative fecundity from 0.45 to 200 eggs per g body weight. These values are lower than the estimated data for pelagic and benthic spawning fishes. Clearly, the semelparous cephalopods allocate much less for reproduction. Not surprisingly, they invest two–eight times more on body growth. Though more data are required to confirm these generalizations, it is an important consideration for enterprising cephalopod aquaculturists. Incidentally, for a given unit of food, the cephalopods convert ~ 50% of the food into its body (O'Dor and Wells, 1987), as compared to ~ 20–30% by fishes (Pandian, 1987). Within a period of 300 days, the cephalopods grow to 5.0 kg in *Octopus vulgaris* and *O. maya* (Forsythe and Van Heukelem, 1987) and 6.5 kg in *O. cyanea* (Van Heukelem, 1983).

Egg size is another decisively important factor in recruitment. In opisthobranch *A. californica*, egg diameter remains constant at ~ 85 µm in the sea hares weighing 20 to 140 g (Yusa, 1994). However, many studies have shown that the egg size is positively correlated with maternal body size (e.g. nudibranch: *Adalaria proxima*, Lambert et al., 2000; sacoglossa: *Elysia stylifera*, Allen et al., 2009). In the bobtail squid *S. atlantica* too, both egg volume and possibly egg weight do not increase with body weight (Rodriques et al., 2011). Still others have found no correlation between these parameters (e.g. *Crepidula*, Collin and Salazar, 2010; bivalve: *Mytilus californianus*, Phillips, 2007). In fact, a non significant negative relation between egg size and body size has been reported for four planar spawning *Chromodoris*. In nudibranch *C. magnifica*, egg size decreases from 125 to 150 µm in females of 40 and 60 mm body length (Trickey et al., 2013). It is likely that scope for intra-specific variation in egg size is limited (i.e. in others than poecilogonous species), as both opisthobranchs and caenogastropods receive the required nutrients from extra embryonic sources, capsular fluid and nurse eggs as well as extra embryonic yolk granules.

4

Regeneration and Asexual Reproduction

Introduction

A taxonomic survey on the modes of reproduction in aquatic invertebrates reveals that the hemocoelomates namely Arthropoda and Mollusca do not reproduce asexually. A new hypothesis was proposed by Pandian (2016) that these hemocoelomates may not have retained Embryonic Stem Cells (ESCs) to agametically clone the whole animal and reproduce asexually. However, *Clio pyramidata*, a pteropod is claimed to reproduce asexually (van der Spoel, 1979). With ongoing differentiation during early embryonic development, the stemness of stem cells progressively decreases from totipotency to pluripotency, multipotency, oligopotency and finally to unipotency (Pandian, 2016, p 122). This account attempts to explain that the claim of asexual reproduction in the pteropod is more an exceptional regeneration involving Multipotent Stem Cells (MSCs).

Studies on regeneration of molluscs have a long history. Many molluscs seem to have retained MSCs, as they are able to completely regenerate one or more organs (including brain) derived from ectoderm, mesoderm and endoderm. In general, most of these organs are either broken, when predating the prey or predated partly or wholly. At an age dependent rate, gastropods can completely regenerate tentacles including an eye and a large olfactory organ as well as proboscis; they may regenerate gonad and penis but not after sexual maturity. In both young and old deposit-feeding lamellibranchs, the inhalant and exhalant siphons are completely regenerated, irrespective of age of the bivalves, from which they are nipped by predators. Arms of cephalopods are also regenerated until sexual maturity, but the heterocotylized arm used to transfer spermatophore is regenerated in mature adults also.

Runham (1993) recognized two types of regeneration and named them as (i) physiological and (ii) restorative. The continuous replacement of cells in expanding and replacing tissues is named physiological regeneration. For example, the radula is continuously produced by gastropods, as it is worned-out due to frequent use in mastigating food. Its production requires multiplication of secretary epithelial cells of the radular gland. Radula is produced at the rates exceeding five rows of teeth a day. But the restorative regeneration involves a static population of cells that do not divide after embryonic development, and if these cells are lost, they are not replaced. This chapter is limited to restorative regeneration.

Regeneration commences with wound healing to blastema formation and then to recovery. Immediately following, siphon nipping, muscular contractions constrict the wound area to arrest loss of hemocoel. Epidermal cells aggregate to form a layer over the wound. Amoebocytes migrate to the wound area to engulf the necrotic tissues. When wound involves a small area of the outer skin of partly retracted tentacle, regeneration does not proceed beyond wound healing in some gastropods. To induce artificial pearl formation, the mantle of bivalve is excised to insert nacre, over which a pearl is formed. The mantle excision causes a large wound and severes the pallial artery that necessitates rapid wound healing to avoid death by bleeding. Following mantle excision in the Akoya pearl oyster *Pinctada fucata*, muscular contractions effectively reduce the size of the wound within 1 hour after mantle excision. During the next 3–6 hours, accumulation of hemocytes and connective tissue is followed and the wound is plugged within 6 hours after excision (Acosta-Salmon and Southgate, 2006). Wound healing is followed by the formation of semispherical or conical blastema. With accelerated mitosis, differentiated muscles (mesodermal derivative) and nerve cells (ectodermal derivative) begin to appear. This is followed by recovery of form and function of the wounded or lost organ.

4.1 Gastropods

Shell: Following removal of a small piece of shell from the whorl adjacent to the aperture without injuring the underlying membrane in *Neritina virginea*, the exposed surface is soon bound by a membrane. Into this membrane, calcareous crystals of varied shapes are deposited to form the periostracum first. Then, a membrane is gradually formed along the remaining layers in such a manner that the solid calcareous layer is continuous with the existing shell. In the regenerated piece, shell ornamentation with colored spiral bands appear, as it has been in the original shell. The regeneration of a piece of shell may last for 17 months. When a larger piece of shell involving one or more whorls far away from the aperture is removed, the removed shell is, however, not regenerated (Andrews, 1935). Injection of ecdysterone α-ecdysone significantly increases calcium deposition into regenerating shell of *Biomphalaria glabrata* (Whitehead, 1977). Understandably, regenerative capacity progressively decreases from the growing edge of the aperture toward spire and apex. Incidentally, holes in *Haliotis* shell are closed from the distal end of the aperture (Fig. 2.2G).

Proboscis: In muricid gastropods, the proboscis and associated organs play an important role in drilling the prey's shell. Drilling and consuming the prey may last for 2.5 days (see p 55). During this long period, the snail runs the risk of losing its proboscis either to a predator or when its irritated prey suddenly clamps shut its valves. Following its loss or proboscisectomy, a blastema consisting of mass of loose cells including numerous amoebocytes bound the amputated ends of oesophagus, buccal artery, ducts of salivary glands and other tissues to form the stump of the proboscis. Being a relatively hard structure, the radula provides a readily measurable parameter for quantitative determination of regeneration. Figure 4.1 shows that a proboscis capable of drilling is formed within 20–30 days following proboscisectomy. In histological details, regeneration of odontophoral cartilages of *Urosalpinx cinera* is strikingly similar to the regenerating limb of vertebrates such as salamander. In both cases, the regeneration process depends on a unique 'cap cell aggregate' (Carikker et al., 1972).

Gonads: Pulmonates have attracted much attention on regenerative testis, vas deferens and penis. The earliest demonstration is that of Gould (1952), who reported the regeneration of testis and penis in juveniles (8–9 mm) of the protandric *Crepidula plana*. Confirming the earlier reports of Laviolette (1954, 1955) on testis regeneration of the gonadal grafts transplanted into the hemocoelom of arionid (*Arion rufus*) and limacid, Runham (1978) reported that the testis regeneration was possible only prior to the appearance of

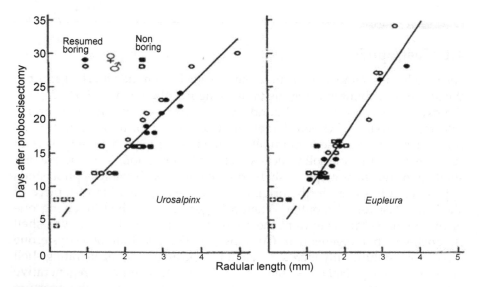

FIGURE 4.1

Extent of radular regeneration in female and male *Urosalpinx cinerea* and *Eupleura caudata etterae* during the 30-d period following proboscisectomy (from Carriker et al., 1972).

spermatids in the testis of *Deroceros reticulatum*. Briefly, the epithelial cells of gonoduct in the proximity of the gonad retain the regenerative ability (Tardy, 1971). These epithelial cells retain the embryonic potency to differentiate into epithelial tube, from which acini develop at first as solid mass and then fill the testis with differentiating gametes—a process very similar to the embryonic testicular differentiation. However, their differentiating potency is lost following their differentiation from ciliary to secretory cells at sexual maturity. In many aeolid nudibranch species, these cells retain the potency to regenerate testis almost throughout the life span following castration, although the rate of regeneration slows down in old animals (Tardy, 1971). Removal of the penis of *Helix pomatia* results in the complete resorption of all other male organs. Hence, it is not clear where exactly the regenerative cells are stored in the reproductive system.

A prelude on stem cells and their role in sexualization is necessary to understand regeneration of the testis in castrated snails. With retention of ESCs, an organism may agametically clone and reproduce asexually. In sexually reproducing animals, Primordial Germ Cells (PGCs) and their derivatives Oogonial Stem Cells (OSCs) and Spermatogonial Stem Cells (SSCs) are responsible for manifestation of sex. In fishes, these stem cells manifest female and male sexes in the presence of OSCs and SSCs, respectively. In the presence of OSCs and SSCs together, simultaneous hermaphrodites are produced. SSCs are lost, when protandrics switch from male to female sex, and the reverse is true of protogynics (Pandian, 2012, p 903–102, 2013, 213–224). Moreover, whether the gonad is developed into ovary or testis, or ovotestis is of paramount importance in the context of regeneration of castrated or ovariectomized gonad. For example, surgical removal of the entire testicular lobe located latero-ventrally in the ovotestis of the protandric black porgy *Acanthopagrus schlegeli* results in the development of the centrally located ovary in the ovotestis and sex change from male to female. On the other hand, the ovariectomized protogynic bluehead wrasse *Thalassoma bifasciatum*, which has presumably retained PGCs/OSCs only in the ovary alone, was unable to regenerate the ovary (Pandain, 2013, p 174).

With regard to gastropods, all the investigations on the effects of castration were made either on protandric hermaphrodites (*C. fornicata*, aeolid nudibranchs, Hyman, 1967, p 293, 482) or simultaneous hermaphrodites (pulmonates). These gastropods have a single ovotestis. However, the ovotestis is not lobular, as in *A. schlegeli* but is either follicular or tubular. In *C. onyx*, the testicular tube is a highly branched organ (Fig. 4.2A) and the large ovary is also dispersed within the visceral mass (Fig. 4.2B). In nudibranchs, pteropods, onchidiceans, the testicular follicles are located centrally but the ovarian follicles/ductules peripherally. The locations of these follicles are reversed in shelled pteropod *Hyalocylix striata*. In the ovotestis, the testicular follicles are located anteriorily in *Archidoris* but posteriorly in *Rhodope* and on the right side in *Sapha*. The ovarian and testicular components of the ovotestis are completely separated from each other (delimited type) in

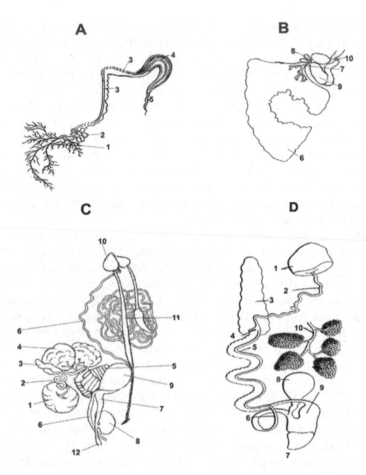

FIGURE 4.2

Reproductive system: protandric prosobranch *Crepidula* A. male 1 = testis, 2 = seminal vesicles, 3 = sperm groove, 4 = penis, 5 = flagella. B. female 6 = overy, 7 = oviduct, 8 = seminal receptacles, 9 = pallial oviduct, 10 = female gonopore. C. hermaphroditic nudibranch *Onchidium*, 1 = ovotestis, 2 = hermaphroditic duct, 3 = ampulla, 4 = albumin gland, 5 = mucous gland, 6 = sperm duct, 7 = oviduct, 8 = spermatocyst, 9 = prostrate, 10 = penial sac, 11 = penis, 12 = vagina and D. hermaphroditic pulmonate *Arion*, 1 = ovotestis, 2 = hermaphroditic gland, 3 = albumin gland, 4 = fertilization chamber, 5 = sperm oviduct, 6 = sperm duct, 7 = penial apparatus, 8 = spermatheca, 9 = vagina, 10 = enlarged ovotestis (Free hand drawings from Hyman, 1967).

Aceton, Tylodina, Pleurobranchaea, Lobiger, Archidoris (Hyman, 1967, p 482). Pulmonates have an ovotestis deeply embedded in the mid gut (Hyman, 1967, p 592), rendering the surgical removal of the testicular follicles/tubules as the most difficult and delicate task. The ovotestis consists of variable number of follicles ranging from five in *Ferrissia tarda* (see Fig. 4.2D for

Arion sp) to 100 in *Lymnaea stagnalis*. It is difficult to know whether Gould (1952), Laviolette (1954, 1955) and Tardy (1971) have correctly identified the testicular lobes from the experimental gastropods *C. fornicata*, aeolids and *Arion rufus*, respectively and entirely removed all the testicular lobes during the years, when suitable surgical instruments may not have been available. The forthcoming description suggests that they have not completely removed all testicular lobes in these gastropods.

In *Crepidula*, sex change from male to female involves (i) penial resorption, (ii) degeneration of testicular follicles, (iii) development of a few oogonia (OSCs) that have were present in the ovarian follicles of the ovotestis from the beginning and (iv) alter reproductive duct into female type (Hyman, 1967, p 295). However, female organs may not redifferentiate so long the spermatogonial stem cells (SSCs) are present in the follicles that have not been removed. As all the testicular follicles may not have been correctly identified and entirely removed, the castrated males of *Crepidula*, aeolids and *A. rufus* could have resumed, and regenerated testis from existing unremoved testicular tubes/follicles. It is likely that all the testicular tubes/follicles were not correctly identified and completely removed by Gould (1952), Laviolette (1954, 1955) and Tardy (1971) in the castrated *Crepidula*, aeolids and *A. rufus*, respectively. Hence, the castrated gonads resume and regenerate testis only. Incidentally, chemical means of castration by feeding on busulfan-supplemented diet (see Pandain, 2011, p 150–151) may eliminate the highly branched (as in *Crepidula*), unidentifiable (as in aeolids), deeply embedded (as in pulmonates) testicular lobes of these gastropods.

Brain: The mammalian brain has the most limited regenerative capacity following injury or damage. However, robust recoverability of the molluscan neurons has long been recognized. Some molluscan species display remarkable ability to regenerate the brain, and can achieve complete functional recovery after injury (Matsuo and Ito, 2011). Following surgical ablation of unilateral cerebrum in *Melampus bidentatus*, commissural connections were spontaneously re-established between the cerebro-lateral cerebral ganglion and the cut ends of the remaining nerves. And the newly formed junction point was slowly enlarged to form a ganglion within a period of 11 months (Moffet, 1995). On implantation of cerebral ganglion into the hemocoelom of *M. bidentatus*, 50% of the grafts developed an eye on the surface of the ganglion and 25% induced supernumerary tentacle often with an eye. A similar observation has also been recorded in another ellobiid *Ovatella mysotis*. Hence, molluscan neurons readily regenerate axons and establish specific connection with target organs. Whereas the severed nerve was regenerated in *Helix pomatia* the dart sac, mucous glands and flagellum failed to regenerate following their surgical removal (Jeppersen, 1976). Experimental studies of Matsuo and Ito (2011) have revealed that the cerebral ganglion is equipped with procerebra (PC) in the terrestrial pulmonates like *Helix* but that of aquatic pulmonates are not (Van Mol, 1967,

Chase, 2000). Unique to the gastropod nervous system, the PC grows by mitotic division from 30–40 day post-hatching (dph) to 20,000–50,000 small uniform sized neurons, even with advancing age and increasing size (e.g. *Limas maximus*, Watanabe et al., 2008, *H. aspersa*, Longley, 2011). The PC plays both olfactory and mnemonic role. Hence, it is analogous to the olfactory bulb and hippocampus of mammals. An intriguing feature is that the PC continues to renew their constituent neurons through neurogenesis almost throughout the life of the terrestrial pulmonates (Matsuo and Ito, 2011).

The bimodal breather *Lymnaea* acquires the required oxygen via cutaneous and/or aerial respiration. The snail is shown to associatively learn not to perform aerial respiration under unfavorable environmental condition (Scheibenstock et al., 2002). The neuron RPeD1, one of the three neurons of Central Pattern Generator (CPG), is known to initiate the aerial breathing activity (Spencer, 2002). Aerial respiratory behavior can be abolished, when the axon of RPeD1 is crushed (Haque, 1999). However, the neurite is regenerated within 10 days after axotomy. Confirming the earlier findings of Scheibenstock et al. (2002), Lukowiak et al. (2003) have reported that (i) the regenerated nervous system is competent to mediate associate memory, i.e. LTM, (ii) LTM survives axotomy and the subsequent regeneration process, (iii) but not the already established memory (ITM).

Tentacles: Readily amenable to amputation, the stylommatophoran pulmonates have served as ideal model for extensive investigations on tentacle regeneration. In the slug *Arion alter* and *Deroceros reticulatum*, a normal functional tentacle including an eye and a large olfactory gland is regenerated within 60 days (Chetail, 1963). In general, regeneration is faster in *Nassarius mutabilis* than in *Viviparus viviparus* (Hyman, 1967, p 321). However, removal of tentacle together with cerebral ganglian in *Patella vulgata* has led to wound healing but not to differentiation of muscle and nerve cells (Choquet and Lemaire, 1968). In *C. fornicata*, Le Gall (1979) has demonstrated that the tentacle abalation without involving the cerebral ganglion results in complete regeneration. He has also shown that the grafts containing cerebral ganglion of *Thais lapillus* (gastropod) and *Mytilus edulis* (bivalve), on transplantation into a decerebrated *C. fornicata* has led to accelerated mitosis. Obviously, a neurosecretory hormone/mytogenic factor is responsible for initiation of tentacular regeneration in *C. fornicata*. Notably, the mitogenic factor is not species specific. Miles (1961) has reported that the regeneration rate is affected by age and hibernation. Briefly, tentacular regeneration involves the formation of new sensory organs and nerve ganglia in a process very similar to embryonic differentiation. In fact, tentacles, eyes, adjacent frontal lobes, operculum, parts or all of the metapodium, entire foot and siphons are all regenerated perfectly (Hyman, 1967, p 321). Runham (1993) suggested that the differentiation potency for regeneration may be related to the frequency, with which these organs are damaged or removed by predators.

Following amputation of an arm of a cuttlefish *Sepia officinalis*, Feral (1988) observed the migration of hemocytes towards the wound, massive synthesis of collagen and spreading of epithelial cells over the wound. Within 3 months following amputation, a thin but fully functional arm was completely regenerated. However, regeneration of an amputated arm from an aged/senescent cuttlefish is limited to wound healing (Feral, 1979). In cephalopods, castration is not followed by regeneration (see Runham, 1993).

4.2 *Clio pyramidata*

In 1979, van der Spoel startled the malacological world by claiming the occurrence of asexual reproduction by strobilization in a pteropod *Clio pyramidata*. His evidence is based on histological observation of 100 specimens. However, no author has thus far checked the claim by van der Spoel. During the so called strobilization in *C. pyramidata*, the columnar muscle running from apex to mouth marked the embryonic axis of the species. The fission between two upper and lower daughter clones ran perpendicular to the long axis (Fig. 4.3). Each daughter clone received a branch of the columnar muscle and septal organs like the penis, accessory

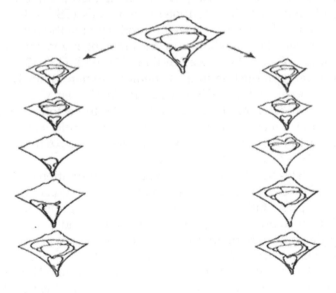

FIGURE 4.3

Regeneration in *Clio pyramidata*. Figures in left and right sides show regeneration from posterior and anterior direction, respectively (modified and redrawn from van der Spoel, 1979).

sexual gland and a lobe of hepatoprancreas glands. The clones were fully developed, after the development of a new gut in the lower clone and a shell in the upper clone together with regeneration of all other missing organs. The following may be inferred from the brief description of van der Spoel: 1. The columnar muscle, a mesodermal derivative, possesses stem cells, from which, all the missing somatic organs are regenerated, 2. the hepatopancreas gland, an endodermal derivative, also possesses stem cells responsible for the development of all missing organs of the alimentary canal system and 3. The septum, a mesodermal derivative, has Primordial Germ Cells (PGCs)/Oogonial Stem Cells (OSCs)/Spermatogonial Stem Cells (SSCs) responsible for manifestation of sex and gonadal development. Notably, the stem cells required for regeneration do not arise from a single source of Pleuripotent Stem Cells (PSCs), as expected in asexual reproduction but from three different sources, each of which is capable of regenerating only specific organ systems and not the whole animal. Hence, the stem cells arising from these three sources are more of Multipotent Stem Cells (MSCs) rather than PSCs. Briefly, *C. pyramidata* seems to possess adequate number of MSCs in the columnar muscle and hepatopancreas gland to regenerate the missing somatic organs and the septum has adequate PGCs/OSCs/SSCs to regenerate the missing gonads in the daughter clones. In fact, the word 'regeneration' is used more frequently by van der Spoel than asexual reproduction. Hence, the claimed (but not yet confirmed by others) 'asexual reproduction' in *C. pyramidata* is more of regeneration involving MSCs rather than asexual reproduction involving pluripotent (PSCs)/totipotent stem cells (TSCs). Incidentally, it may be noted that the claimed asexual reproduction in the parasitic colonial rhizocephalan female was shown to be more of regeneration involving Oligopotent Stem Cells (OSCs) (see Pandian, 2016, p 122–123, 130). Further, the claimed asexual reproduction in these hemocoelomates namely Arthopoda and Mollusca occurs in three of 52,000 species (0.006%) in Crustacea and one of 100,000 (0.0001%) in Mollusca. Therefore, the three–four crustacean and gastropod species may be more an exception than rule. Hence, Pandian's hypothesis that hemocoelomates do not reproduce asexually, as they have not retained ESCs/TSCs/PSCs, may hold true.

Table 4.1 collates bits and pieces of available information on the presence and source of PGCs/OSCs/SSCs, by which the testis/gonad is regenerated. Fortunately, representative observations have been reported for the gonadal regenaration in a protandric (*Crepidula fornicata*) and simultaneous (e.g. aeolids, pulmonates) hermaphrodites and major taxons of gastropods namely prosobranchs: *C. fornicata*, opisthobranchs: *Eubranchus doriae* and pulmonates: *Biomphalaria glabrata*. From the available limited information, the following may be generalized: 1. Regeneration of gonad is possible only in immature gastropods and pulmonates. The PGCs/OSCs/SSCs, responsible for manifestation of gonads, are present in the gonoduct. Differentiation of the gonoductal epithelial cells from ciliary to secretory lining marks the loss

TABLE 4.1

Effect of castration/gonodectomy on regeneration of testis/ovotestis in gastropods

Species/Reference	Reported Observations
Crepidula fornicata Gould (1952)	Castrated immature males regenerated testis and penis
Biomphalaria glabrata Vianey-Liaud (1972)	Young immature snail regenerated gonad (ovotestis) but not old mature snail
Vianey-Liaud and Dussart (2002)	Following surgical removal of ovotestis from mature snails, neither testis nor ovary was regenerated, indicating the loss of PGCs/OSCs/SSCs from their entire body of the snail
Deroceros reticulatum Runham (1978)	Testis was regenerated until the appearance of spermatids in the gonad
Hogg and Wijdenes (1979)	Gonad was regenerated until the ciliary epithelial cells lining of the gonoduct was differentiated into secretory cells lining
Eubranchus doriae Tardy (1971), Tardy and Dufrenne (1976)	Castrated aeolid regenerated testis from the epithelial cells of gonoduct in the proximity of the gonad
Arion rufus Laviolette (1951)	Gonad was regenerated in the body cavity, when sections are gonoduct were transplanted
Aeolids Tardy (1971)	Castrated aeolid nudibranchs regenerated gonad almost throughout their lifetime
Clio pyramidata Van der Spoel (1979)	Gonad originated from the septum, a mesodermal derivative in this 'asexually' dividing pteropod

of these stem cells; the gonoduct loses the potential for gonadal regeneration, i.e. following sexual maturity. 2. However, the capacity to regenerate gonad is retained throughout the life time of opisthobranchs (e.g. aeolids, Tardy, 1971), which seem to retain these stem cells in the septum, a mesodermal derivative. Not surprisingly, the pteropod *C. pyramidata*, has also retained this capacity in mature individuals, which divide to reproduce 'asexually'.

5

Aestivation and Reproduction

Introduction

Also called 'diapause' and 'hibernation' (Hyman, 1967, p 613), aestivation refers to the phenomenon of 'deep sleep' undertaken by animals with no access for water and food to overcome unfavorable environmental conditions. The duration of aestivation may be as short as few hours/days, but regularly recurs, as in bivalves inhabiting intertidal zone and as long as from 1 month (e.g. *Pila globosa*, see Table 5.2) to 2–6 years in freshwater pulmonate *Planorbarius corneus* (Table 5.1) and as long as 23 years in a terrestrial snail *Oxystyla capax* (Hyman, 1967, p 613). In this chapter, the former is named as tidal aestivation and the latter as seasonal aestivation. Being an unusual phenomenon but ubiquitously occurring in animals from freshwater sponges to polar bears, aestivation has attracted much attention from the point of sub-cellular and molecular biological aspects (e.g. Navas et al., 2010). However, the present chapter is limited to physiological and ecological aspects of aestivation in aquatic molluscs from the context of reproduction.

5.1 Seasonal Aestivation

In molluscs, aestivation is prevalent among pulmonates inhabiting freshwater and terrestrial habitats. The tenacity to withstand long durations of aestivation seems to be more common among terrestrial than freshwater pulmonates. The terrestrial snail *Oxystyla capax* can withstand 23 years duration of aestivation and subsequently emerge successfully, in comparison to 2–6 years aestivation duration in freshwater pulmonate *Planorbarius corneus* (Table 5.1). Freshwater pulmonates *Fossaria parva*, *Lymnaea caperata*, *L. natalensis*, *L. peregra*, *L. tomentosa*, *L. truncatula* and *Stagnicola bulimoides techella* aestivate on or beneath the soil but *L. reflexa* and *Stagnicola exilis* do it on plants or tree branches (Jokinen, 1978). As may be seen, burying at greater depths facilitates survival of aestivating snails.

TABLE 5.1

Duration of aestivation sustained by some freshwater pulmonates (compiled from Hyman, 1967, p 613 and others)

Species	Duration of Aestivation
*Pila globosa**	1–5 months, > 2 years
*P. virens****	> 2 years
*Lymnaea elodes***	3–4 months
L. columella	5 months
L. palustris	9 months
L. stagnalis	14 months
L. peregra	15 months
Bulinus trancatus	13 months
Planorbis albus	16 months
P. boissyi	18 months
Ancylus fluviatilis	12 months
Physa gyrina	12 months–24 months†
Planorbarius corneus	2–6 years
Oxystyla capax†	23 years††

*Haniffa (1978b), **Jokinen (1978), ***Meenakshi (1956), †in laboratory, ††terrestrial pulmonate

From a pioneering study at Annamalai University, Tamil Nadu, India, Meenakshi (1956, 1964) reported that a 6 months long aestivation in *Pila virens* involved anaerobic metabolism accompanied by the following biochemical changes: (i) drop in glycogen content by 80–85 mg/snail (ii) a 30-fold increased accumulation of lactate (40–50 mg/snail) in tissues and consequent drop in blood pH 4.5 (iii) increase in magnesium content to 6 mg/100 ml blood, (iv) rise in glutamate to twice the normal level in cerebral ganglian (v) a 10-time elevation of asparagine in the non-saponifiable fraction of the ether extract of the cerebral ganglian. The unusual accumulation of lactate and reduced blood pH dissolves calcium content resulting in decreased shell thickness. Her seminal publications showed perhaps for the first time the scope for survival of aestivating animals under anoxic anaerobiosis for longer duration than what was presumably considered earlier. Many of her findings have been confirmed by subsequent studies. For example, Haniffa (1978b) found that the reductions in shell weight of *P. globosa* were 2.8, 3.3, 3.8, 4.6 and 5.3% following aestivation for 1, 2, 3, 4 and 5 months, respectively.

Haniffa (1978a, b) and Haniffa and Pandian (1978) are the only series of publications, which report almost complete information on active and aestivating population dynamics of *Pila globosa* in a typical southeast Indian

TABLE 5.2

Population dynamics of *Pila globosa* in Idumban Pond, Tamil Nadu, India during 1973. Density values are given in no./m². Column 5 indicates mortality due to predation during the period from May to October and net natality due to recruitment minus predation mortality during remaining period from November to April, the data in column 5 represent net natality due to recruitment minus predation mortality (recalculated and compiled from, Haniffa, 1978a, b)

Month	Surviving Snails (no./m²)			Mortality/ Natality	Aestivation Area (ha)
	Active	Aestivating	Active + Aestivating		
May	13.3	0.5	13.8	−1.9	5.28
June	11.1	0.9	12.0	−1.8	4.48
July	8.7	1.0	9.7	−2.3	5.38
August	5.5	1.5	7.0	−2.7	5.06
September	3.1	1.1	4.2	−3.2	3.52
October	3.1	0	3.1	−1.1	0
November	7.3	0	7.3	+4.2	0
December	11.2	0	11.2	+3.9	0
January	14.8	0	14.8	+3.6	0
February	15.6	0	15.6	+0.8	0
March	15.3	0	15.3	−0.3	0
April	15.7	0	15.7	+0.4	0

irrigation tank. Receiving 77% of 78.7 cm precipitation during northeast monsoon, Pond Idumban remains completely filled with water (62.4 ha) from October to December. Due to withdrawal of water for irrigation and evaporation (from April to July, temperature 34–37°C), water spread area begins to shrink from April to July, when the area is decreased to 15.4 ha (Fig. 5.1). Macrophytes *Chara gracilis*, *Hydrilla verticellata* and *Ceratophyllum demersum* are the major producers, on which *P. globosa* population thrives. The snail grows to ~ 35 g in 64 months in laboratory but within 48 months in the pond. Hence, the life span of the snail is < 4 years. However, marking and recapture of the snails have revealed that due to predation and aestivation stress, survival is reduced to < 2.5 and 0.3% (i.e. 12 of 2.5%) at the ages of 1+ and 2+, respectively. With the spell of monsoon in October, the snails begin to spawn and recruitment is continued at the rate of 3–4 snails/m² from November to January (Table 5.2). Studies are required to know whether the surviving active (3.1/m²) snails alone begin to spawn during October–November and the aestivated snails (1.1/m²) spawn during December–January (cf Malleswar et al., 2013), as the recruitment period is prolonged for 4 months. The proportion of snails commencing aestivation is increased from 4% (0.5 of 13.3 snails per m²) in May to 35% in September (1.1 of 3.1 snails per m²).

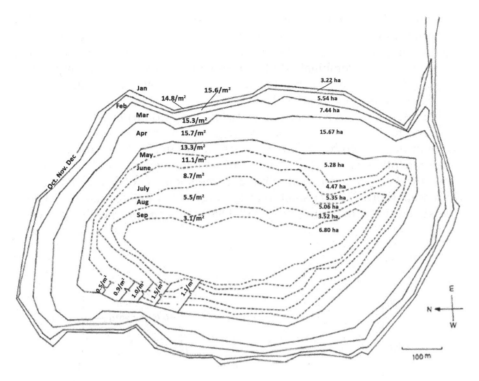

FIGURE 5.1

Morphometric and bathymetric changes in Pond Idumban during 1974 and density of aestivating and surviving *Pila globosa* in littoral snail-inhabiting area. Values given in brackets are density of snails. Values given between two straight lines are the mean density of aestivating snails in the total area (from Haniffa, 1975).

With progressive shrinking of water spread area, more and more snails begin to aestivate: aestivating snail density increases from 0.5 snail/m² in May to 1.5 snails/m² in August, i.e. 0.5 snail/m² aestivates for the ensuing 150 days and 0.4, 0.1 and 0.5 snail/m² aestivates for 120, 90 and 60 days, respectively (Table 5.1); the duration of aestivation is limited to ~ 30 days in 0.4 snail/m² during September. During these variable periods, mortality due to aestivation stress increases from 0.04 snail/m² in May to 1.12 snails/m² in August. With shrinking water spread area from May to September, the snails are easily predated by fishes and aquatic birds. Consequently, cumulative mortality due to predation and aestivation stress is increased from 1.94 snails/m² during May to 2.76 snails/m² during September, i.e. mortality due to predation accounts more than that of aestivation stress. Snails emerging from aestivation add ~ 1.1 individuals/m² to the reported (3.1 snails/m²) active snail population of October. Prior to commencing aestivation, *P. globosa* closes its operculum by mucus epiphragum to reduce desiccation. Soil moisture and temperature combination is a decisive factor in determination of mortality of snails due to aestivation stress. When percentage of surviving snails after aestivation is plotted against soil moisture and temperature, it is apparent that the combination of soil moisture below 7% and temperature above 33°C is critical and the moisture below 5% and temperature above 35°C is lethal to aestivating snails (Fig. 5.2).

The cumulative survival of the snail during the 5 month period of aestivation averaged to 91.31%; however, it progressively decreased to 52 and 64% in snails buried at depths of 15 and 30 cm, respectively. Aestivating snails at greater depth seemed to 'enjoy' relatively low temperature and higher moisture (Fig. 5.3, left panel). Similarly, metabolic level also progressively decreased from 3.17 mg/g snail/d to 2.11 mg/g/day in those that have began to aestivate at 30 cm depth. The parallel trends for the progressive decreases

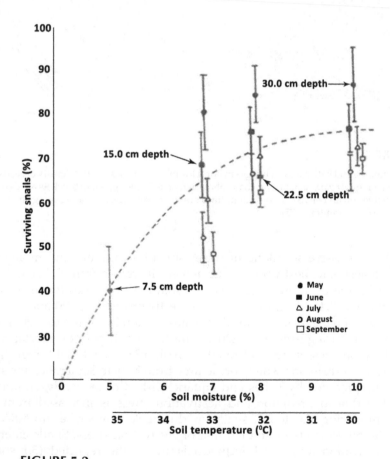

FIGURE 5.2

Number of successfully aestivating *Pila globosa* snail at different depths as function of soil moisture and temperature (from Haniffa, 1975).

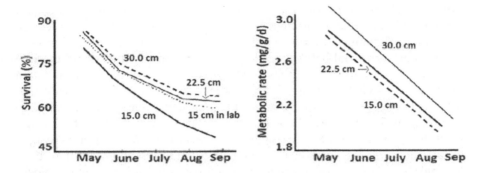

FIGURE 5.3

Calculated changes in survival (left panel) and metabolic level (right panel) of *Pila globosa* which commenced aestivation during different months and at different soil depths (calculated from data reported by Haniffa, 1975).

in metabolic level of aestivating snails at 15.0 and 22.5 cm depth are shown in Fig. 5.3, right panel; remarkably, the depression of metabolic rate increased with increasing depth. Interestingly, metabolic rate of many planorbids, lymnaeids and physids, that were starved but kept in water, dropped to 30 and 20% of the normal level on the 8th and 50th day, respectively. These starving snails could not suppress oxygen uptake any further and died (von Brand et al., 1948). However, snail switching to aestivation (out of water) and anaerobiosis revived successfully even after a period of 5–6 months in *P. virens* (Meenakshi, 1956, 1964 and *P. globosa*, Haniffa, 1988b) and also in others, in which aestivation was prolonged for months and years (Table 5.1). Hence, aestivation is an advantageous phenomenon to tide over unfavorable conditions. Of course, aestivating snails do lose body weight, as they depend on body reserves. Briefly, the weight loss due to aestivation in *P. globosa* ranges from 14 to 50%. Confirming our results that the quantity of weight loss progressively increased with increasing duration of aestivation, Malleswar et al. (2013) reported the losses of 6, 7, 7, 13 and 18 g in *P. globosa* after periods of 7, 15, 30, 60 and 90 days aestivation, respectively. A 50% reduction in body weight of *Helisoma trivolvis* after 4–5 months aestivation was also reported by Russel Hunter and Eversole (1976).

5.2 Tidal Aestivation

The intertidal zone is characterized by dynamic alternative surging waves and submergence, and ebbing waves and regular exposure once every 6 hours 15 minutes. It provides enormous opportunities for feeding, fertilization of gametes and dispersal during submergence but also poses challenges by exposure. Changes in tidal levels inclusive of spring and neap tides, and presence of rocky substratum for sessile molluscs and others render the zone more dynamic. Intertidal animals inhabiting the interface between aquatic and terrestrial habitats are exposed to hypoxia/anoxia and thermal stress during the terrestrial phase. When exposed to hypoxia or anoxia, a mollusc may switch to either (i) anaerobiosis, as in majority of bivalves or (ii) aerial respiration, as in *Littorina* spp, *Monodonta lineata*, *M. turbina* (see Pandian, 1975, p 158). Incidentally, the response of amphibious freshwater snails to hypoxia and/or shrinking water level is to initially suppress oxygen uptake and repay 'oxygen debt' on recovery (von Brand and Mehlman, 1953, Meenakshi, 1956) and/or switch to aestivation accompanied by anaerobiosis, when the adverse condition is prolonged.

Within an animal, the onset of anaerobiosis may be organ-specific (De Zwaan et al., 1991). Also aerobic and anaerobic capacities to sustain an extended aerial exposure differ from species to species (De Zwaan et al., 1991). On aerial exposure, the bivalves close their shell valves and oxygen level inside the closed valves is reduced to < 17% of the normal level. The first response to hypoxia is to 'defend' by suppression of oxygen uptake and 'rescue' it on recovery by repaying oxygen debt within 15 minutes (e.g. *Mytilus galloprovincialis*, Anestis et al., 2010). The second response, which may last for 24 hours, is to draw energy by conversion of aspartate to succinate after transamination to pyruvate yielding alanine (Babarro et al., 2007). The third "phase is characterized by the depletion of glycogen stores, a drop in energy status and a reduction in metabolic rate" (Anestis et al., 2010). In animals, the key glycolytic enzyme Pyruvate Kinase (PK) controls the flux from phosphoenopyruvate (PEP) to succinate. Expectedly, a seven-folded increase in succinate within the first 6 hours aerial exposure has been reported (see Pandian, 1975, p 160, e.g. *M. galloprovincialis*, Babarro et al., 2007). Notably, the PK level is about 45 times higher in a clam *Rangia cuneata*, in comparison to *Littorina irradiata*, which switches to aerial respiration on exposure.

Reproduction: Despite the increased feeding activity by molluscs (e.g. *L. littoria*, Newell et al., 1971, *Lasaea rubra*, Morton et al., 1957) during the limited duration of accessibility to food, the total food energy gained by supralittoral occupants remains far lower than that obtained by sublittoral occupants. Secondly, due to anaerobic or aerial respiration, molluscs also incur heavy losses on metabolism. Consequently, the tidal exposure limits the energy available for growth and reproduction. Expectedly, the supralittoral bivalves have far less energy available for growth and/or reproduction. For example, energy available to supralittoral inhabitants of the black mussel *Choromytilus meridionalis* limits growth to 38 mm, in comparison to 50 mm in sublittoral inhabitants and reproduction to ~ 30% in supralittorals (Fig. 5.4A). However, the reduction in reproduction is limited to < 10% embryos in the ovoviviparous supralittoral clam *L. subviridis* (Fig. 5.4B). Interestingly, Anderson (1975) has

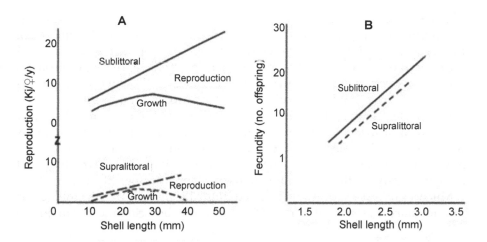

FIGURE 5.4

A. Calculated quantum of energy available for growth and reproduction in supralittoral and sublittoral block mussel *Choromytilus meridionalis* (redrawn from Griffiths, 1977). B. Brood size as a function of shell length in supralittoral and sublittoral ovoviviparous clam *Lasaea subviridis* (redrawn from Beauchamp, 1986).

reported 9 and 5% reduction in sizes of body and gonad, respectively but an increase of 9% in shell size of supralittoral *M. californianus*, in comparison to those of sublittoral occupants. Data reported for Reproductive Efforts (RE) reveals unexpected results (Table 1.12). In comparison to benthic gastropods, the intertidal gastropods invest 5–15 times more energy to insure reproduction. It is not clear whether the aerially respiring (?) intertidal gastropods continue to graze scrapping algae and are able to allocate more resources for reproduction. Secondly, the supralittoral gastropods allocate an unusually higher investment on RE than the midlittorals and sublittorals. More researchers alone can resolve this incomprehensible information.

5.3 Biological Weapon

Schistosomiasis is an ancient scourge of mankind. *Schistosoma* eggs are found in mummies of Egypt and China. Hailing perhaps from Egypt, the debilitating disease causes 200–500 thousand human deaths and is widespread in 70 subtropical countries (Hoffmann and Dunne, 2004) from Brazil (Bavia et al., 1999), Africa, Middle East, China, Japan and the Philippines. More than 200 million people are infected by the disease and an estimated 779 million are at the risk of infection (Protasio et al., 2012). The disease is caused by blood flukes *Schistosoma haematobium, S. mansoni* and *S. japonicum. S. haematobium* is transmitted through urine and snails belonging to the genus *Bulinus,* while that of *S. mansoni* is through feces and *Biomphalaria* spp. Table 5.3 lists the hosts of the diseases schistosomiasis, fasciocoliosis and sanguinicoliasis caused by the flukes to man, livestock and fishes. In the intermediary hosts of *Oncomelania,* a single copulation provides adequate sperm for egg laying up to 385 days (Chi and Wagner, 1957). Life-table parameters from Egyptian fields suggest that *B. truncatus* and *B. alexandrina* double their populations within a fortnight (Dazo et al., 1966) and make many new snails for miracidial infection. This may pose a challenge to those, who intend to control them but an opportunity for those intending to make them a biological weapon.

Much efforts have been made to contain the disease by controlling the intermediary hosts. In this regard, it may not be scientifically unfair to

TABLE 5.3

Parasitic flukes of human, livestock and fish (compiled from Webbe, 1962, Bobek et al., 1986, Bullard and Overstreet, 2002, Brown, 2002, de Kock et al., 2004, Laha et al., 2007, Young et al., 2010, Protasio et al., 2012)

Fluke	Definitive Host	Intermediary Host	Reported Observations
		Schistosomiasis	
S. haematobium	Man	*Bulinus* spp, *B. natus productus*	Gonochoric fluke with female heterogamety, urinary transmission
S. mansoni	Man	*Biomphalaria glabrata, B. pfeifferi, B. tenagophila*	Fecal discharge with 200–300 eggs/d
S. japonicum	Man	*Oncomelania* spp	Fecal discharge with 1000 eggs/d
S. bovis	Livestock	*Bulinus* spp,	1.7 million bovine infected
S. matheei	Livestock	*B. africanus, B. forskali*	
S. leiperi	Ruminants	*B. natalensis, B. tropicus*	
		Fasciolliasis	
Fasciola hepatica	Man		Hermaphrodite. Fecal discharge with 200 eggs/d. Metacercariae stage involving fish as second intermediary host
Clonorchis sinensis	Man		Induces carcinogenesis. 6 million infected in Thailand
Opisthorchis viverrini	Livestock	*Basommatophora*	
		Sanguinicoliasis	
Sanguinicola inermis, Aporocotyle simplex	Fishes	Snails, bivalves, terebellid polychaete, *Artacama proboscidea*	Problems in aquaculture farm, as debilitated fish cannot be treated

consider the scope for employing the infected aestivating snails as weapon for biological warfare. In modern times, a war is made more to destabilize the economy of the enemy country rather than defeating it. Introduction and/ or increasing the incidence of *Schistosoma* through propagation of infected aestivating snails may prove to be a strategy. Firstly, the intermediary host, namely the snails thrive in high density in aquatic bodies at temperatures between 15 and 20°C, an ideal temperature range for the introduction of the infected aestivating snails in southern Europe and North America. However, these snails do not survive at temperatures higher than 25°C, with an exception of *B. africanus* and *B. forskalii*, which may tolerate < 30°C (de Kock et al., 2004). Secondly, the Life Span (LS) of *Bulinus* spp is relatively short. For example, the LS of *B. globosus* is 101 days but on infection by *S. haematobium*, its LS is reduced to 42–71 days (Gracio, 1988). LS of other *Bulinus* species is also less than a year (Brown, 2002) and not many of them undertake aestivation. However, the scope for employing the snail per se does exist with *Nastus productus* with LS of 131 days in Africa. About 48%, i.e. one of two snails is infected and the infected snails survive after aestivation of 98 days within the LS of 131 days (Webbe, 1962). Thirdly, *B. pfeifferi* carries *S. mansoni* infection through a period of aestivation and releases the cercariae within 7–14 days post-aestivation, depending on the age of infection prior to aestivation (Badger and Oyerinde, 2004). Some 23% *B. glabatra* with lamellae survives aestivation in trees for 10 months but *S. mansoni* larvae in them can survive only for 7 months (Richards, 1967). The need for research to employ the infected aestivating snail as a weapon is obvious.

6

Sex Determination

Introduction

In molluscs, sexuality includes gonochorism, parthenogenesis and hermaphroditism; the latter includes Simultaneous Hermaphroditism (SH), protandric, serial and Marian hermaphroditism. This would imply the presence and operation of diverse mechanisms of sex determination. Sex ratio represents the cumulative end product of sex determination and differentiation processes. In the Japanese ampullarid rice field pest snail *Pomacea canaliculata*, Yusa and Suzuki (2003) reported that the ratio was highly variable and ranged from almost exclusive daughters to almost exclusive sons. Further, they also observed that the ratios were constant between successive broods of the same parent. This observation suggests the operation of polygenic sex determination mechanism (Bull, 1983), in which environmental factors like temperature play a role in sex determination. Yusa (2004b) investigated the role of (i) presence of adult males or females, (ii) age of parents, (iii) temperature (20–25°C), (iv) food availability, (v) position in the egg mass, (vi) aquarium size and (vii) indoor (14 L : 10 D) and outdoor (14 L : 10 D to 11 L : 13 D) aquaria. He found that none of these factors had any significant effect on sex ratio. Further, the brood ratios were significantly different among different parents in all these experiments. The results reported in the representative apple snail clearly indicate that environmental sex determination does not occur and both the sex determination and sex ratio variations are under genetic regulation in molluscs.

6.1 Inheritance of Color and Symmetry

The body color of the apple snail *Pomacea canaliculata* can be brown or yellow. Crossing experiments lasting over three generations have revealed the Mendelian pattern of color inheritance (Yusa, 2004a). Apart from it, there are others like the Pacific oyster *Crassostrea gigas*, which displays either continuously varying, color from almost pigment-free white to almost fully black pigmented shells. The pigmentation is a quantitative trait controlled by many genes with additive effects. The dark pigmentation has been identified as foreground color, while the golden or white color is a background color. The locus controlling background color has two alleles with an allele from golden background being dominant to the allele for white background (Ge et al., 2015). Neither the simple color inheritance in *P. canaliculata* nor the complicated color inheritance in *C. gigas* is sex linked. The nudibranch *Dendrodoris nigra* is little different in that the two color variants can be more of sibling species than poecilogonics. All brown gray with white speckles produces 251,000 zygotes/spawn, each zygote measures 75 µm and hatches (after 9.6 days incubation at 23°C) as planktotrophic larva (114 µm) with typically sinistrally coiled shell. The second variant is jet black with red-rimmed mantles and produces 85,000 zygotes per spawn. The zygote and the hatched typical veliger (after 9.6 days incubation at 23°C) measure 129 and 154 µm, respectively. Copulation occurs only between individuals of the same color (Rose, 1985).

Mytilus galloprovincialis is a stable gonochoric but its sex cannot be identified from sex specific morphological characteristics, gonad features or detected from sex chromosomes. Sometimes, the sex of an individual mussel can be identified by the orange colored ovary and pink colored testis (Fig. 6.1A). However, this color sex dimorphism is not dependable, as orange colored mantle occurs not only in females but also in ~ 30% males. On the other hand, ~ 30% females display white and yellow-colored mantle. Exploring the possibility of employing chemical marker to identify sex, Mikhailov et al. (1995) have identified a male specific polypeptide in the mussel's mantle. The presence of the polypeptides 39 kDa (MAP-39) only in the male's mantle has been biochemically and immunochemically confirmed (Fig. 6.1B, C). MAP-39 concentrations, which is associated with testis development, maturity and milting, reaches upto 100% of the total soluble proteins. Notably, the presence of MAP-39 is limited to the mantle alone. Incidentally, the authors have indicated that the gonad color is sex-dependent but not linked.

Shell symmetry: In the early history of Mendelism, the inheritance study on asymmetry of snail has played an important role (Freeman and Lundelius, 1982). However, it is now known that chromosomal locus (see Stone and Bjorklund, 2001), which determines chirality is expressed through maternal cytoplasm so that the direction of coiling in offspring is dictated by maternal genotype, whose expression is delayed by a generation. Schilthuizen and Davison (2005) have brought an acceptable model of delayed inheritance of symmetry in gastropods. The model is explained hereunder with an example. Accordingly, dominant and recessive traits are characterized by large sinistral and dextral, respectively. Due to the delayed maternal effect, all progeny arising from a cross between homozygous sinistral and heterozygous snails are all sinistrals (Fig. 6.2). However, when homozygous recessive sinistral is crossed with another dominant homozygous sinistral, the heterozygous progenies are all dextrals. Clearly, the homozygous but recessive sinistral 'mother' influenced shell symmetry rather than the homozygous dominant sinistral 'father'; the 'mother' delayed the expression of dextral trait by a generation. More evidence was also brought in support of this model. The injection of egg cytoplasm from dextral *Lymnaea peregra* into a sinistral mother generated dextrals alone. Cytological evidence has revealed that the spindle orientation at first cleavage determines the direction of shell coiling in gastropods (see Palmer, 1996). It is likely that the spindle orientation is induced by an unknown cytoplasmic factor in the embryo. However, symmetry inheritance pattern is not sex-linked and may not be associated with sex determination, as both males and females are present in both dextral and sinistral gastropods.

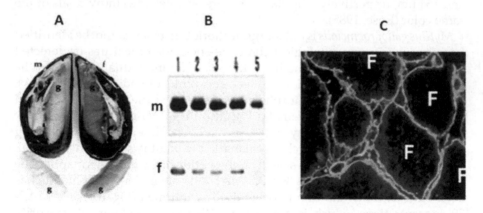

FIGURE 6.1

A. Sex dependent color difference of the gonad tissue of valves 'opened' sexually mature pink testis (m) and orange ovary (f) of *Mytilus galloprovincialis*. B. Immunoblot analysis of MAP39 expression in male (m) and female (f) mantles characterized by different levels of follicles tissues. Note the progressive decreasing levels following spawning/milting from May (Lane 1) to August (Lane 4). C. Immunofluorescent localization of MAP39 in male mantle tissues. Note MAP39 expression on membrane of follicle (F). (A.T. Mikhailov, M. Torrado and J. Méndez. 1995. Sexual differentiation of reproductive tissue in bivalve molluscs: identification of male associated polypeptide in the mantle of *Mytilus galloprovincialis* Lmk. Int J Dev Biol, 39: 545–548. Reproduced with permission from The International Journal of Developmental Biology.)

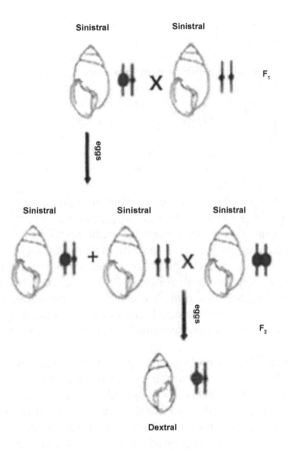

FIGURE 6.2

Inheritance of coiling direction in a model large sinistral and small dextral snails. Note the delayed expression of maternal genotype by a generation (modified and simplified from Schilthuizen and Davison, 2005).

6.2 Karyotypes and Heterogametism

In gastropods, the 2n number of chromosomes ranges from 18 in viviparid *Cipangopaludina chinensis* (Park et al., 1988) to 30 in many pulmonates (Thiriot-Quievreux, 2003), 31 in *Megalobulimus granulosus* (Kawano and Leme, 1994), 32 in lymnaeid *Austropeplea ollula* (Park et al., 1992) and *Conus magus* (Dalet et al., 2015), 34 in bithyniid *Parafossaralus manchouricus* (Chung, 1984) and pleurocerid *Koreanomelania globus* (Park, 1994), 36 in air-breathing shell-less slug *Onchidium esculentum* (Natarajan, 1959), *O. struma* (Shen et al., 2010) and in many pleurocerids and physids (Park, 1994) as well as 38, 42 and

44 in *Succinea urbana, S. campestris* and *S. luteola*, respectively (Laws, 1973). Galante-Oliveira et al. (2012) reported variations in 2n chromosome number in different ecotypes of *Nucella lapillus*. Dalet et al. (2015) employed FISH technique for physical mapping of an 18S rDNA sequence in *C. magus*. An objective of karyotyping is to search and identify heterogametism. However, none of these publications could identify heterogametism in any of the investigated gastropods. An exception may be that of Vitturi and Catalano (1988), who have identified an XO system of male heterogemety in a neretid archeogastropod *Theodoxus meridionalis*. Vitturi et al. (1998) have also reported the operation of XY chromosomes in the neogastropods *Fasciolaria lignaria* and *Pisania striata*.

In bivalves, the 2n chromosome number remains constant at 38 in two species in each of margaritiferids and unionioid ambleminaeids, 12 species belonging to nine genera in unionids, eight species of anodontinaeids, and six species belonging to five genera in lampisilinaeids (see Park and Burch, 1995). In three venerid bivalves *Ruditapes aureus, R. decussatus* and *R. philippinarum* too, the 2n number is 38 (Borsa and Thiriot-Quievreux, 1990). In lampisilinaeid *Epioblasma triquebra*, males have four metacentric, 13 submetacentric and two subterminal chromosomes, against five metacentric, 12 submetacentric and two subterminal chromosomes in females (Park and Burch, 1995).

Hybridization between molluscan species with chromosomes of different sizes and numbers may yield valuable information on heterogemetism. That between snails *Viviparus alter* (2n = 18) and *V. contectus* (2n = 14) produced viable, fertilizable and backcrossable F_1 hybrids. The diploid chromosome component of F_1 hybrid males of both reciprocal crosses consisted of 16 chromosomes, i.e. seven originated from *V. contectus* and nine from *V. alter*. The F_2 hybrid male backcrossed with *V. alter* had more number of chromosomes than that backcrossed with *V. contectus* females. Apparently, the unpaired chromosomes of F_1 hybrid at meiosis were randomly segregated and distributed, when F_1 hybrid male was backcrossed with *V. alter*, in which the sperm of F_1 hybrid with eight chromosomes fathered most of the offspring. But those with *alter*-like nine chromosomes were more successful in backcrosses with *V. contectus*, due to the active selection of the most compatible sperm with *alter*-like nine chromosomes (Sbilordo et al., 2012). Yet the publication could not identify heterogametism in the tested species.

Polymorphic microsatellite locus can also be used to identify heterogamety. A combination of genetic parentage analyses and association studies on microsatellite genotype and sexual phenotype indicated the presence of an XO-like male gamety in the melangerid knobbed whelk *Busycon carica*. At locus *bc* 2.2, 73 out of 139 whelks with penis displayed a single primary DNA band in gels, whereas 62 out of 66 (94%) whelks with no penis showed two bands in gels. Genotype employed by DNA band profiles at *bc* 2.2 revealed the presence of sex linkage. In an egg string of a female, which was heterozygous for alleles 303 and 306 at *bc* 2.2, all 447 assayed progeny displayed either allele 303 or allele 306 and in frequencies not significantly departed from 1 : 1 ratio, i.e. each of the 221 embryos inherited one or the other maternal allele in a model consistent with Mendelian segregation from diploid locus. These embryos also inherited an additional allele from the (XO) father. However, the remaining (203) embryos displayed only one band of maternal (XX) origin. Hence the polymorphic satellite locus *bc* 2.2 is diploid and heterozygous in females but homozygous in males. Its alleles are transmitted from mother (XX) to sons and daughters but from father (XO) to daughters only. Incidentally, four such paternally derived alleles at *bc* 2.2 were detected: allele 285 (120 out of 244, i.e. 50%), allele 330 (7%), allele 309 (22%) and allele 312 (23%). This would imply that at least four different fathers had sired eggs in the strings (Avise et al., 2004).

6.3 Spawning Induction and Cryopreservation

In culturable molluscs, induction of spawning and gamete preservation (Liu and Li, 2015, Hassan et al., 2015) is the first step for selective breeding and ploidy induction. Despite the oft reported long term survival of sperm in seminal receptacle of female gastropods (e.g. *Littorina saxatilis*, see Makinen et al., 2007) and cephalopods (e.g. *Bathyteuthis berryi*, Bush et al., 2012), bivalves have received much attention in recent years for three reasons: (1) their simple reproductive system does not include one or more component organ for long term sperm preservation, (ii) their amenability to ploidy induction and (iii) their importance in aquaculture.

Induction of spawning: The male oyster can be induced to milt by keeping it at 4°C in air and subsequently immersing in warmer water containing algae (Wang and Wang, 2008). Injection of 0.4 ml of 2 mM serotonin into the gonadal tissues induced milting in many bivalves (Table 6.1) in the following descending order: *Argopecten irradians* < *Crassostrea virginica* < *Spisula solidissima* < *Geukensis demissa* < *Mercenaria mercenaria* < *Arctica islantica* (Gibbon and Castagna, 1984). Notably, spawning response of females in all these bivalves is either zero or < 20%. Apparently, female bivalves require higher doses to induce spawning. In fact, it was confirmed by Alvarado-Alvarez et al. (1996), who successfully induced spawning in Pismo clam *Tivela stultorum* with an injection of 0.4 ml of 5 mM serotonin. Further, they also showed that the prophase arrested oocytes scrapped from the ovaries of the clam underwent germinal vesicle breakdown (GVBD) *in vitro*, when incubated at serotonin dose as low as 0.2–20 µM. The *in vitro* oocytes were fertilizable and developed to larvae. Notably, the clam eggs are fertilizable only after GVBD. However, oyster oocytes can be fertilized and they complete normal meiosis and development, when they are inseminated at any stage from prophase to metaphase (see Hassan et

TABLE 6.1

Effect of 0.4 ml of 2 Mm serotonin injection on induced spawning in some bivalves (compiled from Gibbon and Castagna, 1984)

Species	Temperature (°C)	Control/ Serotonin	Mean Spawning (%)	Males (%)	Females (%)
Arctica islandica	15–16	Control	0	0	0
		Serotonin	27	21	6
Spisula solidissima	19	Control	2	0	2
		Serotonin	60	40	20
Argopecten irradians	20–21	Control	6	4	2
		Serotonin	83	83	–
Crassostrea virginica	25	Control	0	0	0
		Serotonin	70	70	0
Geukensis demissa	28	Control	5	5	0
		Serotonin	45	40	5
Mercenaria mercenaria	28–29	Control	0	0	0
		Serotonin	42	36	6

al., 2015). In freshwater mussel *Hyriopsis bialatus*, Meechonkit et al. (2012) successfully induced synchronous release of glochidia with injection of 200 µl of 10^{-3} M serotonin, into the demibranch, where the larvae are brooded. It is known that serotonin accelerates muscular contraction in molluscs (Pavlova, 2001). Understandably, serotonin induces spawning/milting and releases of glochidia by accelerating muscle contractions. Incidentally, Wang and Croll (2006) reported that injections of estradiol into ripe scallop *Placopecten megallanicus* induced spawning in females and milting in males, testosterone injection induced milting in males but not spawning in females and progesterone blocked spawning and milting, as well.

Gamete collection: Following induction, three different procedures are adopted to collect gametes for cryopreservation. Notching is a process of making a hole on the shell by grinding, which ensures collection of required quantity of gametes. As it damages mantle tissue, notching can be lethal. A non-lethal procedure is to anesthetize the animal with Dead Sea salt or Epson salt that helps to open the shells, which are otherwise tightly closed by the adductor muscles. However, due to asynchronous response, the required number of individuals may not response on and at a time. The most widely used procedure is stripping, which is easily practicable and consumes less time. Stripping involves inserting a Pasteur pipette beneath the gonad epithelium and sucking the required amount of sperm from a ripe gonad. For collection of large volume of sperm, dissecting the animal is the only option (see Hassan et al., 2015).

Cryopreservation is an effective and reliable technique, which facilitates reproduction in selective breeding and conservation of endangered species.

As they are larger in size, have lower surface area to volume and a low rate of water and cryoprotectant movement into and out of cells, oocytes are more challenging to cryopreserve than sperm. Not surprisingly, only oocytes of *Crassostrea gigas* (Tervit et al., 2005) and blue mussel *Mytilus galloprovincialis* (Liu and Li, 2015) have been successfully cryopreserved. In comparison to diploid sperm, tetraploid sperm of oyster are more difficult to cryopreserve, as they are more delicate or have different plasma membrane properties (Dong et al., 2006).

Regarding cryopreservation of gametes especially the sperm, the following may be noted: 1. Calcium induces acrosome reaction and agglutination of sperm. Hence sperm can retain motility only in calcium-free water. Therefore, an extender likes the Hank's Balanced Salt Solution (HBSS) or calcium-free HBSS is used (Table 6.2). In liquid nitrogen, bacteria cannot multiply. The use of antibiotics or UV-irradiated sea water reduces multiplication rate of bacteria. 2. "A combination of permeating and non-permeating cryoprotectants can improve sperm viability by (i) regulating membrane fluidity with more cohesive sperm membrane and (ii) increasing

TABLE 6.2

Successful cryopreservation of bivalve gametes. E = Equilibration, FS = Fertilization Success

Species and Reference	Procedure	Reported Results
	Cryopreservation of oocytes	
Mytilus galloprovincialis Liu and Li (2015)	Cryoprotectant consisted of 10% ethyl glycol + 7.5% Ficol PM 70	FS = 38% D-larvae – 14%
	Cryopreservation of sperm	
Crassostrea gigas Dong et al. (2005)	Stripping, calcium-free HBSS + 2% PG + 6% methanol, 10–15 minutes E, 0.5 ml/straw	Fertilizability = 98%
Crassostrea virginica Yang et al. (2012)	Stripping, calcium-free HBSS + 10% PG + 6% methanol, 20 minutes E, 0.5 ml/straw	Fertilizability = 77%
C. tulipa Yankson and Moyse (1991)	Stripping, UV-irradiated sea water + 10–25% DMSO, 30 min E, 1.8 ml cryotubes	Fertilizability = 0–71%
O. iredalei Yankson and Moyse (1991)	Stripping, UV-irradiated sea water + 10–25% DMSO, 30 min E, 1.8 ml cryotubes	Fertilizability = 11–35%
O. edulis Horvath et al. (2012)	Stripping, calcium-free HBSS + 10% DMSO, 30 minutes E, 0.5 ml/straw	8% mortality
Saccostrea cucullata Yankson and Moyse (1991)	Stripping, UV-irradiated sea water + 10–20% DMSO, 30 minutes E, 1.8 ml cryotubes	50% mortality

the membrane hydrophobicity resulting minimal chance for intracellular ice formation" (Hassan et al., 2015). 3. Among the steps in cryopreservation, cooling and freezing as well as thawing are the most stressful events for the gametes. In general, the high viability of oyster sperm in cryopreservation and thawing procedures indicates that they are able to withstand a wider range of cooling and thawing. The procedures thus far developed are more species specific. For more details, Liu and Li (2015) and Hassan et al. (2015) may be consulted. A glance at Table 6.2 shows that combinations of stripping of gametes, use of calcium-free HBSS + 2–10% PG, an equilibration duration of 20 minutes and package of 0.5 ml/straw and 10% ethyl glycol + 7.5 Ficol PM 70 lead to a higher success of cryopreservation of sperm and oocytes, respectively. To ensure a higher Fertilization Success (FS), the reported egg to sperm ratios ranges from 10–15 to 18,000 sperm per egg (Desrosiers et al., 1996). Susceptibility to polyspermy is not uncommon among oysters. For example, Stephano and Gould (1988) found that oocytes of *C. gigas* inseminated immediately after spawning were more polyspermic than oocytes inseminated after an hour.

6.4 Ploidy Induction and Gigantism

Barring about 1,000 species belonging to archeogastropods (e.g. *Haliotis*) and bivalves, reproduction in molluscs is associated with internal fertilization, which is followed by encapsulation of eggs by chorion or jelly. As ploidy induction involves manipulations within a few seconds/minutes before and/or fertilization, only the gamete producing archeogastropods and bivalves are amenable to ploidy induction. In fact, they are amenable to all possible inductions namely gynogenesis and androgenesis, in which ploidy level is not elevated, as well as triploidy and polyploidy, in which ploidy is elevated to 3n, 4n and more. Gynogenesis is induced by activation of egg by genome inactivated (by UV-irradiation) homospecific or heterospecific sperm followed by retention of the second polar body (PB2) (heterozygous meiogynogens) or by arresting cytokinesis of the first mitotic division (homozygous meiogynogens). In fishes, the irradiated heterospecific sperm have also been used to successfully induce gynogenics (Pandian, 2011, p 56). The use of UV-irradiated heterospecific sperm remains to be tested in bivalves. Using genome-inactivated sperm, meiogynogenics alone have been induced in bivalves and archeogastropods by retention of PB1 or PB2 or both PB1 with a chemical shock by cytochalasin B (CB), as in *Crossostrea gigas* (Guo et al., 1993) or thermal (cold) shock, as in *Haliotis discus hannai* (Fugino et al., 1990).

Three procedures have been adopted to induce ploidy in bivalves: 1. Retention of PB1 or PB2 in a fertilized egg (*C. gigas*, Allen and Downing, 1990), 2. Cross between tetraploid female and diploid male, as in *C. gigas*

(Guo and Allen, 1994b, Guo et al., 1996) and 3. At CMFRI, India, Thomas et al. (2004) successfully induced triploidy using safer 6-Dimethylaminopurine (DMAP) in *Crassostrea madrasensis* and the triploid yield increased from 42% for CB to 67% for DMAP. Incidentally, there are concerns on the use of CB induced triploids in edible bivalves. The 'chemical triploid' is also achieved by retention of PB1 or PB2 in a fertilized egg using CB in many bivalves (e.g. *Mya arenaria*, Allen et al., 1986, *C. gigas*, Guo et al., 1996, *Mytilus galloprovincialis*, Kiyomoto et al., 1996). Generation of the chemical triploids is not encouraged for two reasons: firstly, CB is toxic and its possible residual level in these 'chemical triploids' is a public concern. Secondly, induction efficiency of CB is rarely 100%, which complicates hatchery management (Allen et al., 1989). Hence the choice of triploid induction is to cross 4n female with 2n male (Guo and Allen, 1994b, Jouaux et al., 2010). Surprisingly, only a few authors have made a comparative study on growth rates of diploids and triploids. The relatively faster growth of triploid is reported from *C. virginica* (Stanley et al., 1984), *M. edulis* (Beaumont and Kelly, 1989) *Pinctada martensii* (Jiang et al., 1991) and *C. madrasensis* (Mallia et al., 2006, 2007, 2009). Tetraploidy is induced by fertilizing unreduced 3n eggs with haploid sperm from a normal diploid. It was achieved only in *C. gigas* (Guo and Allen, 1994b). 3. Thermal induction inhibits mitotic divisions one and two in *Mulinia lateralis* and has generated viable tetraploids (Yang and Guo, 2006).

Natural androgenics: No mollusc has yet been induced to generate androgenics. However, natural occurrence of diploid androgenics was reported from Simultaneous Hermaphroditic (SH) clam *Corbicula fluminea* from Japanese freshwater (Komaru et al., 1997). In 1999, Komaru and Konishi found both diploid and triploid androgenics of *C. fluminea* from Taiwanese water. In diploid and triploid *C. fluminea* as well as diploid *C. leana*, the mean DNA content of spermatozoa was almost identical to those in somatic cells. DNA microfluorometric study of *C. fluminea* showed that spermatogenesis was non-reductional. Bringing cytological evidence, Komaru et al. (1998) showed that (i) maternal genome was extruded as polar body after first meiosis, (ii) only chromosomes derived from sperm pronucleus constituted the first mitotic cleavage and (iii) DNA content of 7-day old veliger larvae was identical to somatic cells of the parent.

Natural gynogenic is reported from the viviparous crevice-inhabiting clam *Lasaea rubra*. Their F_1 offspring and in one case F_3 offspring displayed identical zymograms at loci Glucose-Phosphotase isomerase (GPi) and Phopho-glucose mutase (Pgm) among themselves and with their respective parents. Based on these observations, the clam was considered as a parthenogenic species (Crisp et al., 1983, Crisp and Standen, 1988). However, the presence of sperm in the clam was repeatedly confirmed (see Beauchamp, 1986). O'Foighil (1987) observed the penetration of an egg by a sperm in *L. cistula*. All the investigated populations of *Lasaea* that lack planktonic larvae are now known to be SH but only with a tracer quantity of allocation to testis (O'Foighil and Eernisse, 1988). However, the presence of odd ploidy numbers and supernumerary chromosomes rendered the description of meiosis impossible. From their study on ploidy and pronuclear interaction of the northeastern Pacific clones of *Lasaea*, O'Foighil and Thriot-Quievreux (1991) reported that the 97 chromosomes of the clam could be arranged in triplicates (Fig. 6.3). Following penetration, the incorporated sperm nucleus disintegrated in the cortex. Clearly, the autosperm only triggered the first cleavage but with the possibility of occasional leakage of sperm chromosome, which were found as supernumerary chromosomes, a phenomenon that also occurs in gynogenesis of some unisexuals fishes (see Pandian, 2013, p 103–104). The 121 out of 123 second polar body examined were diploids. Following the integration of the sperm nucleus, 96 chromosomes were arranged in metaphase of the egg nucleus in preparation of the first cleavage. Briefly, the lineage of the northeastern Pacific clones of *Lasaea* was considered

FIGURE 6.3

Lasaea metaphase karyotype with 97 chromosomes arranged in triplicates (from O'Foighil and Thiriot-Quievreux, 1991).

to be the first bivalve and mollusc, in which gynogenesis naturally occurs. Hence *L. rubra* is not a parthenogenic (O'Foighil and Smith, 1995).

Ploidy induction: Bivalves are characterized by (i) the ability to tolerate chromosomal mosaicism, (ii) prolonged and asynchronized release of polar bodies and (iii) different patterns of segregation. Many bivalves are mosaics for chromosome number. For example, a vast majority of cells (89.3%) in *C. gigas* is normal diploids; others are haploids (0.7%), triploids (1.3%) and aneuploids (8.7%, see Guo et al., 1992a). Rarely but certainly, adductor muscle cells of triploid hermaphroditic *C. gigas* consists of both diploids and triploids (Allen and Downing, 1990). Surprisingly, survival of aneuploids is not uncommon among molluscs (e.g. *C. gigas*, Guo and Allen, 1994a, b). A reason for heavy mortality of molluscan embryos and early larvae is that many are indeed mosaics, which are progressively eliminated during development. For example, the heat shock intended to inhibit meiosis I in dwarf surfclam *Mulinia lateralis* produced larvae that consisted of 41% diploids, 22% triploids and 37% tetraploids on the first day of hatching. However, there were only diploids and tetraploids between the second and sixth day of hatching (Yang and Guo, 2006). In triploid Japanese pearl oyster *Pinctada fucata martensii* too, aneuploids were eliminated during the first 4 months (Komaru and Wada, 1994).

Unlike crustaceans, in which the events of releasing PB occur in quick succession (Pandian, 2016, p 168), meiosis is relatively a prolonged process in molluscs. The PB1 and PB2 are released around 17 and 35 minutes post fertilization (pf), respectively (Guo et al., 1992a). Hence, there is adequate interval for their independent manipulation. However, meiosis in molluscs is not completely synchronized (Fig. 6.4). It is possible that treatment intended for both PBs may effectively inhibit only PB1 (Yang and Guo, 2004). Heat shock used for PB1 inhibition for 10 to 20 minutes post fertilization (PF) in *M. lateralis* induced triploids and pentaploids but not tetraploids (Yang and Guo, 2006).

The following is described to familiarize cytological events that take place in an oocyte. When a sperm penetrates an oocyte, which remains arrested at prophase, is activated and continues subsequent meiotic process. For example, the 10 tetrads (10 pairs of homologous chromosome, 2n = 20) of *C. gigas* go through two successive meiotic divisions and release PB1 and PB2. Hence the PB1 contains 20 diads and PB2 10 chromatids. The remaining 10 chromatids in the oocyte fuse with 10 chromatids from the sperm to form a diploid zygote (Guo et al., 1992b). In bivalves, blocking PB1 is theoretically expected to induce tetraploids along with aneuploids and that of PB2 produces triploids (Gilbert, 1988) and both of PB1 and PB2 may generate pentaploids (Yang et al., 2000). It is also to be expected that meiosis I triploids, involving retention of PB1 have a higher level of heterozygosity and faster growth rate than those of meiosis II triploids, as in *C. virginica* (Stanley et

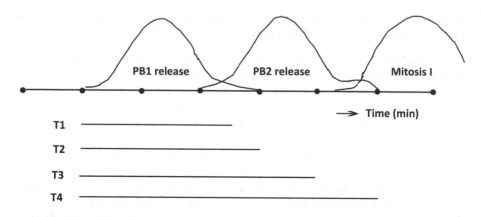

FIGURE 6.4

Schematic representation of temporally overlapping meiotic events during early embryogenesis in bivalves Lines T1–T4 indicate the treatment durations required to achieve the desired ploidy (from Yang and Guo, 2004, redrawn and modified).

al., 1984). This is also true in *C. gigas* (Yamamoto et al., 1988), *Mytilus edulis* (Beaumont and Kelly, 1989) and *P. martensii* (Jiang et al., 1991).

The electron microscopic observation by Longo (1972) on the presence of four groups of maternally derived chromatids in CB-treated eggs of surfclam *Spisula solidissima* suggested the possible entry of two groups of dyad chromosomes from meiosis I through meiosis II indicating bipolar segregations. However, Guo et al. (1992b) found that the bipolar segregations were limited to 12% of CB-treated eggs in *C. gigas*. The majority (68%) of the treated eggs went through tripolar segregation to produce randomized tripolar and unmixed tripolar triploids.

As a consequence of tolerance to mosaicism, asynchronous meiosis and different segregation patterns, induction of ploidy always resulted in simultaneous production of more than one ploidy. For example, the induction of gynogenesis in *M. lateralis* using UV-irradiated sperm and CB-treatment to retain PB2 generated 4% 2n gynogenics, 20% triploids and others (76%), which were not viable up to D-stage perhaps due to aneuploidy (Guo and Allen, 1994a). Similarly, induction of triploidy in *C. gigas* by CB-treatment to retain PB1 produced not only triploids (15.6%), but also haploids (0.7%), diploids (4.5%), tetraploids (19.4%), pentaploids (2.4%) and aneuploids (57.6%, see Guo et al., 1992a). In induction of *C. gigas* tetraploids too, surviving oysters consisted of 67% tetraploids, 23% aneuploids, 3% diploids, 3% triploids and 3% mosaics. However, the ensuing account describes gynogenesis, triploidy and tetraploidy separately to facilitate better understanding.

Induced gynogenics: Despite successful induction in *H. discus hannai* (Fujino et al., 1990) and *C. gigas* (Guo et al., 1993), these gynogenics did not survive long.

Guo and Allen (1994a) is perhaps the only source, which provides adequate information on growth and reproduction in gynogenic *M. lateralis*. Due to homozygosity, survival was reduced to 4% (at D-stage) and fecundity to 80% in gynogenics relative to those of normal female (Table 6.3). Gynogenic Pacific oyster was also significantly smaller than diploids at the age of 10 months (Guo and Gaffney, 1993). However, body growth and egg size were equal in both gynogenic and normal females. Except for isogenic lines, gynogenic bivalves may not be profitable venture for the development of aquaculture.

Triploidy: Natural occurrence of triploids is reported from *C. gigas* (Guo and Allen, 1992a) and *M. lateralis* (Guo and Allen, 1994a). Relevant information indicates that triploids out-grow diploids (Table 6.4). The increase in growth of triploids relative to diploids ranges from 1.09 times in the zhikong scallop

TABLE 6.3

Survival, growth and reproductive characteristics of normal, gynogenic (2n g) and triploid dwarf surfclam *Mulinia lateralis* (compiled from Guo and Allen, 1994a)

Parameter	2n	2n g	3n
Survival (%)	62.2	3.8	20.1
Sex ratio ($♀ : ♂$)	1 : 1	1 : 0	1 : 1
Growth at 3 mo (mg)	372	367	672
Fecundity (no.)	3×10^5	2.4×10^5	1.7×10^5
Egg size (μm)	49.1	48.6	56.6
Sperm count (no.)	2.5×10^9	–	2.0×10^9

TABLE 6.4

Ploidy induction methods, survival and growth of triploid bivalves

Species	Induction Method	Age (mo)	Growth (mg)	Growth in Times Relative to 2n	Reference
Mulinia lateralis	PB2 by CB	3	672	1.80	Guo and Allen (1994a)
Chlamys farreri	PB1 by CB	15	2200	1.09	Yang et al. (2000)
C. nobilis	PB2 by CB	14	8100	1.32†	Komaru and Wada (1989)
Pinctada martensii	by CB*?	24	2250	1.44	Jiang et al. (1991)
P. martensii†	PB1 by CB	9	2247	1.22	He et al. (2004)
C. gigas	4n ♀ X 2n ♂	3	92	1.42	Guo and Allen (1994b)
C. gigas	PB1 by CB	7	28,000	1.39	Allen and Downing (1986)
C. gigas	4n ♀ X 2n ♂	10	4,8007	1.35	Guo et al. (1996)
C. madrasensis	DMAP	12	1450	1.85	Mallia et al. (2006)

† adductor muscle weight of 3n is 1.37 times bigger than that of 2n, * pressure shock also

Chlamys farreri to 1.80 times in dwarf surfclam *M. lateralis*. Firstly, seven of eight publications compare growth in weight of diploids and triploids. Komaru and Wada (1989) have reported more precise estimates on growth of 14 months-old *Ch. nobilis*; these values for 'meat weight' and 'shell weight' are 1.32 and 1.08 times more in triploids than those of diploids. Dry meat weight of triploid *C. madrasensis* is 126% higher than diploid (Mallia et al., 2006). Notably, triploids grow not only faster but also have more meat than diploids. Differences in amounts of water held in the mantle cavity of closed bivalves can considerably alter the actual body weight, especially when estimates for diploids and triploids are made on different dates. For example, the increases in body weight and adductor muscle weight of 3n *Ch. farreri* are 1.09 and 1.31 times, respectively (Yang et al., 2000). Considering the mean values of 1.38 times for the eight 3n bivalves, it may be generalized that growth in 3n bivalves is 1.4 times faster than that of diploids. Irrespective of the difference in production methods, i.e. CB-treated or 4n ♀ X 2n ♂ hybridization, the growth of 4n ♀ X 2n ♂ hybridized *C. gigas* was also 1.4 times faster than diploid. In the PB2-retained triploid *M. lateralis*, increase in growth was faster by 1.8 times. Stanley et al. (1984) considered that retention of PB1 alone induced more heterozygous and faster growth in *C. virginica*. However, this may not be true in other bivalves. Further, the growth rate in *C. gigas* is consistently faster from 3 months of age to 10 months of age (Table 6.4). This faster growth rate also corresponds approximately with an increase in chromosome number from 20 in diploid to 30 in triploid, suggesting the presence of three alleles per gene(s) is responsible for faster growth (cf Jouaux et al., 2014).

The accelerated growth rate of triploids differs significantly between progeny arising from different combination of crosses. In *C. gigas*, triploids were induced using different combination of crosses: (i) 2n (D) ♀ X 2n ♂, (ii) 4n (T) ♀ X 2n ♂, (iii) 2n ♀ X 4n ♂ and (iv) 4n ♀ X 4n ♂. The progenies arising from tetraploid cross did not survive beyond 8th post fertilization day (pfd) (Fig. 6.5). The other progenies were reared for longer durations and their growth estimates were made on the 8th and 10th month. Remarkably, the growth rate of 4n ♀ X 2n ♂ progeny was faster than that of 2n ♀ X 4n ♂.

Table 6.5 summarizes available information on retarded gametogenesis in gynogenic and triploid bivalves. In three of them, sex ratio is skewed toward males; consequently, female ratio in them ranges from 0.0 in *M. galloprovincialis* to 0.2 in Pacific oyster and 0.3 in dwarf surfclam. In others, the ratio is as high as 0.77, i.e. 77% females + 0.007% female-like individuals in softshell clam *Mya arenaria*, in which the gonad remains underdeveloped. Conversely, in all other investigated bivalves with higher male ratio, both oogenesis and spermatogenesis occur, albeit retarded to different levels. An exception is the Japanese pearl oyster, in which spermatogenesis alone occurs. Reasons for successful gametogenesis in these bivalves with higher male ratio have to be explored. It is not clear why the clams with higher female ratio remain partially sterile; it is also not known why sex ratio of the Pacific oyster, which remains as 0.37 ♀ : 0.62 ♂ : 0.1 ⚥ in diploids, is shifted

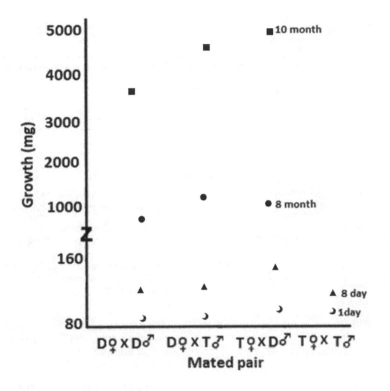

FIGURE 6.5

Growth of F_1 progenies generated by different combinations of crosses 2n ♀ X 2n ♂, 2n (D) ♀ X 4n (T) ♂, 4n ♀ X 2n ♂ and 4n ♀ X 4n ♂ in *Crassostrea gigas*. D = Diploid, T = Tetraploid (drawn from data reported by Guo et al., 1996).

toward hermaphrodite as 0.23 ♀ : 0.48 ♂ : 0.29 ⚥ in triploids. Apparently, hermaphrodites are increased at the cost of about 14% females and 11% males. Allen and Downing (1990) found that some of these triploid hermaphrodites were female-like and male-like individuals.

Interestingly, resource allocation for gametogenesis in the Pacific oyster is reduced by 45% in male and 23% in female (Table 6.5). Contrastingly, the reduced allocation in dwarf surfclam is 20% in male and 28% in female, apart from skewing male ratio to 0.7. Evidently, the resource from the reduced allocation for gametogenesis is diverted to accelerate faster somatic growth. In this context, a comparative study on growth of diploid and triploid in noble clam and dwarf surfclam may be rewarding. Incidentally, from their histological and immuno-histochemical study, Jouaux et al. (2014) have identified α and β individuals in *C. gigas*, in which germinal cells remain unlocked and locked, respectively. Perhaps, it is the difference in proportions of α and β individuals within a triploid group may account for the different levels of partial or complete sterility in the investigated triploid bivalves.

TABLE 6.5

Sex ratio and gametogenesis in induced gynogenic and triploid bivalves

Species	Sex Ratio ♀ : ♂	Heterogamety	Gametogenesis	Reference
		Gynogenic		
Dwarf surfclam *Mulinia lateralis*	1.00 : 0.00	XX–XY	Oogenesis reduced to 79%; 2.4 X 10^5 eggs in 2n g, but 3.3 X 10^5 eggs in 2n	Guo and Allen (1994a)
		Triploids		
Noble scallop *Chlamys nobilis*†	0.50 : 0.50	–	Completely sterile, 26% reduced gonad weight	Komaru and Wada (1989)
Softshell clam *Mya arenaria*	0.77 : 0.16 : ♀ like 0.007	XX–XY	Underdeveloped gonad	Allen et al. (1986)
Mussel, *Mytilus galloprovincialus*	0.00 : 0.90 : 0.10 ⚥	ZW–ZO	For follicular area of 10000 μm, only 9 sperm in 3n but 1072 sperm in 2n	Kiyomoto et al. (1996)
Dwarf surfclam *Mulinia lateralis*	0.30 : 0.70	XX–XY	Only 1.7 X 10^5 eggs in 3n ♀, 2.0 X 10^9 sperm in 3n ♂ but 2.5 X 10^9 sperm in 2n male	Guo and Allen (1994a)
Japanese pearl oyster *Pinctada martensii*	–	–	Reduced but fertile sperm produced, 2n eggs are also produced	Komaru and Wada (1990, 1994)
Pacific oyster *Crassostrea gigas*	0.23 : 0.48 : 0.29 ⚥	ZW–ZO	21% reduction in oogenesis, 23% reduction in follicular area, 2.54 X 10^5 eggs in 3n ♀; 45% reduction in spermatogenesis in 3n ♂	Allen and Downing (1986, 1990), Guo and Allen (1994b)

† maturing stage I at age 1+

Another advantage of triploid bivalves is their suitability to markets. Marketing demands supply of tastier oyster almost throughout the year. Triploid Pacific oyster readily meets these demands, as their meat contains more carbohydrate and their availability during off-season. Following natural spawning from mid July to mid August, diploid oyster incurs heavy loss of body weight and carbohydrate content (Fig. 6.6). Conversely, triploid oyster grows and stores higher carbohydrate content due to retarded gametogenesis. Glycogen provides the essential substrates for lipogenesis, i.e. vitellogenesis in eggs and pentose for nucleic acids. Its storage and utilizations is intimately related with the reproductive cycle. Indeed, its content reflects the extent of

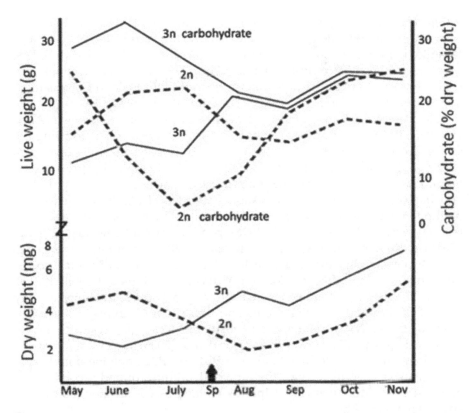

FIGURE 6.6

Dry weight (lower panel), live weight and carbohydrate content (upper panel) of diploid and triploid *Crassostrea gigas* during successive Calendar months intervened by spawning (Sp, indicated by a vertical arrow) (compiled and redrawn from Allen and Downing, 1986).

gametogenesis in *C. gigas* (Perdue, 1983). Hence, diploid and triploid oysters display distinctly different patterns of glycogen (carbohydrate) utilization (Fig. 6.6). The carbohydrate content of triploids remains higher from May to November than that of diploids. This is also true of *C. rivularis* (Perdue and Erickson, 1984). In *C. madrasensis* too, carbohydrate content increases from ~ 13% in 2n to ~ 20% in 3n (Mallia et al., 2009). Interestingly, explants of the oyster in isolated culture with cerebropleural ganglia showed normal gametogenic activity coupled with degeneration and loss of storage tissues. Conversely, explants cultures with visceral ganglia led to retarded gametogenesis and maintenance of storage tissues (Lubet and Mathieu, 1978). Hence, glycogen utilization/gametogenesis and glycogen storage are antagonistically regulated. In *M. galloprovincialis*, Danton et al. (1996) have also demonstrated that the annual cycles of germinal and storage tissue volume are also antagonistically related to the levels of cerebral insulin-like peptide.

Tetraploidy: Natural occurrence of tetraploids and hexaploids is reported from many freshwater snails (e.g. *Melanoidis tuberculata*, Jacob, 1954). To induce tetraploids, three methods have been adopted. The first and successful method involves the retention of PB1 by CB (0.67 mg/l) in unreduced triploid eggs (see below) that are fertilized by sperm form diploid male in *C. gigas* (Guo and Allen, 1994b, Guo et al., 1998). This Guo-Allen method has been used to induce tetraploids in *C. gigas* (Guo et al., 2002a, b) and *C. ariakensis* (Allen et al., 2003). However, 26% tetraploids induced in *Ch. farreri* employing the Guo-Allen method did not survive beyond 60 post-fertilization days (pfd) (Yang et al., 2000). It must also be noted that the tetraploid pearl oyster *P. fucata martensii* do not produce unreduced triploid eggs but only diploid and haploid eggs (Komaru and Wada, 1994, He et al., 2004). Hence, the method may not work in species, in which triploid females produce no eggs (e.g. *M. arenaria*, Kiyomoto et al., 1996), produce diploid eggs (e.g. *P. fucata martensii*, Komaru and Wada, 1994, He et al., 2004) and produce a few eggs (e.g. *M. lateralis*, Guo and Allen, 1994a). Hence, the second direct method of inducing tetraploidy by retention of PB1 has to be adopted. Using CB (0.67 mg/l) to block PB1, tetraploids were produced in *M. lateralis*. However, only 0.6% of the induced were tetraploids and they did not survive to 30 days pf (Yang and Guo, 2004). Evidently, a majority of *M. lateralis* eggs were not yet tetraploids (see p 165). However, Yang and Guo (2006) successfully induced tetraploids in *M. lateralis* by retention of PB1 and PB2 applying double heat (35°C) shocks to fertilize eggs, first at 35–37 minutes pf and second at 50–57 minutes pf. More than 44% triploids were induced but they too did not survive beyond 24 hpf.

A reason adduced for the low viability of tetraploid produced by the direct method involving retention of either PB1 alone or both PB1 and PB2 is that haploid eggs hold too little cytoplasm to sustain tetraploid nucleus. The larger unreduced triploid *C. gigas* eggs provide the required quantum of cytoplasm and enabled the survival of tetraploids. Unfortunately, Yang and Guo (2004, 2006) have chosen eggs from diploid *M. lateralis* to induce tetraploids. Had they used unreduced triploid eggs of *M. lateralis*, they could have been as successful as Guo and Allen (1994b). In *C. gigas*, egg size is 47.8 and 55.2 μm in haploid and triploid, respectively, i.e. the 3n egg is 1.155 times larger than that of a haploid egg. With sizes of 48.6 and 55.6 μm in haploid and triploid eggs of *M. lateralis* (Guo and Allen, 1994a), the 3n egg is 1.165 times larger than that of the haploid egg. Hence, long living and perhaps fast growing tetraploids are certainly achievable using triploid eggs, even in *Tapes phillippinarum* and *M. galloprovincialis* with their haploid eggs of 55–60 μm and 66–70 sizes, respectively (see Allen et al., 1994). Information on growth of tetraploids is available for *C. gigas* only. At the age of 3 months, its reported growth is 65, 92 and 284 mg for diploid, triploid and tetraploid, respectively (Guo and Allen, 1994b), i.e. in comparison to diploid, the growth of triploid and tetraploid is 1.4 and 4.4 times faster.

Polyploid gigantism, a common feature in plants, has also been observed in some aquatic invertebrates like *Artemia salina*. However, the increased cell volume in aquatic vertebrate polyploids like fishes (Pandian, 2011, p 124) and amphibians (Fankhauser, 1945) is compensated by a reduction in cell number. Consequently, the overall size of an individual remains unchanged. The cell number of tetraploid fishes is reduced to maintain equal size with diploids. For example, 54% reduction in cumulative cell number of tetraploid grass carp *Ctenopharyngodon idella* occurs in 5 hour-old alevin (see Pandian, 2011, p 115). In triploid and tetraploid bivalves, the number of cells is not reduced but the size of each cell is increased along with increasing ploidy level. The gigantism is often noted in storage organs like the adductor muscles. In comparison to respective diploid adductor muscle cells, the increased triploid cell size is 73% more in *Argopecten irradians*, 96% in *Ch. farreri* and as much as 198% in *C. gigas* (see Guo and Allen, 1994a). Hence, development and application of triploidy and tetraploidy shall dimensionally increase aquaculture production of bivalves in developing countries like India.

Incidentally, Haley (1977) reported the growth of individual diploid *C. virginica* varied from family to family so much that a fast growing male from a family was 1.33 times more than a slow growing male of another family; this value was as high as 1.71 times more in fast growing females. The need for experiments on family-based polyploid is obvious.

In all, XX/XO-like heterogamety was recognizable in a single gastropod *Busycon carica* (see p 158–159). In bivalves, both XX/XY and ZW/ZO heterogametisms have been recognized in eight species (Table 6.5). The presence of heterogamety in < a dozen species for molluscan phyla with > one million species may be more an exception than a rule. Pieces of *Sepia* gonad cultured in a non-hormonal medium develop into a testis or ovary at ratio of 1 : 1 (see Mangold, 1987), clearly indicating that sex determination has already been made following fertilization in cephalopods. Hence it is likely that the molluscs employ a hitherto unknown mechanism of sex determination.

6.5 Mitochondrial Genome

Genomics: Haploid chromosome number ranges from nine (Park et al., 1988) to 22 (Laws, 1973) in gastropods and from 10 to 23 in bivalves (Thiriot-Quievreux, 1994). The DNA content of haploid bivalve is in the range of 0.65–5.4 pg (Gregory, 2005). The genome size ranges from ~ 550 bp in *Crassostrea virginica* to ~ 2000 bp in *Pecten maximus* (Saavedra and Bachere, 2006). In terms of bivalve species, DNA databases are dominated by oysters (73%) followed by mitochondrial (23%) and microsatellites (4%). Understandably, the Oyster Genome Consortium (OGC), composed of 70 scientists from 11 countries intends to sequence the genome of *C. gigas* (Hedgecock et al., 2005).

Mitochondrial genome: With the discovery of 'doubly uniparental inheritance' (DUI) (Zouros, 2001), the study of mitochondrial genome has become a hot area of research in bivalves during the last two decades. Bivalvia is the only known taxon to comprise species, in which two mitochondrial genomes stably coexist. Species with DUI carry two mitochondrial genomes, the first one called the M genome, is transmitted to sons only and the second one namely the F genome is transmitted to both sons and daughters. In other words, sons and daughters inherit the F genome from the mother but the M genome is inherited from father to sons only. Interestingly, hermaphrodites arising from crossings have M genome in the testicular tissues and F genome in the ovarian tissues of the ovotestis in the Mediterranean mussel *Mytilus galloprovincialis* (Fig. 6.7). This particular pattern of mitochondrial inheritance is called doubly uniparental inheritance (Zouros et al., 1994a, b). Until 2000, DUI has been detected in species from three families of marine (Mytilidae, Skibinski, 1994, Veneridae, Passamonti and Scali, 2001) and freshwater (unionids, Hoeh et al., 1997) bivalves. Presently, DUI is known to be present in 36 species belonging to seven families of bivalves (see Kenchington et al., 2009). It is likely that DUI is a characteristic of bivalves.

Many oyster species are not stable gonochorics (e.g. *Crassostrea, Mytilus,* Haley, 1977) and in some freshwater bivalves, sex is more a population specific trait than species specific trait (see p 88). It must be noted that the sex determination process is immediately followed by sex differentiation and the reported sex ratios are the cumulative end product of these successive

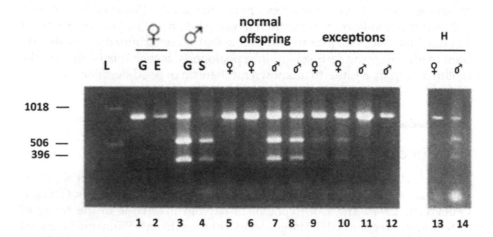

FIGURE 6.7

Normal doubly uniparental inheritance (DUI) in gonochoric (Lanes 5–8) and hermaphroditic (Lanes 13–14) and exceptions (Lanes 9–12) to DUI in pair matings of *Mytilus galloprovinciallis.* Lanes 1 and 2 represents female (G) and egg (E) and lanes 3 and 4 male and sperm, respectively (from Saavedra et al., 1997).

yet separable processes. In the Manila clam *Ruditapes philippinarum*, the overall sex ratio is balanced but male ratio per family ranges from 0.08 to 0.83 (Fig. 6.8). The findings of Rawson and Hilbish (1995) also suggest that sex determination in bivalves is more a family trait rather than species trait. Returning to DUI, pair mating, in which females and males were multiply crossed, Saavedra et al. (1997) found that male ratio ranged from an extreme of 0.0 to another extreme of 0.97 in *M. galloprovincialis* (Table 6.6). Confirming their findings and extending it to *M. edulis* and *M. trossus*, Kenchington et al. (2002) reported that the mating of a female with different males produced the same sex ratio but those of the same male with different females produced different sex ratios. Hence, sex determination in these mussels is independent of male and a property of female parent. Incidentally, *M. edulis*, like *M. galloprovincialis* (Table 6.5) may be female heterogametic. If so, it is to be expected that the heterogametic ZW female determines the sex and the homogametic ZZ male may not alter it. In *M. galloprovincialis*, of 834 progeny generated by crossing five females each with five males, M70 male was found to produce M-negative sons, female F31 was highly male-biased and F19 more female-biased (Table 6.6). Hence, the females were grouped into three classes namely those females that (i) produced only females, (ii) high proportion of males and (iii) almost equal proportion of males and females (Saavedra et al., 1997). Incidentally, Ghiselli et al. (2012) identified 1,575 sex-biased genes that differed in overall expression levels between males and females of *R. Philippinarum*. Among the sex-biased genes, 935 were male-biased and 640 female-biased. Sex-biased genes were more highly expressed in males than females; of them, 165 genes were associated with family-biased expression.

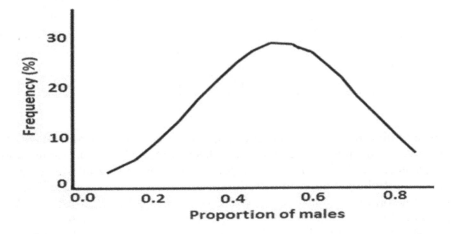

FIGURE 6.8

Frequency of proportion of males in *Ruditapes philippinarum* (modified and simplified from Ghiselli et al., 2012).

TABLE 6.6

Sex distribution and mitotypes in pair mating of *Mytilus galloprovincialis* (from Saavedra et al., 1997, simplified)

Cross	Male Ratio (%)	Presence of M mtDNA			
		Females		Males	
		Present	Absent	Present	Absent
F19 x M27	0.83	0	6	30	0
F19 x M28	0.97	0	1	28	0
F19 x M46	0.68	0	10	21	0
F19 x M54	0.81	0	6	23	0
F19 x M70	0.69	0	9	0	20
F20 x M27	0.85	1	4	5	0
F20 x M28	0.79	0	5	5	0
F20 x M46	0.69	0	5	5	0
F20 x M54	0.62	0	14	23	0
F31 x M27	0.03	0	29	1	0
F31 x M28	0.00	0	32	0	0
F31 x M46	0.03	0	28	0	1
F31 x M54	0.07	0	27	0	2
F31 x M70	0.04	0	27	0	1
F53 x M27	0.26	0	5	5	0
F53 x M28	0.26	0	6	3	0
F53 x M46	0.33	0	5	9	0
F53 x M54	0.08	0	34	2	1
F53 x M70	0.27	0	22	0	8
F66 x M27	0.61	0	5	5	0
F66 x M28	0.61	0	5	5	0
F66 x M46	0.64	1	4	4	1
F66 x M54	0.47	0	17	14	1
F66 x M70	0.50	0	16	1	15

In addition to F and M genome, the existence of a third mitochondrial genome designated as C has also been detected in *M. edulis* and *M. galloprovincialis*. Genome C has a primary sequence like that in F genome but its control region has an insertion of three copies of M genome and its transmission route is also paternal. Hence, it is called the 'masculinized' F genome (see Kenchington et al., 2009). It is not clear whether the presence of C genome is responsible for M-negative sons. The occasional presence of M genome in females may represent a byproduct of the dispersed pattern of sperm mitochondria in eggs that are already committed to develop into females (see Kenchington et al., 2009).

Prior to the description of the fate of sperm mitochondria in embryos, it is necessary to visit the origin and development of Primordial Germ Cells (PGCs) responsible for manifestation of sex in bivalves. In animals, the *vasa* gene is used as a molecular marker to confirm the presence and location of PGCs. In *C. gigas*, the oyster *vasa*-like gene (*Oyvl*) transcript, localized on the vegetal pole of unfertilized oocyte, is maternally inherited, as in fishes (see Pandian, 2011, p 131–132) and crustaceans (see Pandian, 2016, p 161–162). The transcript is segregated into a single blastomere at 2-cell stage and then to 4d mesentoblast of blastula (Fig. 6.9). From the late blastula stage, the mesentoblast divides into cell clumps and the clumps migrate to both sides of the larval body. The clumps, rather PGCs differentiate into germinal stem cells in the juvenile oyster (Fabioux et al., 2004). Remarkably, there are differences in location and direction of migration of PGCs between molluscs on one hand and fishes and crustaceans on the other. In fishes, the maternally supplied mRNAs are located on the animal pole. Subsequently, the developed PGCs migrate dorso-laterally toward posterior direction (e.g. *Carassius auratus*, see Pandian, 2011, p 130–132). In crustaceans too, the cluster

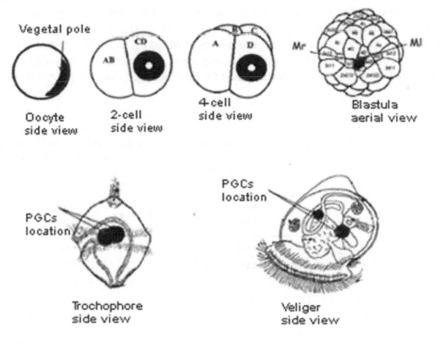

FIGURE 6.9

Schematic representation of the location and migration of germline during the embryogenesis of *Crassostrea gigas* (modified and redrawn from Fabioux et al., 2004).

of PGCs located at the base of antennule, antenna and cerebral lobe migrate dorso-laterally toward posterior direction (e.g. *Fenneropenaeus chinensis* see Pandian, 2016, p 161–162). Conversely, the presumptive PGCs are located at the vegetal pole in unfertilized egg of *C. gigas* and subsequently migrate dorso-laterally toward anterior direction.

Staining the sperm mitochondria of *M. edulis* with the florescent dye MitoTracker Green FM, Cao et al. (2004) and Cogswell et al. (2006) traced their fate in eggs of mothers known to produce female-biased and male-biased progeny. In female-biased eggs, all or most of sperm mitochondria form an aggregate and are attached to 4d blastomere, which is known to give rise to cell line that develops into germinal stem cells. In female-biased eggs, the sperm mitochondria are randomly segregated among the blastomeres. However, only 8.5% of the sperm mitochondria have been found in the 4d cell of the embryos from female-biased mothers but as much as 63% of the mitochondria in male-biased embryos. Clearly, "the cellular mechanism that directs the fate of sperm mitochondria in the fertilized eggs differs among embryos that are hatched by mothers, which produce exclusively daughters, and embryos that are hatched by mothers, which produce mostly sons" Cogswell et al. (2006). Hence, this 'intimate' relationship between DUI and sex determination is more of causative (i.e. has an active role in sex determination), than associative (i.e. DUI is a byproduct of sex determination).

However, hybridization (see also Wood et al., 2003) and/or triploidization almost completely disrupted sex ratios in mytilids. From Table 6.7, the following may be inferred: in female-biased *M. edulis* mother, a homospecific cross produce daughters only, as expected. Surprisingly, triploids and hybrids between *M. edulis* female and *O. trossus* male produced almost equal

TABLE 6.7

Summary of sex ratio and mitotype of progeny pooled from different cross types of (E) *Mytilus edulis*. Values given within brackets are percentages. P = Pure species, H = Hybrid between *M. edulis* and *Ostrea trossus*, D = Diploid, T = Triploid (from Kenchington et al., 2009, modified)

Sex Ratio Bias of Mother	Cross	Progeny	♀ Progeny (no.)	♂ Progeny (no.)	⚥ Progeny (no.)	F Progeny (no.)	F + M Progeny (no.)
Female	E X E	PD	138 (98)	0 (0)	3 (2)	141 (100)	0 (0)
Female	E X T/H	HD	112 (41)	103 (38)	57 (21)	272 (100)	0 (0)
Female	E X E	PT	0	107 (100)	0 (0)	107 (100)	0 (0)
Female	E X H	HT	0	57 (98)	1 (12)	58 (100)	0 (0)
Male	E X E	PD	9 (10)	79 (84)	6 (6)	16 (17)	78 (83)
Male	E X T/H	HD	0	44 (98)	1 (2)	7 (16)	38 (84)
Male	E X E	PT	0	30 (94)	2 (6)	6 (23)	26 (77)
Male	E X H	HT	0	25 (100)	0 (0)	9 (36)	16 (64)

numbers of males and females along with 21% of hermaphrodites. More surprisingly, when she was crossed with a homospecific or hybrid male followed by triploidization, the crosses produced male progenies only and all of them carried only F mitotypes. When a male-biased mother was paired with homospecific male, the cross produced 84% male progeny but 17 and 83% of them bearing F or F + M mitotypes, respectively. Hybridization and triploidization in crosses between male-biased mother and homospecific or hybrid male also produced 94–100 males but 16–36 of them carrying F mitotype. These experimental evidences showed that female "mussels have two separate properties: (i) produce only females and (ii) prevent the transmission of the sperm's mtDNA to their progeny. Hybridization and triploidization interfere with the first property but leave the second unaffected" (Kenchington et al., 2009). Yet, these findings may have to be considered in the light of the following: 1. Disruption of sex ratio by hybridization and/or triploidization is limited to female-biased mothers only. 2. Transcription factors present in hybrids may not function correctly and effectively due to regulatory incompatibility and dosage effect of increasing maleness in triploids.

Females and males share the same nuclear genome. Hence the phenotypic divergence in sex requires different alleles of gene(s), their expression and sex specific regulation. It is likely that transcription analyses provide a better insight into the relationship between DUI and sex determination. A comparison between families with female-biased mother and male-biased mother led to the identification of about 200 genes in *Tapes philippinarum* (Ghiselli et al., 2012). Of these genes, eight were male-biased transcription and their expression was 7.6 times more in mothers of male-biased family than in mothers of female-biased family. Hence, the expression of these genes could bias the development toward maleness. Ubiquitination is a process, in which sperm mitochondria are degraded. The family-biased expression of ubiquitination genes were also linked to mitochondrial degradation in mothers of female-biased family. Briefly, these findings suggest a causative relationship between DUI and sex determination.

6.6 Sex Determination Models

While the relationship between DUI and sex determination in bivalves is being debated as causative or associative, separate models have been proposed in support each of them. The associative model proposed by Kenchington et al. (2009) includes a nuclear locus S with alternative alleles S_1 and S_2 with different 'dosages' ($S_1 = 1, S_2 = 2$) (Fig. 6.10). For determination of male sex dosage S_2 is required. Only maternal allelic is expressed, when sex is determined. The paternal allele contributes a dosage of zero. Paternal mtDNA transmission

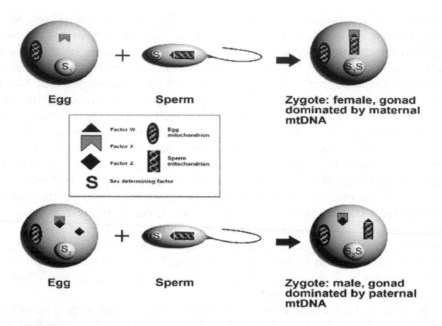

FIGURE 6.10

DUI associative model for sex determination and paternal mtDNA inheritance in mytilid mussels, as proposed by Kenchington et al. (2009). The model is based on their experiments on hybridization and triploidization. For description see text.

is affected by three maternal nuclear genes; of them, locus Z is important. It segregates for an active Z and an inactive z. It supplies the egg with factor an active Z, which suppresses factor X and allows sperm mitochondria to aggregate and attaches to PGCs. There is a tight linkage between Z and S_2 as well as between z and S_1 so that only the combinations of Z S_2 and z S_1 may occur in eggs. Hence DUI is more an associative than causative. This model has many assumptions. For example, Kenchington et al. (2002) have concluded that sex determination in the mussels is independent of male and is a property of female parent. This conclusion may have to be reconsidered in the light of the following: Firstly, this is applicable only for female (ZW) heterogametic but not for male (XY) heterogametic bivalve species (see Table 6.5). Secondly, the female pronuclear genome is disintegrated in the natural androgenic clam of *Corbicula* (see p 163). Thirdly, matings between female-biased mother of *Mytilus edulis* with a male produce 100% females, as expected, indicating that sex determination is solely a property of female parent. If this is correct, it is difficult to comprehend how sex ratio of 1 ♀ : 1 ♂ was obtained, when the female-biased mother of *M. edulis* was crossed with *Ostrea trossus* male, as paternal allele contributes a dosage of zero (see Table 6.7). Hence, sex is determined at fertilization in relation to nuclear and

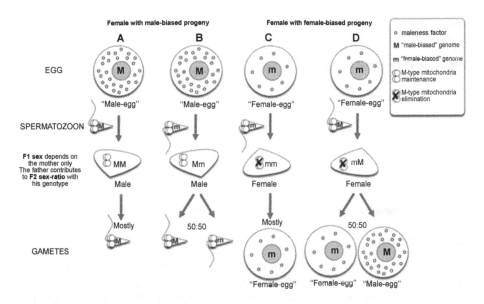

FIGURE 6.11

DUI causative model for sex determination and paternal mt DNA inheritance proposed by Ghiselli et al. (2012) based on their transcriptomoe expression study in *Ruditapes phillippinarum*. Personal communication by Dr. F. Ghiselli of their modified figure. For details see text.

perhaps mitochondrial genomic contributions made by both female and male in bivalves.

The causative model proposed by Ghiselli et al. (2012) is based on transcription expression. Transcription factors like ubiquitination genes, stored in an oocyte committed for female sex, activate sex gene expression in early embryonic stages (Fig. 6.11). Male development requires the crossing of a critical threshold of masculinizing transcripts. The sperm genotype contributes to sex bias in F_2. 'Female eggs' produce F_1 females, regardless of the genotype 'G' or 'g' of the spermatozoan. When the 'female oocyte' is fertilized by a spermatozoan with male-biased genotype 'g', the F_1 produces mostly F_2 'female eggs'. But when it is fertilized by a male-biased genotype 'G' sperm, the F_1 females produce female and male eggs in 0.5 : 0.5 ratios. A male egg produces males only, regardless of being fertilized by 'G' or 'g' spermatozoan. F_1 males homozygous for GG genotype produce only sperm bearing 'G' genotype. But the heterozygous F_1 males produce sperm bearing 'G' or 'g' genotype in 0.5 : 0.5 ratios. The model is simple and supported by evidences like sequence polymorphism and so on.

7

Sex Differentiation

Introduction

In mollusca, the hormone system is perhaps most diverse of all invertebrate phyla, as it differs not only between classes but also within classes reflecting extreme differences in sexuality and life history patterns (e.g. Oekten et al., 2004). Remarkably, 25% of molluscs are simultaneous and sequential hermaphrodites; some of these Simultaneous Hermaphrodites (SH) are capable of selfing. However, hardly a few publications are available on endocrine differentiation in protandric (e.g. *Crepidula fornicata*, Le Gall et al., 1981), unilateral SH (e.g. *Aplysia californica*, Zhang et al., 2000, 2008, *A. depilans*, Lupo di Prisco and Fugheri, 1975) and reciprocal SH (e.g. *Lymnaea stagnalis*, De Jong-Brink et al., 1981). Surprisingly, a host of vertebrate-types steroids is present in molluscs but their endogenic biosynthesis and role in reproductive cycle is questioned (Scott, 2013). Only, a single publication by Wang and Croll (2004) reports sex reversal in scallop *Placopecten magellanicus* treated with one or other steroid. Whereas pollutants like tributyltin (TBT) masculinize (female) fishes, it only superimposes masculine traits on female gastropods. Many molluscs serve as intermediary host to trematode parasites, which either partially or completely sterilize them.

7.1 Neurohormones

In multicellular organisms, the endocrine system employs chemical messengers to coordinate and integrate various mechanisms and physiological processes. Considering their chemical properties and mode of action, neurohormones are divided into two main groups namely neuropeptides and neuroamines (Table 7.1). They act directly on target tissues. In molluscs, the sex differentiation process is accomplished solely by neuropeptides. The neuropeptides act on membrane receptors to trigger a cascade of intracellular events, which result in gene transcription (Ketata et al., 2008). Among them,

TABLE 7.1

Neurohormones of molluscs (compiled from Ketata et al., 2008)

Name	Abbreviation
Neuropeptides	
Caudodorsal cell hormone	CDCH
Dorsal body hormone	DBH
Egg-laying hormone	ELH
Molluscan-insulin like peptide	MIP
Neuroamines	
Dopamine	
Serotonin	
Noradrenalin	

peptidic messengers are most common. In vertebrates, the release and levels of steroids are controlled by Gonadotropin Releasing Hormone (GnRH). In molluscs too, the presence of GnRH-like peptides from the optical glands of *Octopus vulgaris* has been detected using heterologous GnRH antibodies (Di Cosmo and Di Cristo, 1998). In gastropods, reports are also available for the presence of GnRH-like neurons in freshwater snails *Helisoma trivolvis* and *L. stagnalis* (Young et al., 1999) and GnRH-like peptides in the Pacific abalone *Haliotis discus hannai* (Amano et al., 2010). The presence of GnRH-like factor in the Central Nervous System (CNS) has been detected from the extracts from cerebral ganglion and hemolymph of *Mytilus edulis* (Amano et al., 2010). The GnRH is under the ultimate control of CNS (e.g. *O. vulgaris*, Di Cristo, 2013). Like the presence of hypothalamus ⟹ pituitary ⟹ gonad axis in vertebrates, the CNS ⟹ GnRH ⟹ gonad axis seems to exist in molluscs.

Besides ganglia, the optical glands are known to produce neurohormones that control feeding, body growth and sexual maturation in cephalopods. The gland is a small rounded body located on each of the optical stalk and is innervated from the subpeduncle lobe of the brain of octopus (Frosch, 1974). Surgical removal of the glands results in precocious sexual maturity and enlarged ovary and testis (Wells and Wells, 1972a). Blinding followed by removal of the glands has also led to maturity. This observation suggests the sequence of hormonal control from light ⟹ brain ⟹ optical gland ⟹ gonad. The neurohormone is required for follicular proliferation and oogonial multiplication in female, and multiplication of spermatogonia in male *Sepia officinalis* (Richard, 1970). Successful functioning of the implanted gland from opposite sex or species from other families and order also suggests that the hormone(s) arising from the optical gland is common to all cephalopods (Wells and Wells, 1975). Removal of the glands, testis and gonoduct does not alter sexual behavior of *O. vulgaris* and *O. cyanea* (Wells and Wells, 1972b). However, the optical gland abalated *O. hummellincki* females cease to brood. Normal and sham operated female *O. hummellincki* survives for 43–45 days after egg laying (Table 7.2). However, the post-egg laying period is extended to 77 and 175 days, following unilateral and bilateral abalation of the optical glands, respectively. Possibly, the light, through its effect on the brain and optical glands, may play a decisive role on maximum body size and duration of life after egg laying and semelparity.

In crustaceans, the process of sex differentiation is under the control of neurohormones synthesized and released by androgenic glands. But that of reproductive cycle is regulated either individually or jointly by neuropeptides of sinus glands as well as methyl farnesoate and ecdysone synthesized and released by Y-organs and mandibular organs, respectively (see Pandian, 2016, Chapter 7). In daphnids alone, ecdysone and its analogues (the juvenoids) induce sex reversal from female to male. Despite germ cells being developed in these 'chemically induced' embryos, the juvenoids only induced masculine morphological differentiation but not the production of sperm to fertilize eggs (Pandian, 2016, section 7.2.2). The fact that demonstration of the juvenoid's role is limited to daphnids alone indicates the sex differentiation process is entirely under the control of androgenic glands in malacastracans and its equivalent in entomostracans. An array of responses displayed by crustaceans evoked by exposure/injection of estrogen, progesterone, testosterone or other vertebrate-type steroids are entirely experimental and not natural, as none of these steroids is shown to be synthesized in any of the said nervous glands/organs of crustaceans. In molluscs too, these two processes are decoupled, i.e. sex differentiation is under the control of neurohormones and the reproductive cycle may also be regulated by neurohormones/neuroamines or may jointly be regulated by steroids gained from the surrounding water. That sex differentiation is controlled solely by neurohormones/amines is supported by the following evidences: 1. As indicated earlier, pieces of *Sepia* gonad cultured in non-horomonal medium developed into ovary and testis at ratio of 1 : 1 (see Mangold, 1987). Clearly, this finding has the following implications (i) the presence of genetic sex determination mechanism, (ii) the entire process of differentiation of ovary and testis is accomplished by endogenously synthesized neurohormones and (iii) the need for exogenously arising steroids has no role in sex differentiation. 2. However, there are claims

TABLE 7.2

Effect of unilateral and bilateral removal of optical glands of *Octopus hummelincki* (simplified from Wodinsky, 1977)

Optical Glands	Maximum Body Weight (g)	Body Weight at Death (g)	Duration from Egg Laying to Death (d)
Normal	204	188	43
Sham-operated	192	145	45
Unilateral removal	196	105	77
Bilateral removal	211	224	175

that the genetic sex of an undifferentiated bivalve gonad can be reversed by steroid exposure/injection. Wang and Croll (2004) injected 30 μl of steroid yielding 1000 μg/ml of E_2, progesterone (P_4), T or dehydroepiandrosterone (DHEA) into juvenile (7.9 g) scallop *Placopecten magellanicus* and the scallops were observed for sex differentiation during the period of 3 months following the injection. No reason was given for the choice of the steroid especially for DHEA and the selected dose. Injections of androgens T and DHEA resulted in the production of only about 50% of males and did not induce sex reversal in the remaining 50% scallops (Table 7.3). The development of only 8% females in E_2- and P_4-injected scallops was not comprehensible. Instead, the injections of these four steroids produced 42–44% undifferentiated scallops. Clearly, the injection of these steroids did neither influence sex differentiation nor induce sex reversal, i.e. the experiment by Wang and Croll failed to convincingly demonstrate the presence of a steroid mechanism for sex differentiation process and reversing sex in genetic males and females. 3. Imposex, associated with superimposition of masculine reproductive structures over that of females, is induced by a neurohormone Penis Morphogenic Factor (PMF) and may not involve T, as was earlier proposed (see Ketata et al., 2008). Hence sex differentiation process in gonochorics, protandric males and simultaneous hermaphrodites is almost solely under the control of neurohormones. This process is genetically determined and is not amenable to environmental influence (see Yusa, 2004b). In hermaphrodites, selfing and/ or reciprocal or unilateral insemination is a phenotypic behavioural trait. As indicated elsewhere (see p 91–92), the differentiation process leading to sex change from male to female in protandrics is labile and protracted, and can be modulated by environmental factors like temperature, food availability and so on.

Mori et al. (1969) claimed that sex change in protandric *Crossastrea gigas* could be accelerated by cumulative injections of estradiol-3-benzoate directly into the gonad of the three experimental series, the first one (0.4 mg/oyster) was administred from March to May, the second (0.2 mg/oyster) from April to June and the third (0.3 mg/oyster) from May to July. Table 7.4 shows that the

TABLE 7.3

Effects of steroids on sex differentiation in the sea scallop *Placopecten magellanicus* (compiled from Wang and Croll, 2004)

Steroid	Male (%)	Female (%)	Hermaphrodite (%)	Undifferentiated (%)
Testosterone	52.9	7.8	0	39.2
DHEA	43.8	8.3	4.2	43.8
Estrodiol	50.0	8.3	0	41.7
Progesterone	49.0	5.7	1.8	43.3
Vehicle control	26.2	21.4	0	52.4
Blank control	20.8	16.7	0	62.5

TABLE 7.4

Direct injection estradiol-3-benzoate into the gonad of protandric *Crassostrea gigas* and its effect on advancing sex change (compiled from Mori et al., 1969)

Duration of Injection	Raise in Temperature (°C)	Cumulative Dose (mg/oyster)	Female Ratio	
			Natural	Experimental
March–May	7 to 11	0.4	0.70	0.78
April–June	10 to 15	0.2	0.65	0.66
May–July	12 to 31	0.3	0.72	0.75

second and third injections hardly increased female ratio, when the gonads were exposed to 5 and 19°C rises in temperature, respectively. However, the first injection increased female ratio, when the gonad was exposed to a raise in temperature from 7 to 11°C. Notably, the sample size for natural oyster (10 only) against experimental (36) was small and no statistical analysis was made. Hence it is rather difficult to accept the claim of Mori et al. (see also Scott, 2013).

Remarkably, neurohormones are also known to play a role in the reproductive cycle. For example, it activates gametogenesis and vitellogenesis as well as controls the energy storage and spawning process in many molluscs (Mathieu, 1994). They stimulate sexual maturation in *Mytilus edulis* (Mathieu et al., 1991) and *C. gigas* (Pazos and Mathieu, 1999). In *M. edulis*, cerebropleural ganglia is reported to regulate gametogenic activity coupled with degeneration of storage organ (Lubet and Mathieu, 1994). In *Argopecten purpuratus*, serotonin and dopamine levels are always higher in testicular and ovarian tissues, respectively (Martinez and Rivera, 1994). Hence, neurohormones also play an important role in regulation of reproductive cycle. However, the present status of information available on the role of steroids in molluscs is nebulous and confusing. Hence, the ensuing account may be taken with 'a pinch of salt'.

7.2 Steroid Hormones

Steroids are widespread molecules derived from cholesterol or related sterol and are employed by organisms to regulate growth (e.g. brassinosteroids) and reproduction (Lafont and Mathieu, 2007). Aquatic organisms, especially fishes are the potential source of steroids in water. Evidences are reported for continuous traffic of steroids over the gills and frequently through urine and feces (e.g. Scott and Ellis, 2007). However, the excreted steroids are biologically inactive conjugates. *Escherichia coli,* that produces β-glucoronidase, is known to effectively deconjugate and render them active (see Pandian, 2015, p

104). Molluscs are able to convert the biologically inactive conjugates back to free steroids. For example, preparations from the limpet *Patella vulgata*, glucoronidase and sulfatase are commercially available (see Scott and Ellis, 2007).

Among the protostomes, Mollusca are the only phylum, in which vertebrate-type of steroid hormones were considered to endogenously synthesize. However, the endogenous origin is questioned for two reasons: 1. The molluscs have great powers to bioconcentrate lipophilic compounds in their tissues from surrounding water (Le Curieux-Belfond et al., 2001, 2005, Peck et al., 2007, Scott, 2012). Prostoglandins, implicated to induce gamete proliferation and spawning in several molluscan species, can also be accumulated (see Ketata et al., 2008). 2. However, the molluscan genome is now considered not to contain the genes for any of the key enzymes that are involved in biosynthesis of vertebrate steroids. 3. To further complicate the subject, Morely (2006) has justifiably brought another dimension from the point of view of parasitic infection. In this regard, the following are notable: (i) A peptide released by parasitized snail interferes with neuroendocrine regulation (e.g. *Lymnaea stagnalis* infected by *Trichobilharzia ocellata*, De Jong-Brink et al., 2001). (ii) The relationship between the level of parasitic infection and imposex induction is negatively correlated (e.g. *Ilyanassa obsoleta*, Curtis, 1994). (iii) A climax can be the snail *Bulinus globosus*, when infected by *Schistosoma hematobium*, ceases to produce eggs (Gracio, 1988). Hence, Morely suggested that all experiments and estimates related to steroid hormones in molluscs must be carried out only on healthy molluscs. He seems to be right, as some contradictory reports on steroid actions can be resolved, when sexuality and health or infection status are considered. For example, a single dose of injection of E_2 (1–1000 n mol/mussel) induced vitellogenin-like proteins in *Elliptio complanata* from Montreal, Canada (Gagne et al., 2001). But from Boston, USA, Won et al. (2005) could not detect any trace of vitellogenin-like protein, following the injection of 250 nmol E_2/mussel of *E. complanata*. Not surprisingly, such contradictory reports intrigued reviewers like Ketata et al. (2008). It may be noted that sexuality of *Elliptio* is more a population trait than a species trait (Downing et al., 1989, see p 89–90). However, vitellogenesis may not occur in an infected mussel, as infected snails cease to lay eggs (cf Gracio, 1988). It is not clear from which population the sentinel mussels were collected for experiments and whether the collected mussels were infected or not.

Off from these, many laboratory experiments have convincingly demonstrated that bivalves and gastropods have a remarkable ability to absorb the vertebrate type sex steroids, testosterone (T) and estradiol-17β (E_2) from surrounding waters and retain them. The mud snail *Ilyanassa obsoleta* required only 8 hours to absorb 80% of ^{14}C-labeled T from 3 ml solution (Gooding and LeBlanc, 2001). Following exposure of male and female *Mytilus edulis* to 200 ng E_2/l for 13 days, the E_2 uptake progressively increased from

0.4 ng/g body tissues on the 0 day to 14.63 ng/g tissues on the 13th day. The absorbed E_2 was esterified into fatty acids. Yet, the free E_2 level remained at 2.6 ng/ l (Fig. 7.1B). In *M. galloprovincialis*, Janer et al. (2005) reported a dose-dependent increase in E_2 level from 2 to 20 ng/g following exposure to 200 ng/l for 7 days. Exposure of freshwater zebra mussel *Dreissena polymorpha* to ^{14}C-E_2 (7.5 ng/l for 13 d) also showed that the uptake of labeled E_2 was to the tune of 6 mg/g in male and female mussels. A depuration period of 10 days did not reduce the E_2 level in the mussels (Fig. 7.1A). This was true also for the mussel exposed to natural clean waters containing < 0.01 ng E_2/l, suggesting the endogenous origin of E_2 (Peck et al., 2007). Two-hours following the injection of labeled E_2 into *Crassostrea gigas*, 58% of the injected E_2 was metabolized into estrone (E_1) and after 48 hours of injection, the E_2 distribution showed the maximum concentration in the gonad (Le Curieux-Belfond et al., 2005). Clearly, gastropods and bivalves do absorb T and E_2 from surrounding waters and retain a major part of it as esterified fatty acids.

Through his thought provoking reviews, Scott (2012, 2013) questioned the veracity of 85% of over 200 publications reporting the presence of one or more vertebrate type of sex steroid hormones and their role in reproduction of molluscs and the positive conclusion made by 18 out of 21 reviewers (between 1970 and 2012). Scott (2013) rejected more than half of these publications, as they did not actually test steroids but only compounds/mixtures that were presumed to behave as steroids. Of the remaining, 40 out of 55 were rejected for not replicating the treatment. The other 25 out of 55 were also rejected, as the data were not subjected to suitable statistical analyses. Further, Scott (2012) pointed out that mere presence of T or E_2 in molluscan tissues including the gonad may not be an evidence, as sex steroids

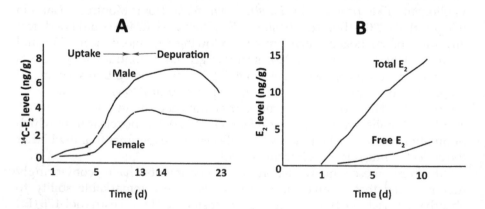

FIGURE 7.1

A. uptake (up to 13th day) and depuration (after 13th day) of E_2 by males and females of the zebra mussel *Dreissena polymorpha* (simplified and redrawn from Peck et al., 2007). B. Levels of free and total absorbed E_2 from water by *Mytilus edulis* (simplified and redrawn from Puinean et al., 2006).

are extracted from almost all organisms (e.g. Janer and Porte, 2007) including plants (Janeczko and Skoczowky, 2005). Reliable evidences are required to show that they are sex specific differences in the levels of T and E_2 in the gonads of males and females and seasonal changes in the levels of these hormones in tune with reproductive cycle in the context food availability and temperature (cf Fig. 7.3). In molluscs, the uptake of T or E_2 from water is in the range of nanogram levels but sex specific and/or seasonal changes in these sex steroids oscillate at dimensionally higher picogram levels. Hence T and E_2 may have to the synthesized in molluscs. From Fig. 7.2, it may be noted that in vertebrates, a series of enzymes are involved in biosynthesis of T and E_2 from cholesterol. CYP11A1, 17β hydroxylsteroid dehydrogenase (17β HSD) and aromatse are key enzymes required for synthesis of pregnenolone, T and E_1/ E_2, respectively. The view of Scott (2012) is discussed in the light of following observations (see also Table 7.5):

1. In molluscs, cholesterol, the precursor for synthesis of T and E_2, is present in adequate quantities (Idler and Wiseman, 1972). The injected ^{14}C-acetate is converted into cholesterol in many molluscs (e.g. *Littorina littorea*, Voogt, 1969), but it is not in carnivorous cephalopods (e.g. *Sepia officinalis*, Carreau and Drosdowsky, 1977) and gastropods (e.g. *Buccinum undatum*, Voogt, 1969). With no evidence for the presence of pregnenolone, the

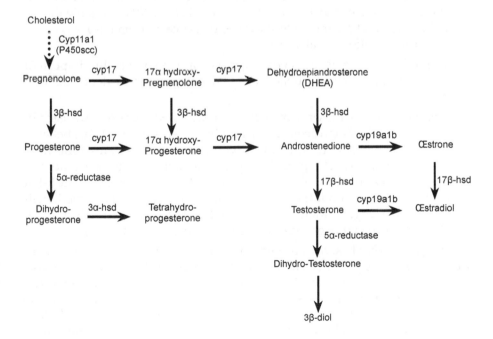

FIGURE 7.2

Steroidogenic pathway in zebrafish (from Diotel et al., 2011).

possibility of these molluscs using steroid other than cholesterol as a starter for the synthesis of steroids, may not exist.

2. mRNA extracted from *Mytilus* spp has been shown to hybridize with a probe based on human CYP11A1 (Wootton et al., 1995). Despite this demonstration, the existence of CYP11A1 in molluscs has to be rejected. For genomic studies indicate that the gene for *CYP11A1* has evolved only in the vertebrate line (Markov et al., 2009). For the likely presence and functions of vertebrate type of sex steroids in many echinoderms, Kohler et al. (2007) may be consulted.

3. Although 13 studies have brought very strong evidences for the presence of 17β HSD and its effective role in conversion of androstenodione (AD) into T and E_1 and vice versa (e.g. Matsumoto et al., 1997) in a few molluscs, Scott (2012) has considered it as a routine in many biochemical pathways and disposed it. However, these evidences may go against the view of Scott.

4. Of 10 publications, which attempted to demonstrate the presence of aromatase in five bivalves, three aquatic gastropods and one cephalopod by adding labeled T or tritium labeled T, the expected conversion to from T to E_2 did not occur in four cases and in others, it occurred at abysmally low level from 0.01% in *L. littorea* (Ronis and Mason, 1996) to 7% in *Ruditapes desscatus* (Morcillo and Porte, 1999), in comparison to bovine tissues. Incidentally, *aromatase* gene first appeared only in chordates (Reitzel and Tarrant, 2010, Callard et al., 2011). Notably the last publication that attempted to show the presence of aromatase was that of Le Curieux-Belfond et al. (2001).

5. The other enzymes in steroidogenesis are also equally important. However, the relevant publications did not provide evidence for their presence.

6. The claims for the presence of nuclear receptors (*nER* or *nAR*) are listed by Kohler et al. (2007) and Scott (2012). Firstly, the receptors (*nER*) are believed to have evolved only after the divergence of vertebrates (Thornton, 2001). No one has yet reported the presence of any molluscan genes that match nuclear receptor of other vertebrate steroids. In molluscs, nER-like proteins identified but they are structurally similar but not functionally (e.g. *Thais clavigera*, Kajiwara et al., 2006, *Octopus vulgaris*, Keay et al., 2006). They are more of constitutive receptors (see Scott, 2012).

Briefly, gastropod larvae are known to acquire amino acids and sugars from surrounding water (Welborn and Manahan, 1990). Similarly, molluscs may treat steroids as a convenient nutrient source. The presence of individual steps of biosynthetic pathway has been shown in some species with very poor yield. Till date, a complete sequence of biosynthesis from cholesterol to E_2 in any one mollusc remains to be demonstrated. Crucially, many genes responsible for key enzymes in biosynthetic pathway of vertebrate type of sex steroidal hormones are considered to be missing in molluscs. In vertebrates, the sex steroids control and regulate both sex differentiation and gonadal maturation. However, sex differentiation in molluscs is completely under the control of neuroendocrines. Similarly, the gonadal maturation and reproductive cycle may also be regulated by neuroendocrines. Yet, the fact that seasonal changes in steroids in both male and female in all 10 species of molluscs, for which information is available, raise a big question on rejecting the role of vertebrate steroids in molluscs. Regarding the statements of Scott (2012, 2013) that the genes responsible for vertebrate type of steroidogenesis, the following may be noted: 1. E_2 is a ligand of ER in Unilateria, the common ancestor of deuterosomes and protosomes, which include schizocoelomata comprising of molluscs and others. 2. "There is no reason.... to imagine that these vertebrate steroids—if present—would be more likely to act as hormones" (Markov et al., 2006). 3a. The selection of *Aplysia californica* by Markov et al. may not be a good representative for Mollusca, as *A. californica* is a simultaneous hermaphrodite and a continuous breeder (see Fig. 3.14). Hence, sex specific and the season's specific changes in steroid level may hardly occur. The choice of *Sinonovacula constricta* may be an ideal representative (see below). 3b. By the yard stick of Scott himself, the results arising from a single species for a speciose phylum Mollusca comprising more than 100,000 species may not be adequate. It is for the molecular biologists to check with more molluscan representative species and to know whether the molluscs, as a taxon of protosomes are able to synthesize T and E_2 by an unknown steroidogenic pathway or not.

In female and male molluscs, seasonal changes of E_2 and T play an important role as endogenous modulators of gonad maturation and reproductive cycle. For example, E_2 level in *Mya arenaria* increased from 150 pg/g live weight to 195 and 334 pg/g during vitellogenesis and ripening of the ovary, respectively. During this period, T level decreased from 29 to 6 pg/g, as if T was converted into E_2. In males, T level, however, remained stable all along testicular maturation between 19 and 32 pg/g (Cauthier-Clare et al., 2006). In cephalopods too, seasonal changes in E_2 and P_4 were accompanied by morphological changes in ovary, oviduct and oviductal gland during reproductive cycle (Di Cosmo et al., 2011). Reporting seasonal changes in P_4 and T throughout the period of reproductive cycle from September to March, Avila-Poveda et al. (2015) reported that levels of P_4 increased from 1 to 611 pg/g and T from 1 to 80 pg/g in female *O. maya*. In males, T increased from 106 pg/g in September to 169 pg/g in March. These season-dependent endogenous modulations clearly show the active role played by steroid hormones in regulation of reproductive cycle (Table 7.5). In the razor clam *S. constricta*, Yan et al. (2011) reported a comprehensive evidence for seasonal changes in steroid hormones level in male and female. In this clam the steroid levels are modulated in response to temperature and food production. Particularly, seasonal modulation in E_2 level of female was parallel to the increase in oocyte diameter (Fig. 7.3). The

TABLE 7.5

Evidences against and support of the presence and functions of vertebrate type sex steroids in molluscs

Opposing Evidences	Supporting Evidences
Absorption of T and E_2	
Absorbed quantity of E_2 or T is equal in males and females, i.e. the uptake from water is more a 'passive' than sex-specific activity. About 85% of the observed E_2 is esterified (Gooding and LeBlanc, 2001, Puinean et al., 2006, Peck et al., 2007)	Free E_2 level was ~ 18% of the total uptake in *M. edulis* (Puinean et al., 2006). E_2 level was high in zebra mussel exposed to natural waters containing almost no E_2, suggesting its endogenous origin (Peck et al., 2007). In *C. gigas*, ~ 58% of labeled E_2 was metabolized into E_1 after 2 hours injection of labeled E_2 (Le Curieux-Belfond et al., 2005)
Sex specific differences	
Of relevant publications related to 20 aquatic molluscan species considered by Scott (2012), there were no sex specific differences in T and E_2 levels in 10 species. For 2 species the reported data were contradictory. Sex specific differences were recognizable only in 2–3 species. Scott disposed them stating that being 'fatty', gonads tend to concentrate steroids	In the razor clam *Sinonovacula constricta*, Yan et al. (2011), which was missed by Scott, T level (343 pg/g) was higher in male than female (81 pg/g) and E_2 higher (770 pg/g) in female than in male. Avila-Poveda et al. (2015), obviously not consulted by Scott, also reported higher T level (169 pg/g) *Octopus maya* male than in female (80 pg/g)
Seasonal differences	
Of 20 species considered by Scott (2012), relevant information was not available for seasonal changes in steroid levels for 9 species. Positive reports for ~ 10 species were disposed by Scott (2013) stating that the experimental design was wrong, samples were inadequate and statistics were not applied. Further, E_2 levels are highly variable and median values can differ by ~ 6-fold factor (Stanczyk et al., 2007)	Seasonal changes in steroid levels were reported for *S. constricta* and *O. maya*. In *S. constricta*, E_2 levels increased from 422 to 770 pg/g and after spawning decreased to 241 pg/g and T increased from 81 to 343 (Yan et al., 2011). In *O. maya* male, T increased from 1 to 80 pg/g during the period from September to March, when the octopus milted. Progesterone (P_4) increased from 1–611 pg/g during September just prior to spawning. Evidence reported for *Mya arenaria* by Cauthier-Clare et al. (2006) cannot be ignored. In females, E_2 level increased from 150 pg/g to 195 and 334 pg/g during vitellogenesis and ripening of the ovary; during the corresponding period, T level decreased from 29 to 6 pg/g ovary, as if T was converted into E_2. In males, T level increased from 9 to 32 pg/g during testicular maturation. More importantly, Yan et al., reported a positive correlation (r = 0.664) between E_2 level and oocyte diameter during ovarian maturation of *S. constricta* (Fig. 7.3)

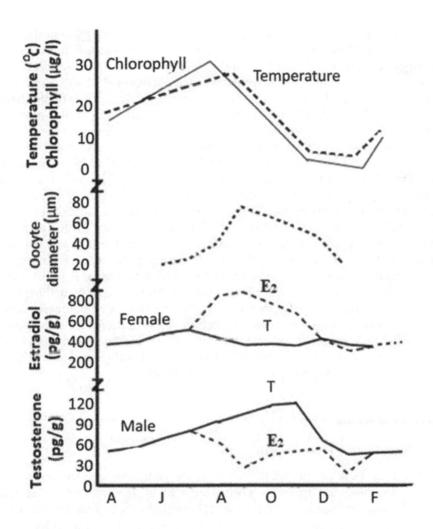

FIGURE 7.3

Seasonal changes in temperature and chlorophyll IIα content of the Yellow River delta, China. Monthly changes in oocyte diameter and sex steroid levels in female and males of the razor clam *Sinonovacula constricta* during 2007 and 2008 (simplified and compiled from Yan et al., 2011).

observations of Yan et al., strongly speak for the role of steroid hormones in regulation of the reproductive cycle.

In gastropods, there are, however, disturbing reports on both the steroid level and steroid binding capacity of estrogen receptors. Goto et al. (2012) reported equal levels of aromatase (40 p mol/g tissue protein/h) in testis and ovary of *Thais clavigera*. Another publication also confirmed the equal (~ 3.5 n M) levels of T in testes and E_2 in ovaries of *T. clavigera* (Lu et al., 2002). It is not clear whether these reflect the equal requirement of steroids in both males and females. Moreover, the binding capacity of E_2 receptor is not strong in investigated gastropods like *T. clavigera* (Kajiwara et al., 2006), *Nucella lapillus* (Castro et al., 2007) and *Marisa cornuarietis* (Bannister et al., 2007). Hence, it is likely that the steroids in gastropods may not play an active role in regulation of reproductive cycle. On the other hand, the steroids seem to play an active role in bivalves and cephalopods, as not only the endogenous modulations in steroids are correlated with seasonal changes but also their estrogen receptors bind E_2 strongly. For example, Kanda et al. (2006) have found that oct-GnRHR is an authentic receptor for oct-GnRH. In *O. vulgaris*, the steroid specific binding molecules at affinity levels (0.5–5.0 p mol/g protein) are comparable to vertebrate steroid receptors (D'Aniello et al., 1996). Croll and Wang (2007) have produced Northernblot-based evidence for the existence of estrogen receptor-like gene expression in the gonad and hepatopancreas of the scallop *Placopecten magellanicus*. Examining the relative abundance of estrogen binding in the cytosolic and nuclear fractions of *P. magellanicus*, Wang and Croll (2007) reported that the ratio of binding capacity in the cytosol to that in the nuclei rapidly increased with initiation of gametogenesis and subsequently decreased with on completion of maturation. These changes in cytosol to nuclear ratio reflect that the receptors are in dynamic status. Hence, steroids seem to regulate the reproductive cycle of bivalves and cephalopods but not that of gastropods (however, see earlier also).

7.3 Endocrine Disruption

"More than 60% of the 100,000 man-made chemicals are in routine use worldwide since 1990s. Every year 200 to 1,000 new synthetic chemicals enter the market (Shane, 1994). Over 900 of these chemicals are identified as established or potential endocrine disrupters (Soffker and Tyler, 2012). The chemicals that either mimic or antagonize the actions of endogenous hormones are known as endocrine disrupters or Endocrine Disrupting Chemicals (EDCs) (Hiramatsu et al., 2006). After use in domestic (into sewage), agricultural (e.g. pesticides, fungicides, herbicides) and industrial (e.g. 4-Nonylphenol) sectors, hundreds of estrogens and their mimics are discharged into aquatic habitats" (Pandian, 2015, p 93). "A casual observation by Sweeting (1981) on 5% incidence of intersex roach *Rutilus rutilus* in River Lee, Southeast England triggered the discovery of the occurrence of sexual disrupting estrogens in Sewage Treatment Effluents (STE) water discharged into rivers" (Pandian, 2014, p 102). For detailed information from both field and experimental observations on endocrine disruption, Pandian (2015) may be consulted.

Unlike vertebrates, invertebrates in general and molluscs in particular have limited ability to metabolize exogenous chemicals and to eliminate pollutants like tributyltin (TBT) through the excretory system (e.g. Ronis and Mason, 1994). With unlimited use of TBT containing anti-fouling paints, the vulnerability of prosobranchs to EDCs became acute; for example, of 303 molluscan species in Germany, 204 were included in the 'Red list' and rated as 'threatened' or already extinct at least regionally (Jungbluth and von Knorre, 1995). Although publications on studies performed on EDC effects impressively increased from 1998, information available for molluscs (> 100,000 species) is less, in comparison to that for fishes (26,000 species). Incidentally, the following must also be noted: 1. Except for a few publications reporting the TBT effects on shell deformations and spatfall declines in oysters like *Crassostrea gigas*, availability of publications per se are limited to gastropods and that too to prosobranchs alone. It is not known why archaeogastropods, and other sessile prosobranchs and bivalves, whose gills with larger surface area are necessarily more exposed to acquire both dissolved oxygen and food, were not considered for EDC studies. It may also be revealing to study the EDC effects on simultaneous hermaphrodites like sea hares and other shell-less opisthobranch. 2. In prosobranchs too, the hitherto investigated studies are all on endocrine disruption with compounds acting as estrogens or androgens in vertebrate test systems (Oehlmann et al., 2007), though these steroids are exogenous to molluscs. 3. As in fishes (Pandian, 2014, Chapter 7, 2015, Chapter 4), the exposure of molluscs to one or other EDC during post-labile adult stage interferes only with secondary sex differentiation and produces intersexes but not sex reversal in females or males. However, the exposure during the labile period may reverse sex, as it is indicated by skewed ratios (toward male) in clam *Mya arenaria* (Gagne et al., 2003) and *Mytilus edulis* (Hellou et al., 2003) exposed to contaminated rivers.

Estrogens: Following long term (12 months) exposure covering the complete life cycle of ramshorn snail *Marisa cornuarietis* to 1–100 µg/l of bisphenyl A (BPA) or octylphenyl (OC), the 'super female' snails were characterized by formation of additional feminine organs, enlarged accessory sex glands, malformed pallial oviduct, increased egg masses and consequent death (Oehlmann et al., 2000, see also Duft et al., 2006). These observations were also confirmed in *Potamopygrus antipodarum* (Duft et al., 2003) and *Nucella lapillus* (Oehlmann et al., 2000). Tillmann et al. (2001) made a remarkable observation that *M. cornuarietis*, on exposure to 500 ng ethylnylestrodiol (EE_2)/l, did not increase egg production but induced time-dependent 'imposex' phenomenon, an effect which was reverted by an addition of 1.25 mg cyproterone acetate (CPA)/l.

Androgens: In prosobranchs, TBT effects are perhaps the best documented example of the impact of an EDC on aquatic invertebrates and imposex in molluscs (Oehlmann et al., 2007). "Imposex is defined as a virilization phenomenon or a pseudohermaphroditic condition characterized by development and superimposition of non-functional male sexual characteristics like penis, vas deferns and/or seminal tubules" (Ketata et al., 2008). On the basis of morphological structures of the pseudohermaphrodites in 69 species belonging to 43 genera, Fioroni et al. (1991) classified the affected snails into seven stages: the classification is advantageously adopted to other affected species (e.g. Ramasamy and Murugan, 2002). Females are sterilized in stage 5–6, when the opening of the vulva is occluded by proliferating tissue of vas deferns. Imposex is a widespread phenomenon from different regions (e.g. Argentina, Penchaszadeh et al., 2001, Canada, Horiguchi et al., 2003, South Africa, Marshall and Rajkumar, 2003, India, Ramasamy and Murugan, 2002). Today, it is reported from as many as 160 prosobranch species (Shi et al., 2005). A number of potent androgen receptor (AR) agonists (e.g. 17α-Methyltestosterone, MT), aromatase inhibitor (e.g. Fedrozole) and anti-androgens (e.g. CPA, Vinclozolin, VZ) are reported to induce imposex. Fenarimol (FEN), a fungicide (Tillmann, 2004) and organochlorine contaminants like polychlorinated bisphenyls (PCBs, pesticides, e.g. Delgado et al., 2013) and polycyclic aromatic hydrocarbons (PAHs) (Maran et al., 2006) can also induce imposex. Tributyltin chloride (TBT) and its (triphenyltin) oxide (TPT) are endocrine disrupting organotin compounds, and potent inducers of imposex and adipogenesis (Iguchi et al., 2007). The induction is effected at levels as low as 2 ng/l (Oberdorster and Mc Clellan-Green, 2002).

Many hypotheses have attempted to explain the mechanism responsible for manifestation of imposex. Table 7.6 lists some of these hypotheses and suggests the need for further research for acceptance of one or another hypothesis. The aromatase hypothesis states that the inhibition of aromatase activity results in increased T level and decreased in E_2 (Oberdorster and Mc Clellan-Green, 2002). The following evidences support it: 1a. Following laboratory administration, T induced imposex. 1b. The TBT-exposed snails were characterized by dose- and time-dependent increase in T levels. 2. A positive correlation between imposex stages (cf Fioroni et al., 1991) and T levels in the field-caught *N. lapillus* and *N. reticulatus* was also established (Bettin et al., 1996). 3. From their ELISA and GC/MS estimates of T levels in gonads of *T. clavigera*, Lu et al. (2002) showed that the T level in imposex-affected female was significantly higher than in snails from uncontaminated sites. 4. Barroso et al. (2005) also found a negative correlation between aromatase activity and organotin content of *Nassarius reticulatus*. 5. The determination of steroid level in female *N. lapillus* revealed that TBT induced an elevation in free T level but not the total (free + esterified) T level. Incidentally, the elevation in T level did not alter the free E_2 level in TBT-exposed females (Santos et al., 2005). The objections against the aromatase inhibition hypothesis are: 1. That the increasing T levels are not necessarily accompanied by decreasing E_2 level. 2. That the increase of T levels in TBT-exposed snails is too low to explain the rapid increase in imposex level (Oehlmann et al., 2007).

TABLE 7.6

Imposex induction in prosobranchs: hypotheses and their status

Hypothesis	Comments
Neuropeptides hypothesis: Feral and Le Gall (1982)	
TBT inhibited the release of the neuroendocrine factor PMF from pleural ganglia, known to suppress penis formation in females. Hence imposex was induced in *Ocenebra erinacea*	Oberdoster and Mc Clellan-Green (2002) have proposed that APG Wamide represents PMF. But its level is equal in control and treated *Ilyanassa obsoleta*
Inhibited T excretion hypothesis: Ronis and Mason (1996)	
Inhibition of T sulfation in periwinkle *Littorina littorea* resulted in reduced elimination of T. Consequently, TBT-exposed periwinkle retained more T from an administred ^{14}C-T dose than control periwinkle	This finding is not confirmed in other studies at environmentally relevant TBT concentrations
T esterification hypothesis: Gooding et al. (2003)	
The hypothesis postulated that modulation of free versus fatty acids-bound T level induced imposex. TBT interfered with esterification of T resulting in elevated free T levels in *I. obsoleta*	The results of Janer et al. (2005c) for FEN, a potent aromatase inhibitor, indicates that imposex induction does not necessarily affected by fatty acids-bound T levels
Retinoid X receptor (RXR) hypothesis: Nishikawa et al. (2004)	
An organotin compound interacted with RXR. TBT and TPT exhibit high affinity to RXR in human and *Thais clavigera*. The injection of its suspected natural ligand 9-*cis* retinic acid (RA) into female snail induced imposex. Analyses of the cloned RXR homologs from the snail revealed that the ligand binding domain of the snail RXR was similar to that of human RXR. Hence RXR induced imposex	In European snails, Di Benedetto (2005) has not been able to confirm the findings of Nishikawa et al. However, Sternberg et al. (2008) have shown that in *I. obsoleta*, RXR mRNA expression remains refractory during non-reproductive season but is primed during reproductive season. *Columbella rustica*, on exposure to high TPT level, does not develop imposex, although other species of the genus do it (Gibbs et al., 1997). It is not clear whether the differences in findings of Nishikawa et al., and Di Benedetto are due to species specific response and/or the reproductive season, in which the specimens were collected for experiments

7.4 Parasitic Disruption

Parasitic trematodes are ubiquitously present in both marine and freshwater habitats. From an egg of the trematode, a free-living miracidium larva emerges. Guided by light and chemical stimuli, it infects an intermediary molluscan host; in the host it undergoes a series of polyembryonic reproductive cycles (see Pandian, 2016, p 9); the complexity of the cycle varies from species to species (see Morley, 2006). Eventually, the embryos develop into infective cercariae and are released from the molluscan host. The prevalence of trematodes may vary from habitat to habitat but 4 to 65% of molluscs are infected in a locality (Esch et al., 2001). That the trematodes castrate/sterilize the molluscan hosts has already been narrated earlier (p 52). In *Biomphalaria glabrata*, *Schistosoma mansoni* reduces the levels of monamine transmitter serotonin and dopamine by utilizing them for its own development. The developing parasitic *Trichobilharzia ocellata* directly inhibits mitotic division in reproductive organs of *Lymnaea stagnalis* (De Jong-Brink et al., 2001). Interestingly, Coustau et al. (1993) have reported that the extract of *Prosorhynchus squamatus* inhibits mitosis in gonad of not only in its specific host *Mytilus edulis* but also additional three bivalve species *M. galloprovincialis*, *Ostrea edulis* and *Pecten maximus*. Clearly, the toxin released by *P. squamatus* is not species specific. *In vitro* studies indicate that the extract of *Zoogonius lastius* inhibits spermatogenesis in *Ilyanassa obsoleta* (Pearson and Cheng, 1985). Fecundity of *B. pfeifferi* infected by *S. mansoni* is significantly reduced from 7–10 days post-infection onwards. From the 13th day after infection, oviposition is completely arrested (Meuleman, 1972). In the *S. hematobium*-infected *Bulinus globosus* too, egg production is markedly reduced with progress of infection and ceased entirely at the onset of cercarial emergences from the snail. The longevity of the host is also reduced from 101 days in a healthy snail to 42–71 days in an infected snail (Gracio, 1988).

In view of the medical importance, molecular biologists have made enormous efforts to understand the ability of *Schistosoma* spp to immunosuppress the definitive human host and intermediary snail hosts. Genome sequencing of *S. japonicum* (The *Schistosoma japonicum* Functional Analysis Consortium, 2009), related genomics studies in *S. mansoni* (Bobek et al., 1986, Hoffmann and Dunne, 2003, Geyer et al., 2011, Protasio et al., 2012), liver fluke *Opisthorchis viverrini* (Laha et al., 2007) and *Fasciola hepatica* (Young et al., 2010). Duvaux-Miret et al. (1992) have produced evidences for the release of at least two (immunoreactive prooplomeanocortin) POMC-derived peptides ACTH (corticotrophin) and β-endorphin by *S. mansoni* to regulate the host's immuno function. These peptides mask the fluke from the host immunosurveillance and the fluke is able to suppress immunoresponsiveness of the host. Hence, the fluke is tolerated by the human host for decades; ACTH also suppresses interferon (IFN) production by T lymphocytes and inhibits IFN activation of macrophages (see Duvaux-Miret et al., 1992).

Efforts have also been made to control snails serving as intermediary hosts. At present, niclosamide (Bayluscide) is the only molluscicide applied on large scale. However, the molluscicide is toxic to fishes also and not affordable in schistosomasis-endemic areas. An alternative is the extract of *Jatropha curcas* seeds. The plant *J. curcas* is widespread in the tropics and planted as a live fence. It is easily grown. Oil from its seeds can be used as a substitute for

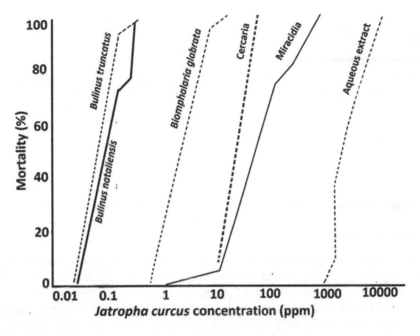

FIGURE 7.4

Mortality of the snail hosts, miracidia and cercaria of *Schistosoma mansoni* as function of extracts of *Jatropha curcus*. Note the high concentration of aqueous extract required to affect same level of mortality as against methanol extract. All other trends are for methanol extracts (compiled from Rug and Ruppel, 2000).

diesel and for producing soap. It is traditionally known for its antimiracidial property. The seeds contain phorbo esters. The esters activate protein kinase C (PKC), which destroys the surface integrity of miracidia and cercariae. It may also induce osmolar instability, surface vesiculation and subsequent death of snails and parasitic larvae. Figure 7.4 shows the comparative mortality of the snails and parasitic larvae as a function of toxicity of the aqueous and methanol extracts of jatropha seeds. Methanol extracts are about 10,000 times more potent than aqueous extract. This is due to the inhibition by hydrophilic compounds like safonins, curcins, phytates and potease inhibitors present in the seeds. *Bulinus* spp are more sensitive to the toxin than *Biomphalaria glabrata*. Miracidia are at least 10 times more sensitive than cercaria. On exposure of both the snails and cercariae to 2.5 ppm methanol extract, 100% mortality occurs within a short time. The methanol extract of *J. curcas* seeds is about two-times more potent than the commercial Bayluscide (Rug and Ruppel, 2000).

8

Molluscs and Ocean Acidification

Introduction

During recent years, research on the unprecedented increase in atmospheric carbon dioxide (CO_2) concentration and consequent global warming and ocean acidification have become the most burgeoning and hottest area of research. This is evidenced by the fact that citation rates of publications in this area run to hundreds and thousands (e.g. Orr et al., 2005 cited 2,746 times as on 24.08.2016). Due to progressively increasing anthropogenic activity, the atmospheric CO_2 concentration has increased from 280 ppm in 1950 to 385 ppm in 2010 (IPCC, 2013). This increase is 100 times faster than that prior to some 650,000 years (Siegenthaler et al., 2005). Atmospheric CO_2 concentration, especially due to combustion of fossil fuels, has increased at the rate of 1%/year during the 20th century and is now increasing ~ 3%/year (see Talmage and Gobler, 2011). As a consequence, global mean temperature has increased by 0.2°C/decade over the last 30 years (see Pandian, 2014). Based on CO_2 emissions and circulation models, The Intergovernmental Panel on Climate Change (IPCC) has indicated that the atmospheric CO_2 concentration is expected to reach 550 and 750 ppm by the middle and end of this century, respectively.

8.1 Chemistry of Seawater

Carbon dioxide combines with water chemically. Oceans cover 70% of the earth's surface and hold 97% of its water (see Pandian, 2011, p 1) and serve to buffer CO_2. Consequently and thankfully, the daily uptake of atmospheric CO_2 by the oceans is 22 million metric tons (mmt) (Feely et al., 2004). Since the advent of the industrial era, the oceans have absorbed 127 ± 18 billion metric tons (bmt) of carbon as CO_2 from atmosphere. The CO_2 absorbed by the marine realm ranges between 25 and 40%, i.e. ~ a third of the atmospheric carbon emissions (Sabine et al., 2004, Zeebe et al., 2008). Without this 'ocean sink', the atmospheric CO_2 concentration would have by now increased to ~ 450 ppm (Sabine and Feely, 2007) and consequent increase in temperature on land.

Hydrolysis of CO_2 increases the hydrogen ion (H^+) concentration with concomitant reductions in pH and carbonate ion (CO_3^{2-}) concentration. This process of reducing sea water pH and carbonate ion concentration is called 'ocean acidification' (Caldeira and Wickett, 2003). Consequent to the acidification process, the mean pH levels of the world oceans have declined by 0.1 unit and another 0.3–0.4 unit reduction is expected by 2010 (Orr et al., 2005). The decrease in sea water pH and carbonate ion concentration is one of the most persuasive environmental changes in the oceans (Feely et al., 2008). The progressive reduction in availability of carbonate ion (CO_3^{2-}) renders the acquisition of biogenic calcium carbonate ($CaCO_3$) by calcifying organisms more and more difficult and energetically costlier (Wood et al., 2008a). Not surprisingly, the reductions in pH are more critical for the calcifying poikilothermic organisms than the increase in sea water temperature (see Denman et al., 2011).

Another consequence of ocean acidification is the extent, to which the calcifying organisms depend on the $CaCO_3$ saturation state (Ω). The availability of carbonates is determined by the concentrations of Ca^{2+} and CO_3^{2-} divided by solubility product for either aragonite or calcite, that are required for calcification of shells and skeletons of aquatic organisms. In regions, where the values for Ω_{arag} and Ω_{cal} are > 1.0, the formation of shells and skeletons is favored. However, when or where these values fall below 1.0, the water becomes corrosive and dissolution of aragonite shells begins to occur (Feely et al., 2008).

Off from the global level ocean acidification, there are also regional level magnifications of acidification. They are related to (i) upwelling and (ii) latitudes. Upwelling can introduce sea water with high CO_2 concentrations up to 800–1,100 ppm along large sections of the continental shelf. The seasonal wind driven upwelling process transports CO_2-rich corrosive deep water into the coastal regions. The central and southern coastal region of western North America is strongly influenced by the wind-driven seasonal upwelling, which transports CO_2-rich corrosive waters into the mid-shelf depths of 40–120 m (Feely et al., 2008). In the tropics, commercial level shell capture is widely practiced from the east coast of south India along the Gulf of Mannar (see Lipton et al., 2013) but is not extended to the west coast, which annually experiences southwest monsoon-driven upwelling. In the western Baltic Sea, characterized by low salinity (10–20‰), Ω_{arag} (> 0.35 but < 1.0) and Ω_{cal} (> 0.58 but < 1.0), pCO_2 exceeds by a factor of three to five times. In Kiel Fjord, upwelling of CO_2-rich waters renders the Fjord more acidic (pH 7.5) with pCO_2 ~ 700 µatm during summer and autumn (Thomsen et al., 2010). Therefore, the upwelling habitats like the Kiel Bay, west coast of North

America and south west coast of India are 'natural analogs' to study how ecosystem may be influenced by the predicted ocean acidification.

At broad geographic scales, molluscan shell thickness decreases with decreasing Sea Surface Temperature (SST) and increasing latitudes (e.g. Briones et al., 2014). With decreasing water temperature, $CaCO_3$ becomes less saturated and the more soluble (see Trussell, 2000) and aragonite also becomes under-saturated (see Orr et al., 2005). Consequently, both deposition and maintenance of shells become more and more difficult in colder waters (see Trussell, 2000, cf Briones et al., 2014). Briefly, colder waters are more acidic and corrosive. Hence Orr et al. (2005) have suggested that some polar and sub-polar surface waters may become totally under-saturated within next 50 years. Alternatively, the subarctic pteropods with generation time of 6 months (e.g. *Limicina retroversa*, Dadon and de Cidre, 1992) or 18 months may have to adapt to the corrosive sea water within the next 50 or 150 generations (Fabry et al., 2008).

To study the effects of ocean acidification on molluscs, two different approaches have been made: 1. Field oriented investigations (including time series) to estimate acidification effects and 2. Experiments involving exposure of molluscs to selected (pre-industrial, present and predicted) levels of reduced carbonate saturation, decreased pH and/or elevated pCO_2. Barring a few studies on pteropods and paper nautilus, almost all investigations are related to bivalves. No investigation on the acidification effects is yet available for freshwater molluscs (however, see Rundle et al., 2004). In view of relatively limited availability of calcium in freshwaters (Dillon, 2000), acidification effects on freshwater molluscs may be critically important.

The uptake of atmospheric CO_2 is pronounced in ocean surface waters (< 250 m depth), where nearly 50% of anthropogenic CO_2 is absorbed, rendering near shore habitats including estuaries more vulnerable to ocean acidification (Doney et al., 2009). Not surprisingly, most available publications are on the holoplanktons such as pteropods and paper nautilus as well as pelagic larvae of molluscs. Benthic molluscs generate pelagic larvae, which undertake recruitment transition to become benthic juveniles. Table 8.1 summarizes available information related to acidification effects from fertilization to metamorphosis, and structural and behavioral responses to predation (structure, e.g. shell thickness) and function (e.g. metabolism, immune defense, etc.). However, no long term studies have been undertaken to study the acidification effect on reproductive performance of molluscs (see Kroeker et al., 2010).

8.2 Pelagic Molluscs

Adapted to holopelagic life cycle, thecosomatic pteropods form thinnest (12 μm in *Creseis acicula*, 20 μm in *Diacavolinia longirostris*, Roger et al., 2012)

TABLE 8.1

Molluscan responses to ocean acidification

Species	CO$_2$ Parameters	Responses	Reference
Tivela stultorum	pH 8.5	Decrease in fertilization rate	Alvarado-Alvarez et al. (1996)
Placopecten magellanicus	pH 8.0	Decrease in fertilization and embryo development	Desrosiers et al. (1996)
Pinctada fucata martensii	pH 7.7; pH 7.4	Shell dissolution, reduced growth, Increasing mortality	Knutzen (1981)
Mercenaria mercenaria	$\Omega_{arag} = 0.3$	Juvenile shell dissolution, increased mortality	Green et al. (2004)
Clio pyramidata	$\Omega_{arag} > 1$	Shell dissolution	Feely et al. (2004), Orr et al. (2005), Fabry et al. (2008)
Mytilus edulis	pH 7.1/10 000 ppm pCO$_2$ 740 ppm	Shell dissolution, 10–25% decrease in calcification	Gazeau et al. (2007) Lindinger et al. (1984)
Crassostrea gigas	pCO$_2$ 740 ppmv	10% decrease in calcification	
Haliotis laevigata H. rubra	pH 7.78; pH 7.39 pH 7.93; pH 7.37	5–50% growth reductions 5–50% growth reductions	Harris et al. (1999)
Mytilus galloprovincialis	pH 7.3, 5000 ppm	Reduced metabolism, growth	Michaelidis et al. (2005)
Dosidicus gigas	0.1% CO$_2$, 1000 ppm	Reduced metabolism/ scope for activity	Rosa and Seibel (2008)
Illex illecebrosus	2000 ppm	Impaired oxygen transport	Portner and Reipschläger (1996)
Ostrea lurida	pH 7.8	Juveniles carry-over energy deficit from larvae	Hettinger et al. (2012)
Saccostrea glomerata	pCO$_2$ 856 atm	Exposed reproductive adults produce more robust larvae	Parker et al. (2012)
Mytilus edulis	pCO$_2$ 1994 ppm	No shell dissolution at increased feeding rate	Melzner et al. (2011)
Littornia littorea	pH 6.5	At dual stressors low pH + 'crab cue', switched to predator avoidance	Bibby et al. (2007)
Lymnaea stagnalis	Calcium poor freshwater	At dual stresses of low Ca^{2+} and 'fish cue' switched to predator avoidance	Rundle et al. (2004)

aragonite shells and others like athecate pteropods (e.g. *Clio*) lack shells as adults (Mackas and Galbraith, 2012). Pteropods are the major planktonic producers of aragonite. As counterpart to krill (e.g. Rosa Sea), high latitude pteropods have one or two generations per year (e.g. *Limicina retroversa*, Dadon and de Cidre, 1997, *L. helicina*, Gannefors et al., 2005), form integral component of food webs and are found in densities of hundreds and thousands of individuals/m^3 up to 300 m depth (e.g. Urban-Rich et al., 2001). Adapted to holopelagic life cycle, the nautiloid cephalopods too have thin (225 µm) shells, lacking an outer periostracum cover (e.g. *Argonauta nodosa*). The females construct paper nautilus shells, in which the egg masses are incubated until the juveniles are ready to hatch into plankton (Wolfe et al., 2012).

Pteropods cannot be maintained for longer than 48 hours in the laboratory. Empty pteropod shells from sedimental traps indicate the presence of pitting and partial dissolution (Feely et al., 1988). As if to confirm this evidence, experimental exposure of *Clio pyramidata* to aragonite under-saturation level for 48 hours has also revealed the etched pits from dissolution and peeling of increased surface area exposing aragonite rods (Orr et al., 2005). Clearly, the polar pteropods have been and continue to suffer from pits and dissolution of their aragonite shells due to ocean acidification. However, a study by the Fisheries and Oceans Canada, who have the longest time series (1979 to 2009) of pteropods from the subarctic Northeast Pacific, has not revealed the expected notable decline in population size of the aragonite shelled *L. helicina*, indicating the limited shell dissolution related mortality. The investigation has also reported an unexpected increase of *Clio pyramidata* (Mackas and Galbraith, 2012). Examining 58-years long time series of pteropods belonging to four families at the California Current Ecosystem-LTER site, Ohman and Lavaniegos (2009) have also found no evidence for significant declines in abundance. It is not clear whether *C. pyramidata* and the Californian pteropods had abundant food supply to meet the extra energy cost of shell building. Consequently, they have incurred no shell dissolution, mortality and reduction in their biomass. Nevertheless, another time series investigation has confirmed that thecosomate pteropods do incur shell dissolution due to declining aragonite saturation in the Australian waters. Roger et al. (2012) have estimated the shell characteristics of tropical pteropods *C. acicula* and *D. longirostris* from material collected between 1963 and 2009. A 10% decline in aragonite saturation level (Ω_{arag}) in surface waters was hindcast. Figure 8.1 shows the SEM image of increased porosity incurred by *C. acicula* and *D. longirostris* shells from 1985 to 2009. During this period, the shell thickness has decreased from 12 to 7 µm and porosity index increased from 0.2 to 0.7%/100 nm^2 in *C. acicula*. The corresponding values are from 20 to 13 µm for shell thickness and from 0.1 to 9.0%/100 nm^2 for the porosity index of *D. longirostris*.

Lacking an outer protective periostracum cover, the aragonaut shells of nautilus are radically different from other molluscan shells. Exposing

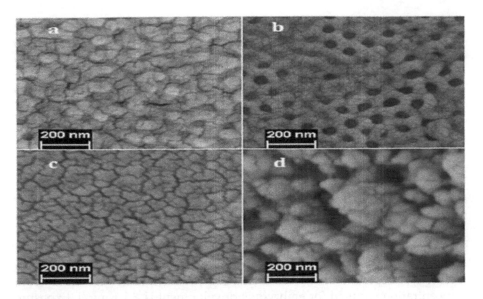

FIGURE 8.1

a. SEM image of *Creseis acicula* shell surface dated 1985—porosity index = 0.21% of 100 nm², b. SEM image of *C. acicula* shell surface dated 2009—porosity index = 0.72% of 100 nm², c. SEM image of *Diacavolinia longirostris* shell surface dated 1985—porosity index = 0.15% of 100 nm², d. *D. longirostris* shell surface dated 2009—porosity index = 8.99% of 100 nm². Magnification: ~ 100 Kx. Personal communication by Dr. L.M. Roger.

the washed ashore (dead) shells of paper nautilus *A. nodosa* to different temperature-pH combinations for 2 weeks, Wolfe et al. (2012) found the dissolution of the paper nautilus shell began to occur at pH 7.8 (projected for 2070). *A. nodosa* lost > 5.3% shell weight on exposure to pH 7.8.

8.3 Benthic Molluscs

Fertilization and embryonic development: Alvarado-Alvarez et al. (1996) noted that fertilization was optimal at pH 8.5 in the Pismo clam *Tivela stultorum* but it required rather higher sperm density at pH 8.0, which facilitated polyspermic fertilization. For the giant scallop *Placopecten magellanicus*, the conditions most favorable for meiotic maturation of the oocyte, fertilization and first zygotic cleavage were in the range of pH 8.0 to 8.5, salinity 25–28‰ at 10°C (Desrosiers et al., 1996). In fishes, maternal genes direct fertilization, egg activation and zygotic cleavage (see Pandian, 2012, p 145). This may be true also for molluscs, although experimental studies using mutants (see Dorsch et al., 2004) are yet to be undertaken for molluscs. Optimal gamete density required for 90–95% monospermic fertilization in *P. magellanicus* is 80,000–150,000 spermatozoa/ml and 5,000 oocytes/ml, which correspond to a ratio of 16–30 spermatozoa per oocyte. For example, the frequencies of monospermic and polyspermic fertilization are 50% for each at pH 7.5. Whereas meiotic maturation cycle is completed within 89–92 minutes, irrespective of changes in pH from 7.0 to 8.5, the duration of the zygotic cleavage requires just 33 minutes at pH 8.5 but as long as 66 minutes at pH 7.5 (Desrosiers et al., 1996). The range of pH required for completion of embryonic development is 7.75 for *Mulinia lateralis* (Sastry, 1979), 7.00–8.75 for *Mercenaria mercenaria* and 6.75–8.75 for *Crassostrea virginica* (Calabrese and Davis, 1966). Apparently, *C. virginica* embryos are more tolerant to pH changes than those of *M. mercenaria*. Examining the relationship between pH and calcium on one hand and reproductive success of the snail *Amnicola limosa* on the other, Shaw and Mackie (1990) have reported that (i) fecundity is positively correlated with buffering capacity in natural habitats, (ii) it is reduced to 44% in lakes with pH ranging between 5.7 and 7.6, (iii) pH 5.5 delays development and reduces hatching success and (iv) critical calcium concentration required for embryonic development is ≥ 1.1 mg/l. Exposure of encapsulated embryos of *Crepidula fornicata* to elevated pCO_2 level of 750 µatm results in the production of 1.5% abnormal embryos (Noisette et al., 2014).

Shell building: The primary function of molluscan shell is defense against predation, although shells may also protect molluscs against physical stress including heat, desiccation and wave force (see Kroeker et al., 2014). Most molluscs are characterized by production and maintenance of shell(s) consisting of organic matrix and calcium carbonate. Besides the cost of organic matrix, calcification involving the accumulation, transportation and precipitation of calcium carbonate is also an energy costing process. Incidentally, from a purely physic-chemical standpoint, $CaCO_3$ should precipitate spontaneously, when $\Omega > 1$ and start to dissolve, when concentration is < 1. However, the process of biogenic $CaCO_3$ deposition is a highly controlled and far more complex process (Nienhuis et al., 2010). Acquiring Ca^{2+} and HCO_3^- from ambient water across cell membrane, elimination of excess proton (H^+) released during mineralization (Allemand et al., 2004) and precipitation of $CaCO_3$ (McConnaughey, 1995) cost energy. At 1–2 J/mg of $CaCO_3$, the calcification costs 5% of the proteinaceous organic fraction of molluscan shells on a per gram basis (Palmer, 1992). Not surprisingly, the body growth of thick shell molluscs is significantly slower than that of thin shell molluscs (Palmer, 1981). In broadcast spawning molluscs, high mortalities (> 98%, see Gosslein and Qian, 1997) occur. Two critical stages are identified, when molluscan larvae encounter relatively high mortalities. They are the stages, when transformations (i) from preceding to ensuing larval stages occur and (ii) transition from pelagic larval stage to demersal juveniles. Transformations from veliger to pediveliger (e.g.

Mercenaria mercenaria, Talmage and Gobbler, 2010)/D-shaped larva (e.g. *Mytilus galloprovincialis*, Kulihara et al., 2009) are the most critical life stages.

Strikingly, mortality of *M. mercenaria*, which was > 5% during veliger stage increased further to 49% in the 10 day-old pediveliger stage. On exposure to 390 ppm CO_2, it increased further to 80% (see Table 8.2). Mortality incurred during transformations from one to ensuing stages was uniformly high at 68–69% for the abalone *Haliotis kamtschatkana* and mussel *M. galloprovincialis* (Table 8.3). Hence, these transformations are recognized as the most critical stage, when extremely high mortalities occur in shelled molluscs that broadcast their gametes. Consequently, any factor like the ocean acidification may inflict even higher mortality. Not surprisingly, many molluscs have chosen to brood their embryos and release juveniles at the advanced shelled stage. The recruitment transition from pelagic to benthic life is another but less critical stage. During the transition, 31% (from 46% normocapnia to 77% at hypercapnia), mortality is encountered by the newly settling juveniles of *Crassostrea virginica* (Table 8.3). Following settlement of *M. mercenaria*, mortality continues to occur but at progressively decreased rate from 11.8%/d

TABLE 8.2

Development of veliger to pediveliger and metamorphosis in *Mercenaria mercenaria* and *Argopecten irradians* reared under pre-industrial, present and predicted levels of CO_2 in seawater (compiled from Talmage and Gobler, 2010, 2011*)

Parameter	CO_2 Levels in Sea Water (ppm)		
	250	390	750
Mercenaria mercenaria			
Survival (%) pediveliger on 12 d	51	20	18
Survival (%) at metamorphosis	40	18	15
Diameter of larva (µm)	523	282	270
Shell thickness (µm)	18	7	3
Shell growth (mg/d)*	–	1.0	–
Tissue growth (mg/d)*	–	0.06	–
Lipid index	0.34	0.34	0.30
Argopecten irradians			
Survival (%) pediveliger on 14 d	87	68	30
Survival (%) at metamorphosis	74	43	18
Diameter of larva (µm)	520	440	350
Shell thickness (µm)	20	12	11
Shell growth (mg/d)*	–	4.6	–
Tissue growth (mg/d)*	–	0.18	–
Lipid index	0.57	0.46	0.38

TABLE 8.3

Representative examples to show effects of decreased pH, Ω_{arag}, Ω_{cal} or elevated pCO_2 on larvae, juveniles and adults of selected molluscs

Species/Reference	Levels of Exposure	Observations
Early larval stage		
Mytilus galloprovincialis Kurihara et al. (2008) Eggs incubated for 144 h at 13°C	pH : 8.13 & 7.42	Delayed development in 70% trochophore on 54th d
	Ω_{cal} : 3.37 & 0.75	Mortality of 69% abnormal larvae
	Ω_{arag} : 2.23 & 0.49	Decreases of 25% in shell length and 18% in shell height
Mytilus californianus Gaylord et al. (2011) 0–9 d larva at 15°C	pH : 8.15 & 8.03 pCO_2 : 380 & 540 ppm	On 8th day decreases were 18% of breaking force, 7% of shell surface, 14% of body mass, 12% of shell thickness
Haliotis kamtschatkana Crim et al. (2011), Abalone larva for 8 d at 12°C	pCO_2 : 400 & 800 ppm	Mortality increases from 43 to 68%, 60% increase of shell deformities and 5% decrease in shell length
Juvenile stage		
Crassostrea virginica Beniash et al. (2010) Transforming juveniles	pH : 8.2 & 7.5 pCO_2 : 380 & 3500 μ atm	Mortality increases from 46 to 77%. Decreases were 40% for shell mass, 25% for soft tissues, 16% for calcite lath thickness, 18% for Vickers hardness number. 100% increase in juvenile O_2 uptake but 15% decrease in adults
Mercenaria mercenaria Green et al. (2004), Post-settled juveniles for 21 d	pH : 7.8 & 7.1 Ω_{arag} : 1.5 & 0.3 in sediments	Mortality rate decreases from 11.8%/day in a 0.2 mm in a newly settled juvenile to 1.1%/day in 2 mm juvenile
Adults		
Mytilus edulis Melzner et al. (2011) Adult mussels for 7 w	pCO_2 : 987 & 1994 ppm, fed at low and high food densities	Shell growth decreases by 10% and 30% at high and low food densities. No dissolution at high food density but 50% dissolution at low food density
Strombus luhuanus Shirayama and Thornton (2005), < 1 y old adults reared for 26 w at < 30°C	pH : 7.94 & 7.90 pCO_2 : 260 & 560 ppm	10% mortality, 33% decrease in cumulative shell height and 45% decrease in live weight growth

in smaller newly settled juveniles (0.2 mm) to 1.1% in larger individuals of 2 mm (Green et al., 2004). The recently settled, smaller juvenile (e.g. 0.2–0.3 mm) has a larger surface area to volume ratio than the older juvenile (e.g. 2 mm) (Morse, 1984). Hence, any factor including decreasing pH that decreases body size, may increase the surface area of the juvenile, rendering it more vulnerable. Exposure of < 1 year-old *Strombus luhuanus* adults for 7 weeks at pCO_2 560 ppm results in just 45% decrease in live weight and 10%

mortality. Though the effects of stress induced by experimental exposures to different levels of reduced pH and/or elevated CO_2 levels on growth rate are not strictly comparable, the emerging picture from this analysis reveals that extreme mortalities occur during the critical transformation from one to ensuing larval stages and transition from pelagic to benthic juvenile stage (Fig. 8.2).

Shell characteristics: Exposing 3 day-old veligers of *M. mercenaria* and *Argopecten irradians* to the pre-industrial (~ 250 ppm), present (~ 390 ppm) and expected (~ 750 ppm) levels of CO_2 for 20 days period of larval development and metamorphosis, Talmage and Gobler (2010) have made an interesting observation. Notably, shell thickness has been reduced from 18–20 μm during the pre-industrial era to 7–12 μm in the present-day bivalves (Table 8.2). Decreases in shell thickness by 12% in *M. edulis* 16% in *C. virginica* and 40% in *C. gigas* are predicted by 2010 (Table 8.3). Besides the decrease in shell thickness, the other expected negative effects in the shell's integrity are (i) decrease in shell size, (ii) loosening of the hinge to valve locking mechanism, as in *M. mercenaria* and increasingly riddled shells with holes, pockmarks and crevices, as in *A. irradians* (Fig. 8.3). From separate estimates of $CaCO_3$ deposition and dissolution rates in shells of the snail *Nucella lamellosa* exposed to different levels of CO_2, Nienhuis et al. (2010) reported important observations namely the elevated CO_2 did not deter $CaCO_3$ deposition rate but the decrease in shell weight was due to faster dissolution rate rather than

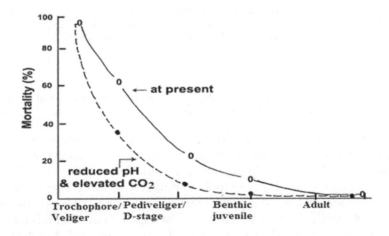

FIGURE 8.2

Trends in mortality at present (0–0) and at reduced pH/elevated CO_2 (•···•) during ontogenetic development of molluscs. Note the sharp decreases in survival, when trochophore/veliger is transformed into pediveliger/D-stage and when pelagic larvae are recruited into benthic juveniles (based on data reported in Table 8.2).

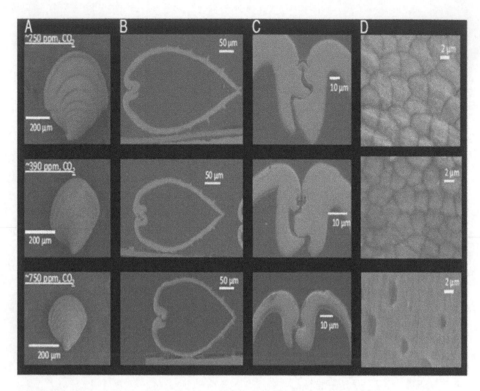

FIGURE 8.3

SEM images of 36 d-old *Mercenaria mercenaria* reared under ~ 250 A. 390 B. and 750 ppm C. levels of CO_2. Note the reduced size and loosening of the hinge to valve locking mechanism with increase in CO_2 levels, D. A magnified SEM image of the outer most shell of 50 d-old *Argopecten irradians* (from Talmage and Gobler, 2010).

reduced deposition rate. Further, shells composed of mostly calcite are less soluble than aragonitic and high-Mg calcitic shells (Andersson et al., 2008).

Feeding and lipid index: At both 250 and 390 ppm CO_2 exposures, the larvae of *M. mercenaria* and *C. virginica* accumulated adequate resources, as indicated by high lipid index (Table 8.2). However, the index was significantly decreased to 0.3–0.4 levels, when exposed to 750 ppm CO_2. In this context, Melzner et al. (2011) made an interesting observation. Following the exposure to high pCO_2 at 987 and 1994 ppm for 7 weeks, shell growth of *M. edulis* adults has decreased at 10 and 30%, respectively. However, no shell dissolution was observed in the mussels receiving abundant food supply. Apparently, the mussel gain adequate energy to rebuild the dissolving shells and reduce shell dissolution-dependent mortality. This important finding has implication to the inconsistent observation reported for *Clio pyramidata* as well as Californian pteropods (see p 204). It is likely that the Californian pteropods and *C.*

pyramidata received abundant food supply of their respective choice, gained adequate energy to rebuilt their shells, and thereby avoided shell dissolution-dependent mortality and maintained equal or even higher biomass at the tested time series.

8.4 Persistent Carry-over Effects

Describing a persistent carry-over effect in planktonic larvae of Olympia oyster *Ostrea lurida*, Hettinger et al. (2012) showed that the larvae reared at pH 7.8 exhibited 15 and 7% decreases in larval shell growth and shell surface area. This kind of stressful larval experience, that reduced lipid index, as in *Mercenaria mercenaria* and/or increased metabolic cost, as in *Crassostrea virginica* (Table 8.2), led the subsequent juveniles to begin with an energy deficit. Acidic pollution and/or poor availability of food (cf *Mytilus edulis*, Table 8.3) may induce a negative 'carryover' deficit. Consequently, juveniles of Olympia oyster, whose larvae were reared under stressful low pH 7.5, exhibited 41% decrease in shell growth rate. Therefore, the adverse effects of early exposure to low pH persisted even after 1.5 month depuration in normal sea water.

On the other hand, the negative effects of elevated CO_2 on the reproductive adults led to the production of more robust and larger larvae also developed rapidly. The lone publication by Parker et al. (2012) is concerned with acidification effects on reproducing adult oyster. The ready to reproducing adults (1.5–2.0 year-old) from wild (7.15 g) and selectively bred (7.13 g) Sydney rock oyster *Saccostrea glomerata* were exposed to ambient (385 µatm) and elevated (856 µatm) CO_2 for a period of 15 days. The larvae of both wild and selected oysters exposed to elevated CO_2 survived more successfully, grew faster and had relatively larger shells than those produced by the oysters reared at ambient CO_2. Notably, *M. edulis* larvae reared under the stress of low pH carried over the negative effects and energy deficit to the juveniles. In contrast, reproductive adults exposed to the stress of elevated CO_2 generated more robust, fast growing larvae. Therefore, there is a need for more research like that of Parker et al. (2012).

Investigating settlement and shell growth of the blue mussel *M. edulis*, Thomsen et al. (2010) recorded peak settlement of *M. edulis* during July-August, when surface pCO_2 levels in Kiel Bay were around 967 µatm. Further, they observed increases in shell growth of the mussel receiving a good food supply in the field to exceed 1 mm/week and thereby confirmed the earlier report by Kossak (2006). There were no significant changes in the thickness of the calcite layers of the newly formed shell parts, when pCO_2 was elevated. Whereas the calcite layer thickness was comparable between pCO_2 385 µatm and 400 µatm, the aragonite platelet layer decreased from 0.6 to 0.4 µm. This observation reveals that (i) the reported decreases in shell thickness are due to the dissolution of aragonite layer (Fig. 8.4). Incidentally, aragonite layers are in direct contact with extrapallial fluid. Notably, aragonite is less stable form of calcium carbonite than calcite (Hyman, 1967, p 157). From their field cum experimental studies, Thomsen et al. (2010) also reported that the mussels grew and did not incur reduction in shell thickness in acidic waters, when fed adequately. Remarkably, the alarm on ocean acidification rung by some scientists may be justified. But the investigations by Thomsen et al. (2010), Melzner et al. (2011) and Parker et al. (2012) indicate that calcifying marine molluscs do adapt to acidic waters, of course, at a higher metabolic cost. Hence, the need for the high pitch alarm may not fully be justified.

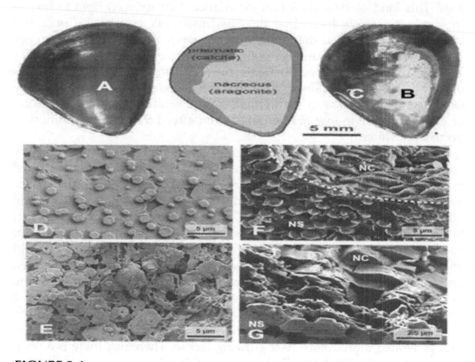

FIGURE 8.4

Mytilus edulis showing internal shell dissolution of the nacreous layer A. Microscopic image of the inner surface of a control, B. Schematic drawing of the inner surface of the control, C. Shows the corroded nacreous layer and unimpacted nacreous surface. SEM images of the control D. and severely corroded E. nacre surface of the shell. F. and G. SEM images of shell cross sections of uncorroded and corroded nacreous layer at different magnifications (from Melzner et al., 2011).

8.5 Effects on Prey-Predators

Sessile molluscs, such as oysters and mussels that cannot move to escape from penetration by drills or their valves forcibly opened by crabs and others, have to rely on shell thickness alone. For example, the drills spent 20% less time to penetrate the oysters from acidified waters. This is attributed to reduced shell thickness (Amaral et al., 2012, Landes and Zimmer, 2012). Larger bivalves have stronger adductor muscles that can powerfully hold shells closed on being attacked by a crab (Kautsky et al., 1990). But smaller bivalves with less powerful adductor muscles are more vulnerable. On the other hand, motile gastropods have a choice of structural increase in shell thickness or behavioral response to escape from predators.

Inducible defenses have been documented for molluscan prey against predation by drills, crabs and fishes (e.g. Appleton and Palmer, 1988). Available publications related to acidification effects on molluscan prey suggest that the responses range from changes in shell shape (axial and spiral sculpture, Appleton and Palmer, 1988) to shell thickness and size. Structural changes demand a higher metabolic cost than behavioral escape. Through an interesting combination of pH and 'crab cue' experiment, Bibby et al. (2007) demonstrated that on exposure to dual stresses of low pH and 'crab cue', the littoral limpet *Littorina littorea* (12 mm) switched to more economic 'predator' avoidance behavior. The 'crab cue' water was prepared by immersing a crab (*Carcinus meanas*, carapace width: 71 mm) in 5 l sea water for a 48 hour period prior to the experiment. The avoidance response was measured by the percentage of time spent by the limpet at or above the water surface. On exposure to the 'crab cue' sea water at pH 7.97, the limpet responded by a 0.05 mm increase in shell thickness. However, on exposure to low pH (6.45) + 'crab cue' water, the limpet switched to (an 18% increase) predator avoidance behavior. More interestingly, oxygen consumption by the limpet increased from 0.040 ml/g/h at normal pH to almost equal levels of 0.047 ml/g/h, when exposed to either low pH alone or normal pH + 'crab cue'; to the latter, the limpet responded by an increase in shell thickness. However, on exposure to dual stress of low pH + 'crab cue', the snail switched to predator avoiding response, costing oxygen uptake of just 0.027 ml/g/h. Clearly, the switch to predator avoiding behavioral response (instead of increasing shell thickness) reduced the metabolic cost by < 50%. Hence, behavioral response to predators is profitable.

In freshwaters, calcium concentrations vary widely between habitats (Dillion, 2000). Expectedly, the structural response to predation by changing shell thickness may depend on Ca^{2+} concentration. Rundle et al. (2004) demonstrated that calcium availability interacts with predation cues to modify growth and form of shells. They estimated the shell characteristics and predator avoidance by the proportions of the freshwater snail *Lymnaea stagnalis* and time required to 'crawl-out' to positions at or above the water surface. Snails, especially the smaller ones grew faster in length in the presence of 'fish cue' but this growth was limited by Ca^{2+} (Table 8.4). Hence, the smaller snails opted for behavioral response to escape from 'predator'. Both the percentages of snails and time required for avoidance of predator were significantly increased in water containing 45 mg Ca^{2+}/l. Remarkably, the changes in aperture ratio indicated that longer and narrower aperture reduced vulnerability by reducing the predator entry width (see also DeWitt et al., 2000). Shell mass increased from 29 to 42 mg/mm in snails at low calcium concentration waters + 'fish cue'. Briefly, the freshwater snail opts to a longer and narrower aperture and escaping behavior, when calcium limits the increase of shell thickness.

Coral reef-inhabiting conch snails have a modified foot and operculum, allowing them to a rapidly jump backwards on an encounter of predator. Watson et al. (2014) reported that elevated CO_2 (~ 961 µatm) reduced the number of jumps by the snail from 24 to 13 and thereby reduced their potential to an escape-making decision. The exposure also delayed the jumping behavior in jumping snails and thus prolonged the time taken to jump and extended the escape duration to predators. It also changed the escape trajectory in such a way that the snail moved on an angle closer to the predator. Hence, the exposure interfered with the function of neurotransmitter receptors and reduced the scope for escape from predators (see also Chivers et al., 2013).

TABLE 8.4

Structural and behavioral responses of the freshwater snail *Lymnaea stagnalis* to 'fish cue' at two calcium concentrations (compiled from Rundle et al., 2004, * approximate values)

Characteristics	Calcium Concentration (mg/l)			
	45		95	
	− cue	+ cue	− cue	+ cue
Shell characteristics				
Shell length increase (%)				
a) Small snail 5 mm	−1.0	3.8	3.8	6.3
b) Large snail 10 mm	3.5	6.5	5.5	5.0
Shell mass (mg/mm)	29.0	42.0	42.0	40.0
Crushing resistance (N)	1.3	1.5	1.7	1.9
Aperture ratio (% change)	−1.0	−0.5	10.0	0.0
Behavioral characteristics				
Avoiding response by snails (%)	42	72	68	74
Time required for avoidance (%)	75	95	80	83

As acidification may affect both the predatory drilling snails and preyed bivalves, Sanford et al. (2014) reared the Olympia oyster *Ostrea lurida* and Atlantic oyster drill *Urosalpinx cinera* at ambient (~ 500 µatm) or elevated (~ 1000 µatm) CO_2 level from the hatching to adult stage. The drills were tolerant to elevated CO_2 but the oysters grew 29% smaller in size and had thicker shells. Two trails of experiments were run. In the first, the drills were offered the oyster reared at the elevated CO_2 level alone but in the second, they were given a choice to predate either oyster raised at ambient or elevated CO_2 level. To compensate the smaller sized oysters, the drills, with no choice, predated 20% more snails. When given a choice, the drills preferred to predate the oyster raised at elevated CO_2 level. Whereas the predatory behavior of the drills was not affected by acidification, it affected other predators like fishes and crabs.

More importantly, the responses to elevated CO_2 vary in different family lines of *Crassostrea gigas* at different durations of exposure (Wright et al., 2014). Radical changes ranging from decrease (e.g. *Ruditapes desscatus*, Range et al., 2011) to increase (e.g. *Crepidula fornicata*, Ries et al., 2009) in calcification have been reported. These changes are attributed to the ability of these molluscs to maintain pH at the local scale in the site of calcification. Incidentally, Wright et al., also found differences in shell compression, thickness and mass among family lines of *C. gigas*. Despite increased strength of the oyster and reduced shell strength of the predatory whelk *Morula marginalba*, previously exposed to elevated CO_2, the whelk consumed significantly more oysters, regardless of whether the oyster had been exposed to ambient or elevated CO_2. This observation is in line with that of Sanford et al. (2014) on prey-predation relation between *O. lurida* and *U. cinera*, both of which were exposed to elevated CO_2.

Briefly, this account shows that 1. Production and maintenance of shells by molluscs requires energy (Palmer, 1981, 1992). 2a. Where or when possible, molluscs acquire additional energy required to maintain calcification by increasing feeding rate (e.g. Melzner et al., 2011, Thomsen et al., 2013), possibly with little to trade off in growth, reproduction and defenses, 2b. Alternatively, net calcification may be maintained at the expense of growth, reproduction and/or defenses. The reflected negative effects measured were (i) decreased shell thickness/strength, as measured by crushing resistance (see Table 8.4) (ii) reduced biomass and lipid/energy index as well as increased energy deficits, (iii) reduced immune response (Bibby et al., 2008, Matozzo et al., 2012), (iv) impaired neurotransmition receptors (e.g. Watson et al., 2014) and (v) chemoreception (e.g. for predators, Munday et al., 2009, Landes and Zimmer, 2012). A notable finding is that with acidification intensity, dual stresses (low pH + predator cue) or calcium limitation (e.g. *L. stagnalis*), gastropods opt for energetically more economic predator avoidance behavior. Molluscan response to acidification of waters differs from species to species, which has been elegantly summarized by Kroeker

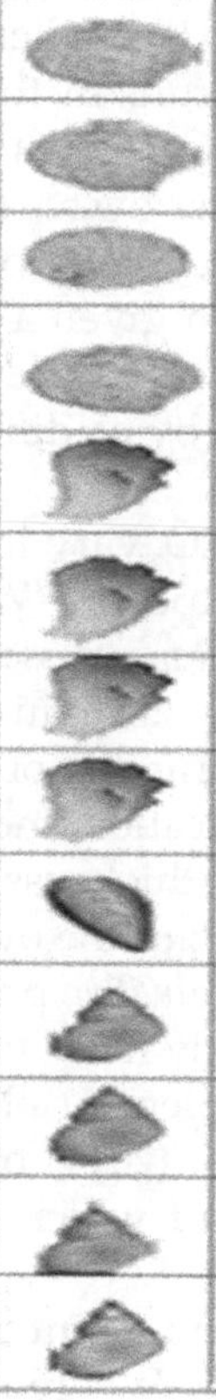

Species	Shell thickness or strength	Shell Size	Energy content	Adaptive behavior	Predation pressure	Reference
Argopecten irradians		=	=			Talmage and Gobler (2010)
Mercenaria mercenaria		=	=			Talmage and Gobler (2010)
Concholepas concholepas	=			⬆		Manriquez et al. (2013)
Mercenaria mercenaria	⬇	⬆	⬆			Dickinson et al. (2013)
Crassostrea virginica		⬇	=			Talmage and Gobler (2010)
Saccostrea glomerata	⬇				⬆	Amaral et al. (2012)
Crassostrea virginica	⬇		⬇			Beniash et al. (2010)
Ostrea lurida	=	⬇			⬆	Sanford et al. (2014)
Mytilus californianus	⬇	⬇	⬇			Gaylord et al. (2011)
Littorina littorea	⬇			⬆		Bibby et al. (2007)
Littorina littorea	⬇	=				Metalunan et al. (2013)
Bembicium auratum	⬇	=				Amaral et al. (2012)
Lymnaea stagnalis	⬆				⬆	Rundle et al. (2004)

FIGURE 8.5

Examples for prey-predator relation as measured by important traits. An equal sign denotes no change, a down-facing arrow denotes reduction in the given parameter and an upward-facing arrow denotes an increase in the parameter (modified and compiled from Kroeker et al., 2014).

et al. (2014). Figure 8.5 is a slightly modified version of Kroeker et al., and includes also representation of a freshwater gastropod. Of 13 species tested against acidification, seven species responded by thickening shell but only three opted for more economic behavioral predator avoidance.

9

Uniqueness of Molluscs

Introduction

Molluscs are unique for the presence of a protective shell, defensive inking, geographic distribution from the depth of 9,050 m to an altitude of 4,300 m (Table 1.6), gamete diversity, the use of nurse eggs/embryos to accelerate the first few mitotic divisions in embryos, the occurrence of natural androgenics, gigantism induced by elevated polyploidy the complementary role of mitochondrial genome in sex determination by nuclear genes and the uptake of steroids from surrounding water.

9.1 Shell and Iteroparity

More than 98% of molluscs carry external protective shell(s). The shell's morphometric traits like size, shape, thickness and weight are phenotypic traits although one or other of these traits is rarely inherited (e.g. freshwater pulmonate *Physa acuta*, Dillon and Jacquemin, 2014). Many authors have estimated the proportion of resource allocation to shell, body growth and reproduction, considering the (dry) shell weight along with live/wet weights of growth and/or reproduction (Table 9.1). Rarely, Komaru and Wada (1989) have reported these values both in dry and wet weights in 14-month old diploid and triploid noble clam *Chlamys nobilis*. On the dry weight bases, the proportion of shell makes up to 87–90% of the clam. However, these values shrink to 53–58%, when proportions of growth and reproduction are considered on wet weight basis. With this limitation for the reported values, the following may be generalized: 1. The values for the shell weight ranges from 31 to 91% for the single valve gastropods but 37 to 90% for bivalves. With a single valve, gastropods too invest on shell weight as much as bivalves. 2. In shelled molluscs, the presence of predators and/or parasites (e.g. *Mytilus californianus*, see also Figs. 2.8, 2.9) alters the investment on

TABLE 9.1

Effects of selected factors on allocation of assimilated energy for shell, growth and reproduction in bivalves and gastropods. * includes for reproduction also, † includes for shell also, †† includes for growth also ** ~ 4.45% energy allocated for regeneration of nipped siphon, ***indicate shell dry weight but growth and reproduction in wet weight

Species/Reference	Remarks	Resource Allocation (%)		
		Shell	Growth	Reproduction
Bivalvia				
Yoldiella valettai	Amundan Sea	37	63*	–
Reed et al. (2013)	Weddell Sea	62	38*	–
Choromytilus meridionalis	Sublittoral	–	22†	78
Griffiths (1981b)	Supralittoral	–	7†	93
Mytilus californianus	Sublittoral, healthy	55	27	18
Anderson (1975)	Supralittoral, healthy	64	20	16
	Sublittoral, infected	63	24	13
*Tellinia tenuis***	4-year old		87.0	8.6
Trevallion (1971)	6-year old		41.7	54.1
Chlamys nobilis	Diploid***	57.5	35.7	6.7
Komaru and Wada (1989)	Triploid***	52.7	42.8	4.5
All by dry weight	Diploid	90.4	8.0	1.6
	Triploid	86.7	12.8	0.5
Gastropoda				
Nucella emarginata	Thick morph	90.9	–	9.1††
Geller (1990)	Thin morph	69.0	–	30.9††
Thais lamellosa	Sublittoral, thin	75	25*	–
Palmer (1981)	Sublittoral, thick	79	21*	–
	Supralittoral, thin	74	26*	–
	Supralittoral, thick	81	19*	–
Littorina obtusata	45.0° N	31.1	68.8*	–
Trussell (2000)	42.5° N	73.8	26.2*	–
	42.5° N to 45.0° N	50.8	49.2*	–
	45.0° N to 42.5°N	62.9	37.1*	–

shell weight. For example, shell weight of the thick shell morph *Nucella emarginata* constitutes as much as 91% of the total weight but is limited to 69% in the thin shell morph. 3. In the intertidal zone, the habitat location significantly alters the investment on shell weight in both gastropods (e.g. *Thais lamellosa*) and bivalves (*Choromytilus meridionalis*). 4. With increasing cost of calcium acquisition at higher latitudes, the investment on shell weight is decreased (e.g. *Littorina obtusata*). 5. Induced triploids invest less on shell weight as well as reproduction and thereby increase 'meat weight'. However, shell size of triploid *Crassostrea madrasensis* increases faster than diploid in all dimensions and meat weight (Fig. 9.1). Evidently, the molluscs allocates fluctuating quantum of assimilated energy between intensely competing protective external shell, body growth and reproductive output.

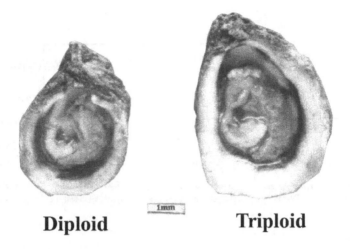

Diploid   Triploid

FIGURE 9.1

One year old diploid and triploid *Crassostrea madrasensis* (from Mallia et al., 2007)

Hence molluscs do stand unique among aquatic invertebrates and render the study of reproduction and development a fascinating field of research. A comparative study on the cost of production and maintenance of external skeleton in the form of chitinous structure in crustaceans, calcium coated dermis in starfishes and shell in molluscs may prove interesting. The regular loss on exiviae following molt in relatively more motile crustaceans ranges from 0.4% in the copepod *Tigriopus brevicornis* to 10% in deep sea euphausiids (Pandian, 1994, p 106) and averages to 4.9%, although the total cost of molting may be more. With sedentary/sessile mode of life, molluscs seem to invest more on the external protective shell.

From a compilation of widely scattered bits and pieces of information on reproductive cycle (monocyclic or polycyclic), a new frame-work has emerged on the protective role of shell and its impact on reproductive cycle and life span as well as sexuality and body size of aquatic molluscs (Fig. 9.2). The following generalizations may hold true for at least 95% of aquatic molluscs.

1. The presence of shell has afforded iteroparity and relatively longer life span in prosobranchs and bivalves but its absence semelparity in opisthobranchs and cephalopods. In cephalopods, the sepiodids and octopodids are characterized by synchronized terminal spawning prior to death. However, there are also monocyclic sepiodids that oviposit in batches but within a spawning season (e.g. *Dosidicus gigas*, Table 1.9). Notably, the shelled nautiloids are iteroparous. With the life span of 16–18 months, many shells-less pulmonates, for example, *Arion lusitanicus*

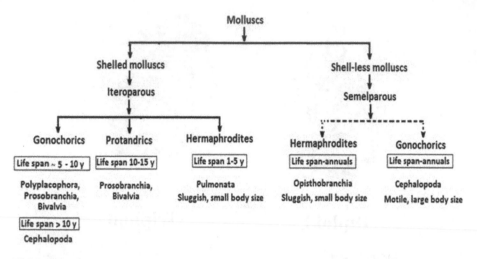

FIGURE 9.2

Potential role of shell on reproductive cycle, sexuality on life span and body size in aquatic molluscs.

is semelparous. It is not clear whether the following two examples can be more of exception to this generalization. Most opisthobranchs are semelparous. However, those belonging to the following four orders Cephalaspidea, Pteropoda, Sacoglossa and Notaspidea may or may not have a shell. Within the life span of 7–8 months, the bivalved sacoglossa *Julia japonica* spawns 40–50 times followed by death (Hyman, 1967, p 502). This is also true of many shelled pteropods (e.g. *Limacina helicina*, Gannefors et al., 2005). The shell-less aplysiads display daily spawning pulses (see Fig. 3.12) buy may be within a monocyclic spawning prior to death. This is also true of many shelled snails belonging to the genera *Bulinus* and *Biomphalaria* (Dazo et al., 1966, Brown, 2002). In fact, it is difficult to designate some of the shell-less slugs as semelparous and shelled snails as iteroparous.

2. Within each of these iteroparous and semelparous groups, sexuality plays a key role in fixing the length of Life Span (LS). In molluscs, hermaphroditism imposes a shorter LS ranging from < 1 year to < 5 years in shelled pulmonates and shelled/shell-less opisthobranchs. In shelled pulmonates and shell-less opisthobranchs, the obligate need to establish and maintain dual sexes and associated reproductive systems in simultaneous hermaphrodites has reduced the LS to < 5 years. In fact, the LS of many shelled pulmonate schistosome-intermediate host snails is so short (< 7 months) that they double the populations within a fortnight (Dazo et al., 1966). The LS of polyplacophoran hermaphrodite *Trachyderma raymondi* is ~ 1 year (Hyman, 1967, p 121), in comparison

to 4+ years and 9+ years for the gonochoric *Chaetoplura apiculata* and *C. tuberculatus* (see Table 1.14), respectively. Of course, there are exceptions to this generalization. The shelled gonochoric prosobranch *Tulotoma magnifica* lives for 2 years only (Christman, 1996). Other exceptions are the freshwater Marian hermaphroditic bivalves (e.g. *Elliptio arca*, 20 years, Haag and Staton, 2003). However, it must be noted that hermaphroditism in these bivalves is not stable and their populations comprise of pure males and females as well as male-like females and female-like males (e.g. *E. complanata*, Downing et al., 1989). On the other hand, the LS of iteroparous gonochorics ranges from 5–10 years in shelled prosobranchs and bivalves. It is extended for longer durations in protandric gastropods and bivalves. For example, the LS of the limpet *Patella vulgata* is 14+ years (Wright and Hartnool, 1981) and black pearl oyster *Pinctada margaritifera* > 12 years (see Fig. 3.7A).

3. Within semelparous cephalopods, gonochorism in combination with high motility facilitates faster growth and larger body size. On the other hand, hermaphroditism coupled with sluggish mode of life has reduced the body size in semelparous shell-less opisthobranchs and iteroparous shelled pulmonates. For example, most opisthobranchs are limited within the size range of 1.5 to 100 mm length; the exceptions are the herbivorus aplysiads; for example, *Aplysia vaccaria* grows to 14 kg body weight (Table 1.13). Contrastingly, the highly motile gonochoric cephalopods grow to large size and *Architeuthis* sp measures 18 m length and weighs 450 kg, the largest known living molluscs.

9.2 Gamete Diversity

With different types of eggs (Table 9.2) and spermatozoa (Table 9.3), molluscs are unique in that no other aquatic invertebrate taxon is known to generate so many types of gametes. Except for a few polychaetes and nemerteans, the presence and prevalence of nurse eggs and embryos are unique to gastropods. There are a large number of publications on the size and proportion of nurse eggs in capsules of gastropods (section 3.8). However, no publication is yet available on the cytogenetic mechanism that arrests the development of nurse eggs at gastrula stage and nurse embryos at a later stage. Research on this aspect by molecular biologists may be rewarding. Secondly, poecilogony and dispersal dimorphism are again unique to opisthobranchs (section 3.9). It is understandable that with abundant food supply during spring enables the opisthobranch *Elysia subornata* mother to produce planktotrophic eggs, lecithotrophic eggs with pelagic larvae during summer; but with limited food supply during fall and winter, the same mother is constrained to produce only lecithotrophic eggs directly developing into juveniles. It is also

TABLE 9.2

Types of eggs produced by molluscs

Egg Types	Example
Nurse eggs and embryos	
Developing eggs and ingestible nurse eggs simultaneously produced by the same mother	Gastropoda: Neogastropoda
Development-arrested nurse eggs and embryos simultaneously produced by the same mother	Gastropoda: Vermetids
Planktotrophic and lecithotrophic eggs	
Planktotrophic and lecithotrophic eggs simultaneously produced by	Opisthobranchs:
(i) The same mother	*Alderia willowi*
(ii) Mothers from different sites	*Elysia chlorotica*
(iiia) Mothers from the same site	*E. pusilla*
(iiib) Different mothers from the same site	Prosobranch: *Calyptrea lichen*
The same mother simultaneously produce eggs within the same egg mass, from which some develop as planktonic, non-feeding veliger and others directly hatching as juveniles	Opisthobranch: *Haminoea japonica*
The same mother oviposits planktotrophic egg masses in spring, lecithotrophic eggs involving pelagic larvae in summer and lecithotrophic eggs directly producing juveniles during fall and winter	Opisthobranch: *Elysia subornata*
Natural and induced ploidy	
Degenerating pronucleus containing eggs resulting in natural androgenesis	Bivalvia: *Corbicula*
Unreduced 2n eggs by hermaphrodite resulting in natural gynogenesis	Bivalvia: *Lasaea*
Reduced 2n eggs produced by natural tetraploid gonochorics	Gastropoda: *Bulinus*
Unreduced 4n eggs by allotetraploid parthenogenics	Gastropoda: *Melanoides tuberculata*
Unreduced 6n eggs by allohexoploid parthenogenics	*M. lineatus*
Unreduced 2n eggs by fertilized by haploid sperm? in parthenogenic	*Pomatogyrgus andipodarum*
Induced tetraploid produce unreduced 3n eggs and others	Bivalvia: *Crassostrea gigas*
Induced tetraploid produce unreduced 2n eggs and others	*Pinctada martensii*

understandable that snails from different sites or different mothers from the same site produce planktotrophic and lecithotrophic eggs in *E. chlorotica* and *E. pusilla*. But it is difficult to comprehend how vitellogenesis is tuned to simultaneously produce planktotrophic and lecithotrophic eggs by the same mother in poecilogonic *Alderia willowi* (Krug, 1988) and dimorphic larvae in *Haminoea japonica* (Gibson and Chia, 1989). It is for the endocrinologists to explain this odd vitellogenic process, in which yolk-less and yolk-laden eggs are simultaneously generated.

TABLE 9.3

Types of sperm produced by molluscs

Sperm Types	Example
Fertilizing eupyrene and non-fertilizing apyrene sperm simultaneously produced by the same male	Gastropoda: *Melanopsis*
Eusperm-transporting modified apyrene spermatozeugma	Gastropoda: *Janthina*
Dimorphic smaller and larger sperm produced by consort and sneaker males, respectively	Cephalopoda: *Loligo bleekeri*
Sperm nucleus does not fuse with egg pronucleus resulting in gynogenesis	Bivalvia: *Lasaea*
Degeneration of egg pronucleus. Unreduced 2n or 3n sperm pronucleus induces natural androgenesis	Bivalvia: *Corbicula leana*: 2n sperm *C. fluminea*: 2n or 3n sperm
Haploid egg fertilized by unreduced 2n sperm? in parthenogenics	Gastropoda: *Pomatogyrgus antipodarum*
Sperm harboring developing mitochondria (M genome) and sperm harboring degenerating mitochondria (F genome) are simultaneously produced by the same male	Bivalvia

Molluscs do produce different types of spermatozoa. The process of spermatogenesis resulting in the production of eupyrene and apyrene sperm has been described. However, molecular mechanism(s) that lead to the disintegration of sperm pronucleus in *Lasea* and that of eggs in *Corbicula* remain to be described.

9.3 Gigantism and Polyploidy

In aquatic animals like fishes (Pandian, 2011, p 124) and higher crustaceans (Pandian, 2016, p 167–170), induction of polyploidy results in increased volume of individual cells but decreased number of cells. Consequently, diploids and polyploids in them are almost equal in size. However, induction of polyploidy in bivalves not only increases the volume of the individual cell but also retains the cell number unchanged. As a result, the size of triploid and tetraploid bivalves is 1.4 and 4.4 times larger than diploids (Guo and Allen, 1994b). Among aquatic invertebrates, bivalves are again unique in that the induced polyploidy leads to the so called polyploid gigantism, a feature that has the potential to be exploited by aquaculturists, especially in developing countries like India. It is not clear whether or not elevation of ploidy due to hybridization in allotetraploid and hexoploid *Melanopsis* led to gigantism in gastropods.

Unusually, the quantity of soft tissues of *Cardium edule* infected by trematode parasite *Labratrema mininus* is increased by 30 and 10% in 2 years and 3–4 year old clams, respectively (see Fig. 2.7A). In *Lepidochitona cinereus* (from Scotland) infected by sporozoan parasite *Haplosporidium chitonis*, the entire body growth (including shells) of the infected chiton was almost doubled (24 mm length), in comparison to uninfected chiton (16 mm) (Fig. 2.7B), i.e. the infected chiton grew 1.5 times faster throughout its life. Trematode parasites may inhibit or accelerate body growth in their respective hosts. It is not clear whether *H. chitonis* induced triploidy, which is known to facilitate ~ 1.4 times faster growth in many bivalves. Studies are required to know whether the induced gigantism in the infected chiton is a result of elevated triploidy.

9.4 Sexuality and Paternity

In molluscs, sexuality includes gonochorism, parthenogenesis and hermaphroditism, which includes simultaneous, Marian, protandry and serial. Remarkably, > 25% of molluscs are hermaphrodites which may be compared with 2% in fishes (Pandian, 2010) and 8% in crustaceans (Pandian, 2016). Uniquely, ~ 23% of molluscs are Simultaneous Hermaphrodites (SH), which may be compared with < one dozen fish species belonging to the genera *Kryptolabias*, *Serranus* and *Hypoplectrus* (Pandian, 2010, Chapter 6) and 4% in crustaceans (Pandian, 2016). Even within this 4% SH crustaceans, the SH is diluted by the presence of dwarf males and androdioecious mating system in Notostraca and Spinicaudata (see Pandian, 2016, p 106–109).

The molluscan SH, especially the pulmonates do not self, unless enforced naturally or experimentally. In the presence of autosperm and allosperm, the snails preferably use allosperm to fertilize their eggs (p 98–100). Most pulmonates and opisthobranchs like *Aplysia* spp are unilaterals. Inbreeding depression is successfully minimized or avoided by pulmonate and opisthobranch SH by opting for unilateral and/or reciprocal inseminations. As a result, the multiple paternity is > 2.0 sires/brood and 2.1 sires/brood in many pulmonates and *A. californica*, respectively (Table 9.4). Notably, with a wide range of structural diversity, on which classification is based, opisthobranchs are accommodated in 12 orders (Hyman, 1967, p 12). But pulmonates, perhaps with limited structural diversity, are included within two orders and their multiple paternity is also equal to that of opisthobranchs. However, pulmonates are more speciose (> 24,000 species) and have a very wide distribution covering sea, freshwater and land. Conversely, opisthobranchs are limited to 2,000 species and are also limited to marine habitat alone. Hence, the pulmonates seem to employ other than multiple paternity to achieve a greater genetic diversity and speciation.

In molluscs, parthenogenesis occurs but limited to a few species belonging to the genera *Melanoides*, *Potamopyrgus* and *Campeloma* within gastropods

TABLE 9.4

Characteristic features of reproduction, development and distribution of major molluscan taxons. Bold letters indicate the relatively larger prevalence

Parameter	Proso-branchs	Opistho-branchs	Pulmonates	Bivalves	Cephalopods
Species (no.)	75,000	2,000	24,000	9,000	760
Sexuality	**Gonochoric/**Protandric	Herma-phrodite	Herma-phrodite	**Gonochoric/**Protandric/Serial	Gonochoric
Fertilization	**Internal/**External	Internal	Internal	**External/**Internal	Internal
Development	**Indirect/**Direct	Indirect	**Direct/**Indirect	**Indirect/**Direct	Direct
Paternity ($\male$/brood)	4.3	2.1	2.0	3.7	2.3
Shell as % body weight	~ 30	–	~ 20	~ 50	–
Habitats	**Marine/**Freshwater	**Marine**	**Land/**Freshwater	**Marine/**Freshwater	**Marine**
Density	Moderate	Moderate	Moderate	**High in marine/**low in freshwater	Low
Distribution	**Benthic,** 4300 m attitude	**Benthic/**Pelagic	Arboreal/Littoral	**Benthic** to 9050 m depth	Pelagic/**Benthic** to 4848 m depth
Largest size	0.3 m	1 m	–	1.4 m	18 m
Largest weight	1.4 kg	14 kg	–	21 kg	450 kg

alone. It is not as prevalent, as in Crustacea, in which it occurs across all the major taxons including the decapod *Procambarus fallax*. As many as 150 and 760 parthenogenic species are known within the genus *Daphnia* and order Cladocera, respectively. In the stable marine habitats too, parthenogenesis occurs in species belonging to five genera. Whereas parthenogenesis originates *de nova* in crustaceans (Pandian, 2016, sections 3.4.1 and 3.4.2), it is induced by (i) hybridization and consequent ploidy elevation (Table 3.2) (ii) conductivity of freshwater (Fig. 3.4A) and/or (iii) parasitic infection (Fig. 3.4B, C). Consequently, males do occur, though at a highly reduced frequency in parthenogenic gastropods. But males are not present but occur sporadically in crustaceans. Hence, parthenogenesis in gastropods may be more an exception than a regular feature.

9.5 Double Uniparental Inheritance (DUI)

In the animal kingdom, the hereditary transmission of sex determination mechanism by chromosomes as carriers of sex determining gene(s) is well known. In mammals, sex is decisively determined by a single *Sry* gene harbored on Y chromosome. However, our understanding on chromosomal sex determination in aquatic animals is limited. Hence, it has become necessary to depend on fishes, for which relatively more information is available. Of 30,000 and odd fish species, cytogenetic and/or genetic analyses of chromosomal inheritance of sex determination mechanism are limited to 1,700 species. The presence of sex chromosomes and heterogamety has been recognized only in 11% of fish species (Pandian, 2013, p 2). Intriguingly, even within the 190 and odd species, in which heterogamety is recognized, 9% have multiple sex chromosome system. Moreover, sex determination by sex chromosomes has repeatedly emerged from autosomes (Pandian, 2011, p 18). Unlike in fishes (2n = 12–500, Pandian, 2011, p 19) and crustaceans (2n = 20–254, Pandian, 2016, p 163), the number of chromosomes in molluscs ranges from 18 to 44 only (see p 157–158). However, heterogamety has been recognized only in three species of gastropods (see p 158) and five species in bivalves (Table 6.5). It is in this context, that the discovery of DUI is of great interest. The presence of DUI is unique to bivalves in the animal kingdom. Its sex-dependent transmission and complementary role with nuclear genome in sex determination is also a unique feature (Zouros et al., 1994a, b, Ghiselli et al., 2012). Sex has evolved as early as 1.6–2.0 billion years ago and has been successfully manifested in a wide range of microbes, plants and animals (see Pandian, 2012, p 17). During the ancient evolutionary history of sex determination mechanism, nuclear DNA that determines sex, may have been complemented by DNA of cytoplasmic organelle like mitochondria, although the complementary role of mitochondrial DNA may have been subsequently eliminated. Interestingly, an unknown oocyte cytoplasmic factor determines the direction of chirality in gastropods. Hence, it is likely that the presence and complementary role of DUI may be more a vestigial representation of the ancient evolutionary history of sex determination mechanism.

9.6 Vertebrate Type Steroids

In aquatic invertebrates like molluscs and crustaceans, sex differentiation is accomplished by, discrete androgenic glands, sinus glands, Y-organ and mandibular organs are established sources of neurohormones and others in crustaceans. These neurohormones have also been a subject of numerous studies (see Pandian, 2016, section 7.1). However, the neurohormones responsible for sex differentiation in molluscs are synthesized and released from ganglia located at the cephalic, visceral and pedal regions. The optical gland is the only organ known to secrete neurohormones in cephalopods.

Uniquely, molluscan genome is now indicated not to contain genes for any enzymes that are involved in biosynthesis of vertebrate type steroids. Molluscs are also suspected not to have functional steroid nuclear receptors (Scott, 2013). For reason not yet known, molluscs also accumulate vertebrate type steroids from surrounding water (Le Curieux-Belfond et al., 2001, 2005, Scott, 2012). The implicated role of these steroids in molluscs is a matter of debate. In crustaceans, responses evoked by exposure to one or other vertebrate type steroid are experimental and not natural. Crustaceans synthesize ecdysteroid, which regulates molt cycle. Both of these aquatic major invertebrate taxons namely Crustacea and Mollusca employ neurohormones for sex differentiation. The implicated role of vertebrate type steroids in reproductive cycle of bivalves and cephalopods merits further research.

9.7 Aestivation and Cysts

Seeds are a wonderful discovery of plants. So are the spores of microbes. Seeds and spores constitute a strategy of plants and microbes for dispersal. They can successfully survive over thousands and millions of years and yet can be revived. For example, the paddy plant can be revived from its seeds after 2000 years. Spores of the microbe *Bacillus sphericus* can be revived after 30 million years (see Pandian, 2002). To tide over unfavorable season, animals too have discovered cysts and the like (Table 9.5). Unlike plants and microbes, tenacity of the cysts to withstand unfavorable conditions does not last long; they survive for a few months/years but rarely for 49 years, as in the protozoan *Monas vestita* (see Pandian, 2002) and 332 years, as in copepod *Diaptomus sanguineus* (see Pandian, 2016, p 150). There are other phenomena like diapause and aestivation, through which animals can withstand an unfavorable season. Dormancy during embryonic and larval stages is named as diapause, which mostly occurs in insects. It is also reported from a few freshwater cyclopoid copepods and ostracods (see Pandian, 2016, p 148–149). Dormancy during summer and winter is called aestivation and hibernation. Polar bear, air-breathing fishes and snails aestivate. Uniquely, freshwater (and also terrestrial) snails are the only aquatic invertebrate taxon known to aestivate in the animal kingdom.

As adult snails aestivate, they require a large quantum of nutrient reserves prior to commencing aestivation. Conversely, cysts are reproductive products produced through parthenogenic (e.g. *Daphnia*) and/or sexual reproduction. They require relatively less nutrient and water reserves; for example, *Artemia*

TABLE 9.5

Contrasting features of cysts and aestivation

Features	Cysts and the like	Aestivation
Occurrence	Gemmules of sponges, cysts of protozoans and parasitic helminthes, cocoons of annelids, cysts and ephippia of crustaceans	Air-breathing snails, fishes and polar bears
Sexuality	Parthenogenesis in cladoceran and ostracodan crustaceans	Hermaphroditism in snails
Life stage	Reproductive product	Mature adults
Strategy	To tide over unfavorable season and for dispersal in free-living animals and infection of new host by parasites	To tide over unfavorable season only
Requirement	Relatively less nutrients and water reserves	Large quantity of nutrient reserves
Survival	Few months/years; rarely up to 332 y	Few months/a season; rarely to 23 y
Metabolism	Can be depressed up to 50,000 times, e.g. *Artemia*	Can be depressed up to < 100 times in aestivating snails

cyst can withstand desiccation with only 0.7 µg residual water/g dried cysts; when revived, they absorb 7 g water/g cyst (see Pandian, 2016, p 150). At this juncture, a comparison between cysts of crustaceans and aestivation in snails may be interesting. Entomostracan cysts have amazing capacity to withstand intense desiccation and freezing up to –18°C (e.g. *Heterocypris incongruens*). The cysts are resistant to digestion by predators. They can be blown (Caceres and Soluk, 2002) and carried by running waters (Caceres and Tessier, 2014) to different habitats. By 'bed-hedging', the cyst of different year classes may be hatched in a season/year. Briefly, the buoyant cysts serve as spatial dispersers, while the sunken cysts function as temporal dispersers. Their ability to survive in transient, hazardous and unpredictable aquatic bodies is phenomenal (see Pandian, 2016, p 133–134).

The astatic freshwater habitats dry out or freeze periodically or are subjected to striking changes in water level. Some of them remain dry for several years; others are inundated for a few weeks or days only. Some others are flooded several times a year; still others have only a single wet phase. With regard to colonization of freshwaters, a comparison of the shortest life span between the hemocoelomate molluscs and crustaceans provides interesting information. In many cladocerans, the life span lasts for only a few days or weeks; for example, it is only 5–6 days in *Moina microrura* (Sarma et al., 2005). With aestivation as a strategy to tide over unfavorable season, freshwater snails could only inhabit in such aquatic bodies, in which water is available for a minimal period of 100–250 days (cf Table 1.14). To colonize all the aquatic bodies, which may range from ephemeral and unpredictable to permanent and predictable, cyst production is the option of crustaceans but aestivation is the option of gastropods; cladoceran and ostrocodan crustaceans have opted for parthenogenesis but gastropods for hermaphroditism. Adoption of hermaphroditism and aestivation strategy has imposed molluscs to limit their distribution to such freahwater bodies, in which water is predictably available for a minimum period of 100–250 days. Adopting parthenogenesis and cyst production, crustaceans have conquered every 'nook and corner' of freshwaters. Notably, metabolism in an *Artemia* cyst is depressed to 50,000 times, but it is to < 100 times in aestivating snails. These alternative strategies of cyst formation and aestivation serve as a decisive factor in geographical distribution of these two major aquatic invertebrate taxons. Incidentally, the snails do not produce cyst nor the entomostracan crustaceans aestivate. They are mutually exclusive.

In the years to come, space travel may become more and more important. With space travel requiring long years of passage, 'cystification' or aestivation has to be resorted. The human body has 200 and odd tissue types and some like bones have < 10% water and others like blood > 90% water. Cystification involving dehydration and rehydration may prove to be an onerous task. But research on aestivation in mammals like polar bear may lead to 'aestivating man' to tide over "the long duration of passage to a desired planet and return home to mother Earth safely" (Pandian, 2016, p 226).

10

References

Acosta-Salmón, H. and P. C. Southgate. 2006. Wound healing after excision of mantle tissue from the Akoya pearl oyster, *Pinctada fucata*. Comp Biochem Physiol Part A, 143: 264–268.

Allemand, D., C. Ferrier-Pagès, P. Furla et al. 2004. Biomineralisation in reef-building corals: from molecular mechanisms to environmental control. Comptes Rendus Palevol, 3: 453–467.

Allen, R. M., P. J. Krug and D. J. Marshall. 2009. Larval size in *Elysia stylifera* is determined by extra-embryonic provisioning but not egg size. Mar Ecol Prog Ser, 389: 127–137.

Allen, S. K. Jr. and S. L. Downing. 1986. Performance of triploid Pacific oyster *Crassostrea gigas* (Thunberg). I. Survival, growth, glycogen content and sexual maturation in yearlings. J Exp Mar Biol Ecol, 102: 197–208.

Allen, S. K. Jr. and S. L. Downing. 1990. Performance of triploid Pacific oysters *Crassostrea gigas*. Gametogenesis. Can J Fish Aquat Sci, 47: 1213–1222.

Allen, S. K., H. Hidu and J. G. Stanley. 1986. Abnormal gametogenesis and sex ratio in triploid soft-shell clams (*Mya arenaria*). Biol Bull, 170: 198–210.

Allen, S. K. Jr., S. L. Downing and K. K. Chew. 1989. *Hatchery Manual for Producing Triploid Oysters*. University of Washington Press, Seattle.

Allen, S. K. Jr., M. Shpigal, S. Uttig and B. Spencer. 1994. Incidental production of tetraploid Manila clams *Tapes philippinarum* (Adams and Reeve). Aquaculture, 128: 13–19.

Allsop, D. J. and S. A. West. 2003. Constant relative age and size at sex change in sequentially hermaphroditic fish. J Evol Biol, 16: 921–929.

Allsop, D. J. and S. A. West. 2004. Sex ratio evolution in sex changing animals. Evolution, 58: 1019–1027.

Alvarado-Alvarez, R., M. C. Gould and J. L. Stephano. 1996. Spawning, *in vitro* maturation, and changes in oocyte electrophysiology induced serotonin in *Tivela stultorum*. Biol Bull, 190: 322–328.

Amano, M., S. Moriyama, K. Okubu et al. 2010. Biochemical and immunohistochemical analyses of a GnRH like peptide in the neural ganglia of the Pacific abalone *Haliotis discus hannai* (Gastropoda). Zool Sci, 27: 656–661.

Amaral, V., H. N. Cabral and M. J. Bishop. 2012. Effects of estuarine acidification on predator-prey interactions. Mar Ecol Prog Ser, 445: 117–127.

Ambrose, H. W. and R. P. Givens. 1979. Distastefulness as a defense mechanism in *Aplysia brasiliana* (Mollusca: Gastropoda). Mar Behav Physiol, 6: 57–64.

Anderson, G. L. 1975. The effect of intertidal height and the parasitic crustacean *Fabia subquadrata* Dana on the nutrition and reproductive capacity of the California sea mussel *Mytilus californianus* Conrad. Veliger, 17: 299–306.

Anderson, M. B. 1994. *Sexual Selection*. Princeton University Press, Princeton.

Andersson, A. J., F. T. Mackenzie and N. R. Bates. 2008. Life on the margin: implications of ocean acidification on Mg-calcite, high latitude and cold-water marine calcifiers. Mar Ecol Prog Ser, 373: 265–273.

Andreu, B. 1960. Un parasite del mejillon. Propagacion del copepode parasite *Mytlicola intestinalis* en el Santander. Boln Inst esp Oceanogr, 22: 1–7.

Andrews, E. 1935. Shell repair in *Neritina*. J Exp Zool, 70: 75–107.

Anestis, A., H. O. Portner and B. Michaelidis. 2010. Anaerobic metabolic patterns related to stress responses in hypoxia exposed mussels *Mytilus galloprovincialis*. J Exp Mar Biol Ecol, 394: 123–133.

Angeoloni, L., J. W. Bradburry and R. S. Burton. 2003. Multiple mating, paternity, and body size in a simultaneous hermaphrodite *Aplysia californica*. Behav Ecol, 14: 554–560.

Ansell, A. D. 1982. Experimental studies of a benthic predator-prey relationship. II. Energetics of growth and reproduction, and food-conversion efficiencies, in long-term cultures of the gastropod drill *Polinices alderi* (Forbes) feeding on the bivalve *Tellina tenuis* (da Costa). J Exp Mar Biol Ecol, 61: 1–29.

Appeltans, W., P. Bouchet, G. A. Boxshell et al. 2011. *World Register of Marine Species*. marinespecies.org.

Appleton, R. D. and A. R. Palmer. 1988. Water borne stimuli released by predatory crabs and damaged prey induce more predator-resistant shells in a marine gastropod. Proc Natl Acad Sci USA, 85: 4387–4391.

Asami, T. 1993. Genetic variation and evolution of coiling chirality in snails. Forma, 8: 263–276.

Asami, T. 2007. Developmental constraint and stabilizing selection against left-right reversal in snails. In: *Evolution of Left-right Asymmetry: Why Isn't the Snail Mirror Flat?* (eds) K. Jordaens, N. van Heute, J. Van Goethem and T. Backeljau. Proc Wld Cong Malacol. Unitas Malacologica, Antwerp. p 13.

Asami, T., R. H. Cowie and K. Ohbayashi. 1998. Evolution of mirror images by sexually asymmetric mating behavior in hermaphroditic snails. Am Nat, 152: 225–236.

Austin, C. R., C. Lutwark-Mann and T. Mann. 1964. Spermatophores and spermatozoa of the squid *Loligo pealii*. Proc R Soc, 161B: 143–152.

Avila-Poveda, O. H., R. C. Montes-Perezy, N. Koueta et al. 2015. Seasonal changes of progesterone and testosterone concentrations throughout gonad maturation stages of the Mexican *Octopus maya* (Octopodidae: *Octopus*). Moll Res, 35: 161–172.

Avise, J. C., A. J. Power and D. Walker. 2004. Genetic sex determination, gender identification and pseudohermaphroditism in the knobbed whelk, *Busycon carica* (Mollusca: Melongenidae). Proc R Soc B, 271: 641–646.

Babarro, J. M. F., V. Labarta and F. M. J. Reiriz. 2007. Energy metabolism and performance of *Mytilus galloprovincialis* under anaerobiosis. J Mar Biol Ass UK, 87: 941–946.

Babarro, J. F., M. J. Fernández Reirez and U. Labarta. 2008. Secretion of byssal thread and attachment strength of *Mytilus galloprovincialis*: the influence of size and food availability. J Mar Biol Ass UK, 88: 783–791.

Babu, A., V. Venkatesan and S. Rajagopal. 2011. Contribution to the knowledge of ornamental molluscs of Parangipettai, Southeast Coast of India. Adv Appl Sci Res, 2: 290–296.

Badger, L. I. and J. P. O. Oyerinde. 2004. Effect of aestivation of *Biomphalaria pfeifferi* on the survival and infectivity of *Schistosoma mansoni* cercariae. British J Biomed Sci, 61: 138–141.

Bai, Z., M. Luo, W. Zhu et al. 2011. Multiple paternity in the freshwater pearl mussel *Hyriopsis cumingii* (Lea, 1852). J Moll Stud, 78: 142–146.

Bannister, R., N. Beresford, D. May et al. 2007. Novel estrogen receptor-related transcripts in *Marisa cornuarietis*: a freshwater snail with reported sensitivity to estrogenic chemicals. Environ Sci Technol, 41: 2643–2650.

Barbieri, E., K. Barry, A. Child and N. Wainwright. 1997. Antimicrobial activity in the microbial community of the accessory nidamental gland and egg cases of *Loligo pealei* (Cephalopoda: Loliginidae). Biol Bull, 193: 275–276.

Barco, A., A. Corso and M. Oliveirio. 2013. Endemicity in the Gulf of Gabes: The small mussel drill *Ocinebrina hispidula* is a distinct species in the *Ocinebrina edwardsii* complex (Muricidae: Ocinebrinae). J Moll Stud, 79: 273–276.

Barratt, I. M., M. P. Johnson and A. L. Allcock. 2007. Fecundity and reproductive strategies in deep-sea incirrate octopuses (Cephalopoda: Octopoda). Mar Biol, 150: 387–398.

Barroso, C. M., M. A. Reis-Henriques, M. Ferreira et al. 2005. Organotin contamination, imposex and androgen/estrogen ratios in natural population of *Nassarius reticulatus* along a ship density gradient. Appl Organometallic Chem, 19: 1141–1148.

Barsby, T., R. G. Linington and R. J. Andersen. 2002. *De novo* terpenoid biosynthesis by the dendronotid nudibranch *Melibe leonina*. Chemoecology, 12: 199–202.

Bartoli, P. 1974. Researches sur les Gymnophallidae F. N. Morozov, 1955 (Digenea) parasites d'oiseaux des cotes de Comargue: Systematigue, biologie et ecologie. These, Univ Aix-Marseille.

Baur, B. 1998. Sperm competition in molluscs. In: *Sperm Competition and Sexual Selection*. (eds) T. R. Bukhead and A. P. Moeller. Academic Press, New York. pp 255–305.

Bavia, M. E., L. F. Hale, J. B. Malone et al. 1999. Geographic information systems and the environmental risk of schistosomiasis in Bahia, Brazil. Am J Trop Med Hyg, 60: 566–572.

Baxter, J. M. 1983. Annual variation in soft-body dry weight, reporductive cycle and sex ratios in populations of *Patella vulgata* at adjacent sites in the Orkney Islands. Mar Biol, 76: 149–157.

Baxter, J. M. and A. M. Jones. 1978. Growth and population structure of *Lepidochitona cinereus* (Mollusca: Polyplacophora) infected by *Murichinia chitonis* (Protozoa: Sporozoa) at Easthaven, Scotland. Mar Biol, 46: 305–313.

Bayne, B. L. and C. Scullard. 1977. An apparent specific dynamic action in *Mytilus edulis* L. J Mar Biol Ass UK, 57: 371–376.

Bayne, B. L. and C. Scullard. 1978. Rates of feeding by *Thais* (*Nucella lapillus*) (L.). J Exp Mar Biol Ecol, 32: 113–129.

Bayne, B. L. and R. C. Newell. 1983. Physiological energetics of marine molluscs. In: *Mollusca: Physiology*. (eds) A. S. M. Saleuddin and K. M. Wilbur. Academic Press, London. 4: 407–571.

Beauchamp, K. A. 1986. Reproductive ecology of the brooding, hermaphroditic clam *Lasaea subviridis*. Mar Biol, 93: 225–235.

Beaumont, A. R. and K. S. Kelly. 1989. Production and growth of triploid *Mytilus edulis* larvae. J Exp Mar Biol Ecol, 132: 69–84.

Bell, G. 1982. *The Masterpiece of Nature: The Evolution and Genetics of Sexuality*. Cambridge University Press, Cambridge.

Ben-Ami, F. and J. Heller. 2005. Spatial and temporal patterns of parthenogenesis and parasitism in the freshwater snail *Melanoides tuberculata*. J Evol Biol, 18: 138–146.

Beniash, E., A. Ivanina, N. S. Lieb and I. I. M. S. Kurochkin. 2010. Elevated levels of carbon dioxide affect metabolism and shell formation in oysters *Crassostrea virginica*. Mar Ecol Prog Ser, 419: 95–108.

Benkendorff, K., A. R. Davis and J. Bremner. 2001. Chemical defense in the egg masses of benthic invertebrates: an assessment of antibacterial activity in 39 mollusks and 4 polychaetes. J Invert Pathol, 78: 109–118.

Bergman, M. J. N. and J. W. van Santbrink. 2000. Mortality of megafaunal benthic populations caused by trawl fisheries on the Dutch continental shelf in the North Sea in 1994. ICES J Mar Sci, 57: 1321–1331.

Bernard, F. 1974. Annual biodeposition and gross energy budget of mature Pacific oysters, *Crassostrea gigas*. J Fish Res Bd Can, 31: 185–190.

Berry, A. J. 1983. Oxygen consumption and aspects of energetics in a Malaysian population of *Natica maculosa* Lamarck (Gastropoda) feeding on the trochacean gastropod *Umbonium vestiarium* (L.). J Exp Mar Biol Ecol, 66: 93–100.

Berry, P. F. 1978. Reproduction, growth and population in the mussel *Perna perna* (Linnaeus) on the east coast of South Africa. Invert Rep Oceanogr Res Inst, Durban, 48: 1–28.

Bettin, C., J. Oehlmann and E. Stroben. 1996. TBT-induced imposex in marine neogastropods is mediated by an increasing androgen level. Helgol Wissens Meeresunters, 50: 299–317.

Bibby, R., P. Cleall-Harding, S. Rundle et al. 2007. Ocean acidification disrupts induced defences in the intertidal gastropod *Littorina littorea*. Biol Lett, 3: 699–701.

Bieler, R. and M. G. Hadfield. 1990. Reproductive biology of the sessile gastropod *Vermicularia spirata* (Cerithioidea: Turritellidae). J Moll Stud, 56: 205–219.

Block, J. D. and S. Rebach. 1998. Correlates of claw strength in the rock crab *Cancer irroratus* (Decapoda: Brachyara). Crustaceana, 71: 468–473.

Boag, D. A. 1986. Dispersal in pond snails: potential role of waterfowl. Can J Zool, 64: 904–909.

Bobek, L., D. M. Rekosh, H. Van Keulen and P. T. LoVerde. 1986. Characterization of a female-specific cDNA derived from a developmentally regulated mRNA in the human blood fluke *Schistosoma mansoni*. Proc Natl Acad Sci USA, 83: 5544–5548.

Borsa, P. and C. Thiriot-Quievreux. 1990. Karyological and allozyme characterization of *Ruditapes philippinarum*, *R. aureus* and *R. decussatus* (Bivalvia, Veneridae). Aquaculture, 90: 209–227.

Borysko, L. and P. M. Ross. 2014. Egg capsules and larval development of *Nassarius burchardi* (Phillipi, 1849) and *Nassarius jonaskii* Dunker, 1846 and comparisons with others Nassaridae (Caenogastropoda). Moll Res, 34: 213–221.

Bouchet, P. 1989. A review of poecilogony in gastropods. J Moll Stud, 55: 67–78.

Bouchet, P., P. Lozouet, P. Maestrati and V. Heros. 2002. Assessing the magnitude of species richness in tropical marine environments: exceptionally high numbers of molluscs at a New Caledonia site. Biol J Linn Soc, 75: 421–436.

Bowers, E. A. 1969. *Cercaria bucephalopsis haimeana* (Lacaze-Duthiers) (Digenea: Bucephalidae) in the cockle *Cardium edule* L. in South Wales. J Nat Hist, 3: 409–422.

Boyle, P. R. 1983. *Chephalopod Life Cycles*. 1. *Species Accounts*. Academic Press, London. p 475.

Boyle, P. R. 1987a. *Cephalopod Life Cycles: 2. Comparative Reviews*. Academic Press, London. p 441.

Boyle, P. R. 1987b. Molluscan comparisons. In: *Cephalopod Life Cycles*. (ed) P. R. Boyle. Academic Press, London. 2: 306–327.

Brahmachary, R. L. 1989. Mollusca. In: *Reproductive Biology of Invertebrates*. (eds) K. G. Adiyodi and R. G. Adiyodi. Oxford & IBH Publishers, New Delhi, Vol IV Part A: 280–348.

Brante, A., M. Fernandezaw and F. Viard. 2011. Microsatellite evidence for sperm storage and multiple paternity in the marine gastropod *Crepidula coquimbensis*. J Exp Mar Biol Ecol, 396: 83–88.

Briones, C., M. M. Rivadeneira, M. Fernández and R. Guiñez. 2014. Geographical variation of shell thickness in the mussel *Perumytilus purpuratus* along the Southeast Pacific coast. Biol Bull, 227: 221–231.

Brown, A. C. 1979. The energy cost and efficiency of burrowing in the sandy beach whelk *Bullia digitalis* (Dilwyn) (Nassariidae). J Exp Mar Biol Ecol, 40: 149–154.

Brown, D. S. 2002. *Freshwater Snails of Africa and their Medical Importance*. CRC Press, Boco Raton, USA. p 608.

Brown, R. J. 2007. Freshwater mollusks survive fish gut passage. Arctic, 60: 124–128.

Browne, R. A. 1978. Growth, mortality, fecundity, biomass and productivity of four lake populations of the prosobranch snail, *Viviparus georgianus*. Ecology, 59: 742–750.

Browne, R. A. and W. D. Russell-Hunter. 1978. Reproductive effort in molluscs. Oecologia, 37: 23–27.

Bruun, A. F. 1957. Deep sea and abyssal depths. In: *Geological Society of America Memoirs*. (ed) J. Hedgpath. 67: 641–673.

Bull, J. J. 1983. *Evolution of Sex Determining Mechanisms*. Benjamin/Cummings Publishing, London. p 316.

Bullard, S. A. and R. M. Overstreet. 2002. Potential pathological effects of blood flukes (Digenea: Sanguinicolidae) on pen-reared marine fishes. 53rd Proc Gulf and Caribbean Fish Inst, Fort Piece, 10–25.

Buresch, K. M., R. T. Hanlon, M. R. Maxwell and S. Ring. 2001. Microsatellite DNA markers indicate a high frequency of multiple paternity within individual field-collected egg capsules of the squid *Loligo pealeii*. Mar Ecol Prog Ser, 210: 161–165.

Bush, S. L., H. J. T. Hoving, C. L. Huffard and L. D. Zeidberg. 2012. Brooding and sperm storage by the deep-sea squid *Bathyteuthis berryi* (Cephalopoda: Decapodiformes). J Mar Biol Ass UK, 92 (07). doi: 10.1017/S0025315411002165.

Butzke, D., N. Machuy, B. Thiede et al. 2004. Hydrogen peroxide produced by *Aplysia* ink toxin kills tumor cells independent of apoptosis *via* peroxiredoxin I sensitive pathways. Cell Death Differ, 11: 608–617.

Caceres, C. E. and D. A. Soluk. 2002. Blowing in the wind: a field test of overland dispersal and colonization by aquatic invertebrates. Oecologia, 131: 402–408.

Caceres, C. E. and A. J. Tessier. 2004. To sink or swim: variable diapause strategies among *Daphnia* species. Limnol and Oceanogr, 49: 1333–1340.

Calabrese, A. and H. C. Davis. 1966. The pH tolerance of embryos and larvae of *Mercenaria mercenaria* and *Crassostrea virginica*. Biol Bull, 131: 427–436.

Caldeira, K. and M. E. Wickett. 2003. Anthropogenic carbon and ocean pH. Nature, 425: 365.

Callard, G. V., A. M. Tarrant, A. Nobvilo et al. 2011. Evolutionary origins of the estrogen signaling system: insights from amphioxus. J Steroid Biochem Mol Biol, 127: 176–188.

Calow, P. 1977. Ecology, evolution and energetic: a study in metabolic adaptation. Adv Ecol Res, 10: 1–62.

Calow, P. 1978. Evolution of life-cycle strategies in freshwater gastropods. Malacologia, 17: 351–364.

Calow, P. 1987. Fact and theory—an overview. In: *Cephalopod Life Cycles: Comparative Reviews.* Academic Press, London. 2: 351–365.

Calvo, J. and E. Morriconi. 1978. Epibionte et protandrie chez *Ostrea puelchana*. Haliotis, 9: 85–88.

Calvo, M. and J. Templado. 2004. Reproduction and development in a vermetid gastropod *Vermetus triquetrus*. Invert Biol, 123: 289–303.

Campos, E. O., D. Vilhena and L. Caldwell. 2012. Pleopod rowing is used to achieve high forward swimming speeds during the escape response of *Odontodactylus havanensis* (Stomatopoda). J Crust Biol, 32: 171–179.

Cao, L., E. Kenchington and E. Zouros. 2004. Differential segregation patterns of sperm mitochondria in embryos of the blue mussel (*Mytilus edulis*). Genetics, 166: 883–894.

Carefoot, T. H. 1967a. Growth and nutrition of *Aplysia punctata* feeding on a variety of marine algae. J Mar Biol Ass UK, 47: 565–589.

Carefoot, T. H. 1967b. Growth and nutrition of three species of opisthobranch molluscs. Comp Biochem Physiol, 21: 627–652.

Carefoot, T. H. 1970. A comparison of absorption and utilization of food energy in two species of tropical *Aplysia*. J Exp Mar Biol Ecol, 5: 47–62.

Carefoot, T. H. 1987. *Aplysia*: its biology and ecology. Oceanogr Mar Biol Annu Rev, 25: 167–284.

Carreau, S. and M. Drosdowsky. 1977. The *in vitro* biosynthesis of steroids by the gonad of the cuttlefish (*Sepia officinalis*). Gen Comp Endocrinol, 33: 554–565.

Carriker, M. R. 1981. Shell penetration and feeding by naticacean and muricacean predatory gastropods: a synthesis. Malacologia, 20: 403–422.

Carriker, M. R. and L. G. Williams. 1978. The chemical mechanisms of shell dissolution by predatory boring gastropods: a review and an hypothesis. Malacologia, 17: 88–111.

Carriker, M. R., P. Person, R. Libbin and D. V. Zandt. 1972. Regeneration of proboscis of muricid gastropods after amputation with emphasis on the radula and cartilages. Biol Bull, 143: 317–333.

Carroll, D. J. and S. C. Kemff. 1990. Laboratory culture of the aeolid nudibranch *Berghis veruciocornis* (Mollusca, Opisthobranchia): some aspects of its development and life history. Biol Bull, 179: 243–253.

Cary, S. G. and S. J. Giovannoni. 1993. Transovarial inheritance of endosymbiotic bacteria in clams inhabiting deep-sea hydrothermal vents and cold seeps. Proc Natl Acad Sci USA, 90: 5695–5699.

Castro, L. F. C., C. Melo, R. Guillot et al. 2007. The estrogen receptor of the gastropod *Nucella lapillus*. Modulation following exposure to an estrogene effluent? Aquat Toxicol, 84: 465–468.

Castro, N. F. and A. Lucas. 1987. Variablity of the frequency of male neotony in *Ostrea puelchana* (Mollusca: Bivalvia). Mar Biol, 96: 359–365.

Cauthier-Clare, S., J. Pellerim and J. C. Amiard. 2006. Estradiol-17β and testosterone concentrations in male and female *Mya arenaria* (Mollusca: Bivalvia) during the reproductive cycle. Gen Comp Endocrinol, 145: 133–139.

Central Marine Fisheries Research Institute (CMFRI). 2012. Green mussel extract (GMe): First nutraceutical produced by an ICAR Institute. CMFRI Newsletter, 135: 5–6.

Central Marine Fisheries Research Institute (CMFRI). 2013. *Handbook of Marine Prawns of India.* Central Marine Fisheries Research Institute, Kochi. p 414.

Chaffee, C. and R. R. Strathmann. 1984. Constraints on egg masses. I Retarded development within thick egg masses. J Exp Mar Biol Ecol, 84: 73–83.

Chaparro, O. R. and K. A. Paschke. 1990. Nurse eggs feeding and energy balance in embryos of *Crepidula dilatata* (Gatropoda: Calyptraeidae) during development. Mar Ecol Prog Ser, 65: 182–191.

Chaparro, O. R., R. F. Oyarzun, A. M. Vergara and R. J. Thompson. 1999. Energy investment in nurse eggs and egg capsules in *Crepidula dilatata* Lamarck (Gastropoda, Calyptraeidae) and its influence on the hatching size of the juvenile. J Exp Mar Biol Ecol, 232: 261–274.

Chapman, A. D. 2009. Number of living species in Australia and the world. Report for the Australian Biological Study. Department of Environment, Government of Australia. p 80.

Charnov, E. L. 1982. *The Theory of Sex Allocation.* Princeton University Press, Princeton.

Charrier, G., A. K. Ring, E. Johansson et al. 2013. Characterization of new EST-linked microsatellites in the rough periwinkle (*Littorina saxatilis*) and application for parentage analysis. J Moll Stud, 79: 369–371.

Chase, R. 2000. Structure and function in the cerebral ganglion. Microsc Res Tech, 49: 511–520.

Chávez-Villalba, J., C. Soyez, A. Huvet et al. 2011. Determination of gender in the pearl oyster *Pinctada margaritifera*. J Shellfish Res, 30: 231–240.

Checa, A. G., F. J. Esteban-Delgado and A. B. Rodríguez-Navarro. 2007. Crystallographic structure of the foliated calcite of bivalves. J Struct Biol, 157: 393–402.

Chellam, A., T. S. Velayudhan, S. Dharmaraj et al. 1983. A note on the predation of pearl oyster *Ptnctada fucata* (Gould) by some gastropods. Indian J Fish, 30: 337–339.

Chétail, M. 1963. Etude de la regeneration du tentacule oculaire chez un Arionidae (*Arion rufus* L.) et un Limacidae (*Agriolimax agrestis* L.). Arch Anat Microsc Morphol Exp, 52: 129–203.

Chi, L. and E. D. Wagner. 1957. Studies on reproduction and growth of *Oncomelania quadrasi*, *O. nosophora* and *O. formosana*, snail host of *Schistosoma japonicum*. Am J Trop Med Hyg, 6: 949–960.

Chia, Fu-S., G. Gibson and P. Y. Qian. 1996. Poecilogony as a reproductive strategy of marine invertebrates. Oceanolog Acta, 19: 203–208.

Chivers, D. P., M. I. McCormick, G. E. Nilsson et al. 2013. Impaired learning of predators and lower prey survival under elevated CO2: a consequence of neurotransmitter interference. Glob Change Biol, 20: 515–522.

Choquet, M. and J. Lemaire. 1969. Contribution a l'etude de la regeneration tentaculaire chez *Patella vulgata* L. (Gasteropode Prosobranche). Arch Zool Exp Gen, 109: 319–337.

Christman, S. P., E. L. Mihalick and F. G. Thompson. 1996. *Tulotoma magnifica* (Conrad, 1834) (Gastropoda, Viviparidae) population status and biology in Coosa River, Alabama. Malacol Rev, 29: 17–63.

Chung, P. R. 1984. A comparative study of three species of Bithyniidae (Mollusca: Prosobranchia) *Parafossarulus manchoricus*, *Gabbia misella* and *Bithynia tentaculata*. Malacol Rev, 17: 1–66.

Clark, K. B. and K. R. Jensen. 1981. A comparison of egg size, capsule size and development patterns in the order Ascoglossa (Sacoglossa) (Mollusca: Opisthobranchia). Int J Invert Reprod, 3: 57–64.

Clark, K. B., M. Busacca and H. Stirts. 1979. Nutritional aspects of development of sacoglossan *Elysia cauze*. In: *Reproductive Ecology of Marine Invertebrates*. (ed) S. Stancyk. University of South Carolina Press, Columbia. pp 11–24.

Clarke, M. R. 1980. Cephalopods in the diet of sperm whales of the southern hemisphere and their bearing on sperm whale biology. Discover Rep, 37: 1–324.

Clemens-Seely, K. and N. E. Phillips. 2011. Effects of temperature on hatching time and hatchling proportions in a poecilogonous population of *Haminoea zelandiae*. Biol Bull, 221: 189–196.

Clewing, C., P. V. Von Oheimb, M. Vinarski et al. 2014. Freshwater mollusc diversity at the roof of the world: phylogenetic and biogeographical affinities of Tibetan Plateau *Valvata*. J Moll Stud, 81: 452–455.

Coe, W. R. 1941. Sexual phase in wood-boring mollusks. Biol Bull, 81: 168–178.

Cogswell, A. T., E. L. R. Kenchington and E. Zouros. 2006. Segregation of sperm mitochondria in two-and four-cell embryos of the blue mussel *Mytilus edulis*: implications for the mechanisms of doubly uniparental inheritance of mitochondria DNA. Genome, 49: 799–807.

Collin, R. 1995. Sex, size and position: a test of models predicting size at sex change in the protandric gastropod *Crepidula fornicata*. Am Nat, 146: 815–831.

Collin, R. 2003. Worldwide patterns in mode of development in calyptraeid gastropods. Mar Ecol Prog Ser, 247: 103–122.

Collin, R. 2006. Sex ratio, life-history invariants, and patterns of sex change in a family of protandrous gastropods. Evolution, 60: 735–745.

Collin, R. 2012. Non-traditional life history sources: what can "intermediates" tell us about evolutionary transitions between modes of invertebrate development. Integ Comp Biol, 52: 128–137.

Collin, R. and M. Z. Salazar. 2010. Temperature-mediated plasticity and genetic differentiation in egg size and hatching size among populations of *Crepidula* (Calyptraeidae: Gastropoda). Biol J Linn Soc, 99: 489–499.

Collin, R., M. McLellan, K. Gruber and C. Bailey-Jourdain. 2005. Effects of conspecific associations on size at sex change in three species of calyptraeid gastropods. Mar Ecol Prog Ser, 293: 89–97.

Collins, M. A., C. Yau, L. Allcock and M. H. Thurston. 2001. Distribution of deep-water benthic and bentho–pelagic cephalopods from the north-east Atlantic. J Mar Biol Ass UK, 81: 105–117.

Constello, D. P. and C. Henley. 1971. *Methods of Obtaining and Handling Marine Eggs and Embryos.* Mar Biol Lab Woods Hole, Mass. p 247.

Coolie, J. S., G. A. Escanero and P. C. Valentine. 1997. Effects of bottom fishing on the benthic megafauna of Georges Bank. Mar Ecol Prog Ser, 155: 159–172.

Cortie, M. B. 1989. Models for mollusc shell shape. South Afri J Sci, 85: 454–460.

Coustau, C., J. Robbins, B. Delay et al. 1993. The parasitic castration in the mussel *Mytilus edulis* by the trematode parasite *Prosorhynchus squamatus*: specifically and partially characterization of endogenous and parasite-induced anti-mitotic activities. Comp Biochem Physiol, 104A: 229–233.

Covich, A. P., T. A. Crowl, J. E. Alexander, Jr. and C. C. Vaugnan. 1994. Predator avoidance responses in freshwater decapods-gastropod interactions mediated by chemical stimuli. J N Am Benthol Soc, 13: 28–290.

Crim, R. N., J. M. Sunday and C. D. G. Harley. 2011. Elevated seawater CO_2 concentrations impair larval development and reduce larval survival in endangered northern abalone (*Haliotis kamtschatkana*). J Exp Mar Biol Ecol, 400: 272–277.

Crisp, D. J. and A. Standen. 1988. *Lasaea rubra* (Montagu) (Bivalvia: Erycinacea), an apomictic crevice-living bivalve with clones separated by the tidal level preference. J Exp Mar Ecol, 117: 27–45.

Crisp, D. J., A. Burfutt, K. Rodriquez and M. D. Budd. 1983. *Lasaea rubra*: an apomictic bivalve. Mar Biol Lett, 4: 127–136.

Croll, R. P. and C. Wang. 2007. Possible roles of sex steroids in the control of reproduction in bivalve molluscs. Aquaculture, 272: 76–86.

Cummins, S. F., A. E. Nichols, A. Amare et al. 2004. Characterization of *Aplysia* enticin and temptin, two novel water-borne protein pheromones that act in concert with attractin to stimulate mate attraction. J Biol Chem, 279: 25614–25622.

Cunha, R. L., R. Castilho, L. Ruber and R. Zardoya. 2005. Patterns of cladogenesis in the venomous marine gastropod genus *Conus* from the Cape Verde Islands. System Biol, 54: 634–650.

Curtis, L. A. 1994. A decade long perspective on a bioindicator of pollutant imposex in *Ilyanassa obsoleta* on Cape Henlopen, Declaware Bay. Mar Environ Res, 38: 291–302.

Dadon, J. R. and L. L. de Cidre. 1992. The reproductive cycle of the thecosomatous pteropod *Limacina retroversa* in the western South Atlantic. Mar Biol, 114: 439–442.

Dalet, J. T., C. P. Saloma, B. M. Olivera and F. M. Heralde. 2015. Karyological analysis and FISH physical mapping of 18S rDNA genes, (GATA)$_n$ centromeric and (TTAGGG)$_n$ telomeric sequences in *Conus magus* Linnaeus, 1758. J Moll Stud, 81: 274–289.

Dalziel, B. and E. G. Boulding. 2005. Water-borne cues from a shell-crushing predator induce a more massive shell in experimental populations of an intertidal snail. J Exp Mar Biol Ecol, 317: 25–35.

Dame, R. F. 1976. Energy flow in an intertidal oyster population. Estuar Coast Mar Sci, 4: 243–253.

D'Aniello, A., A. Di Cosmo, C. Di Cristo et al. 1996. Occurrence of sex steroid hormones and their binding proteins in *Octopus vulgaris* Lam. Biochem Biophys Res Comm, 227: 782–788.

Danton, E., M. Kiyomoto, A. Komaru et al. 1996. Comparative analysis of storage tissue and insulin-like neurosecretion in diploid and triploid mussel *Mytilus galloprovincialis* Link in relation to their gametogenic cycle. Invert Reprod Dev, 29: 37–46.

Darwin, C. 1859. *On the Origin of Species by Means of Natural Selection.* Murray, London. p 247.

Davies, M. S., S. J. Hawkins and H. D. Jones. 1990. Mucus production and physiological energetics in *Patella vulgata* L. J Moll Stud, 56: 499–503.

Davison, A., H. T. Frend, C. Murray et al. 2009. Mating behavior in *Lymnaea stagnalis* is a maternally lateralized trait. Biol Lett, 5: 20–22.

Dazo, B. C., N. G. Haiston and I. K. Dawood. 1996. The ecology of *Bulinus truncatus* and *Biomphalaria alexandrina* in the Egypt 49 project area. Bull Wld Hlth Org, 35: 339–356.

De Goeij, P., P. C. Luthikhuizen, J. van der Meer and T. Piersma. 2001. Facilitation on an intertidal mudflat: the effect of siphon nipping by flatfish on burying depth of the bivalve *Macoma balthica*. Oecologia, 126: 500–506.

De Jong-Brink, M. 1973. The effects of dedication and starvation up on the weight, histology and ultrastructure of the reproductive tract *Biomphalaria glabrata*, an intermediate host of *Schistostoma mansoni*, Z. Zellforsch, 136: 229–262.

De Jong-Brink, M., L. P. C. Scot, H. J. N. Schoemarkers and M. J. M. Bergmin-Sassen. 1981. A biochemical and quantitative electron microscopic study on steroidogenesis in ovotestis and digestive gland of the pulmonate snail *Lymnaea stagnalis*. Gen Comp Endocrinol, 45: 33–38.

De Jong-Brink, M., M. Bergganin-Sussen and M. Solis Soto. 2001. Multiple strategies of schistosomes to meet their requirements in the intermediate snail host. Parasitology, 123: S129–S141.

de Lima, F. D., T. S. Leite, M. Haimoviei and J. E. Oliveira. 2014. Gonadal development and reproductive strategies of the tropical octopus (*Octopus insularis*) in northeast Brazil. Hydrobiologia, 725: 7–21.

de Koch, K. N., C. T. Wolmarans and M. Bornmann. 2004. Distribution and habitats of *Biomphalaria pfeifferi* snail intermediate host of *Schistosoma mansoni* in South Africa. Water South Africa, 30: 29–36.

De Zwaan, A. and R. H. M. Eertman. 1996. Anoxic or aerial survival of bivalves and other euryoxic invertebrates as a useful response to environmental stress-A comprehensive review. Environ Pollut, 113: 299–312.

De Zwaan, A., P. Cortesi, G. van den Thillart et al. 1991. Differential sensitivities to hypoxia by two anoxia-resistant marine molluscs. A biochemical analysis. Mar Biol, 111: 343–351.

Delay, B. 1992. Aphally versus euphally in self-fertile hermaphrodite snails from the species of *Bulinus truncatus* (Gastropoda: Planorbidae). Am Nat, 139: 424–434.

Delgado, G. A., R. A. Glazer and D. Wetzel. 2013. Effects of mosquito control pesticides on competent Queen Conch (*Strombus gigas*) larvae. Biol Bull, 225: 79–84.

Denman, K., J. R. Christian, N. Steiner et al. 2011. Potential impacts of future ocean acidification on marine ecosystems and fisheries: current knowledge and recommendations for future research. ICES J Mar Sci, doi: 10.1093/icesjms/fsr074.

Denny, M. 1980. Locomotion: the cost of gastropod crawling. Science, 208: 1288–1290.

Derbali, A., O. Jarboui, M. Ghorbel and K. Dhieb. 2009. Reproductive biology of the pearl oyster *Pinctada radiata* (Mollusca: Pteriidae) in northern Kerkennah Island (Gulf of Gabes). Cah Biol Mar, 50: 215–222.

Derby, C. D. 2007. Escape by inking and secreting: marine molluscs avoid predators through a rich array of chemicals and mechanisms. Biol Bull, 213: 274–289.

Deroux, G. 1960. Formation, reguliere de males murs de taile et d'organization larviaire chez Eulamellibranche commensal (*Montacuta phascolionis Dautz*). Hebd Seanc Acad Sci, Paris, 250: 2264–2266.

Desrosiers, R. R., J. De´silets and F. Dube. 1996. Early developmental events following fertilization in the giant scallop *Placopecten magellanicus*. Can J Fish Aquat Sci, 53: 1382–1392.

Dewitt, T. J., B. W. Robinson and S. Witson. 2000. Functional diversity among predators of a freshwater snail composes and adaptive trade off for snail morphology. Evol Ecol Res, 2: 129–148.

Dharmaraj, S., A. Chellam and T. S. Velayudhan. 1987. Biofouling, boring and predation of pearl oysters. In: *Pearl Culture*. (ed) K. Alagarswami. Bull Cent Mar Fish Res Inst, 39: 92–99.

Di Benedetto, P. 2005. Investigations in imposex development in marine prosobranchs. Diploma Thesis, Johann Wolfgang Goethe University, Frankfurt.

Di Cosmo, A. and C. Di Cristo. 1998. Neuropeptilergic control of the optic gland of *Octopus vulgaris*. FMRF-Amide and GnRH immunoreactivity. J Comp Neurol, 398: 1–12.

Di Cosmo, A., C. Di Cristo and M. Paolucci. 2001. Sex steroid hormone fluctuations and morphological changes of the reproductive system of the female of *Octopus vulgaris* throughout the annual cycle. J Exp Zool, 289: 33–47.

Di Cosmo, A. N., C. Di Cristo and M. Paolucci. 2002. A estradiol-17β receptor in the reproductive system of the female of *Octopus vulgaris*: characterization and immunolocalization. Mol Reprod Dev, 61: 367–375.

Di Cosmo, A., M. Paolucci, C. Di Cristo et al. 1998. Progesterone receptor in the reproductive system of the female of *Octopus vulgaris*: characterization and immunolocalization. Mol Reprod Dev, 50: 451–460.

Di Cristo, C. 2013. Nervous control of reproduction in *Octopus vulgaris*: a new model. Invert Neurosci, 13: 27–34.

Di Cristo, C., M. Paolucci and A. Di Cosmo. 2008. Progesterone affects vitellogenesis in *Octopus vulgaris*. Open Zool J, 1: 29–36.

Dickson, G. H., O. B. Matoo, R. T. Tourek et al. 2013. Environmental salinity modulates the effects of elevated CO_2. J Exp Biol, 216: 2607–2618.

Dillon, R. T. 2000. *The Ecology of Freshwater Molluscs*. Cambridge University Press.

Dillon, R. T. Jr. and S. J. Jacquemin. 2015. The heritability of shell morphometrics in the freshwater pulmonate gastropod *Physa*. PLoS ONE, doi: 10.1371/journal.pone.0121962.

DiMatteo, T. 1981. The inking behavior of *Aplysia dactylomela* (Gastropoda: Opisthobranchia): evidence for distastefulness. Mar Behav Physiol, 7: 285–290.

Diotel, N., J. L. Do Rego, I. Anglade et al. 2011. Activity and expression of steroidogenic enzymes in the brain of adult zebrafish. Eur J Neurosci, doi: 10.1111/J.1460-9568.2011.07731X.

Distel, D. L. and C. M. Cavanaugh. 1994. Independent phylogenic origins of methanotrophic and chemoautotrophic bacterial endosymbioses in marine bivalves. J Bacteriol, 176: 1932–1938.

Dodson, S. I. 1984. Predation of *Heterocope septentrionalis* on two species of *Daphnia*: morphological defenses and their costs. Ecology, 65: 1249–1257.

Doney, S. C., V. J. Fabry, R. A. Feely and J. A. Kleypas. 2009. Ocean acidification: the other CO_2 problem. Mar Sci, 1: 169–192.

Dong, Q., C. Huang, B. Eudeline and T. R. Tiersch. 2005. Systematic factor optimization of shipped sperm samples of diploid Pacific oyster *Crassostrea gigas*. Cryobiology, 51: 176–197.

Dong, Q., C. Huang, B. Eudeline et al. 2006. Systematic factor optimization for sperm cryopreservation of the tetraploid Pacific oyster *Crassostrea gigas*. Theriogenology, 66: 387–403.

Dorsch, R., D. S. Wagner, K. A. Mintzer et al. 2004. Maternal control of vertebrate development before the midblastula transition. Mutants from zebrafish. Dev Cell, 6: 771–780.

Downing, J. A., J. -P. Amyot, M. Perusse and Y. Rochon. 1989. Visceral sex, hermaphroditism and protandry in a population of the freshwater bivalve *Elliptio complanata*. North Am Benthol Soc, 8: 92–99.

Duft, M., U. Schulte-Oehlmann, M. Tillmann and J. Oehlmann. 2003. Stimulated embryo production as a parameter of estrogenic exposure via sediments in the freshwater mudsnail. *Potamopyrgus antipodarum*. Aquat Toxicol, 64: 437–439.

Duft, M., C. Schmitt, J. Bachmann et al. 2006. Prosobranch snails as test organisms for the assessment of endocrine active chemicals—an overview and a guideline proposal for a reproduction test with the freshwater mudsnail *Potamopyrgus antipodarum*. Ecotoxicology, doi: 10.1007/s10646-006-0106-0.

Duperron, S., C. Bergin, F. Zielinski et al. 2006. A dual symbiosis shared by two mussel species, *Bathymodiolus azoricus* and *Bathymodiolus puteoserpentis* (Bivalvia: Mytilidae), from hydrothermal vents along the northern Mid-Atlantic Ridge. Environ Microbiol, 8: 1441–1447.

Duperron, S., S. Halary, M. Lorion et al. 2008. Unexpected co-occurrence of 6 bacterial symbionts in the gill of the cold seep mussel *Idas* sp (Bivalvia: Mytilidae). Environ Microbiol, 10: 433–445.

Dupont, L., Y. -M. Richard, M. Paulet et al. 2006. Gregariouness and protandry promote reproductive insurance in the invasive gastropod *Crepidula fornicata* evidence: from assignment of larval paternity. Mol Ecol, 15: 3009–3021.

Duvaux-Miret, O., G. B. Stefano, E. M. Smith et al. 1992. Immunosuppression in the definitive and intermediate hosts of the human parasitic *Schistosoma mansoni* by release of immunoactive neuropeptides. Proc Natl Acad Sci USA, 89: 778–781.

Dybdahl, M. F. and C. M. Lively. 1995. Diverse, endemic and polyphyletic clones in mixed populations of freshwater snails *Potamopyrgus antipodarum*. J Evol Biol, 8: 385–398.

Eckelbarger, K. J. and C. M. Young. 1999. Ultrastructure of gametogenesis in a chemosynthetic mytilid bivalve (*Bathymodiolus childressi*) from a bathyal, methane seep environment (northern Gulf of Mexico). Mar Biol, 135: 635–646.

Ehara, T., S. Kitajima, N. Kanzawa et al. 2002. Antimicrobial action of achacin is mediated by L-amino acid activity. FEBS Lett, 531: 509–512.

Ellingson, R. A. and P. J. Krug. 2006. Evolution of poecilogony from planktotrophy: cryptic speciation, phylogeography, and larval development in the gastropod genus *Alderia*. Evolution, 60: 2293–2310.

Emery, A. M., P. W. Shaw, E. C. Greatorex et al. 2000. New microsatellite markers for assessment of paternity in the squid *Loligi forbesi* (Mollusca: Cephalopoda). Mol Ecol, 9: 107–118.

Endow, K. and S. Ohta. 1990. Occurrence of bacteria in the primary oocytes of vesicomyid clam *Calyptogena soyoae*. Mar Ecol Prog Ser, 64: 309–311.

Equardo, J. and A. R. Marian. 2012. A model to explain spermatophore implantation in cephalopods (Mollusca: Cephalopoda) and a discussion on its evolutionary origins and significance. Biol J Linn Soc, 105: 711–726.

Esch, G. W., L. A. Curtis and M. A. Berger. 2001. A perspective on the ecology of trematode communities in snails. Parasitology, 123: S57–S75.

Estebenet, A. L. and N. J. Cazzaniga. 1992. Growth and demography of *Pomacea canaliculata* (Gastropoda: Ampullariidae) under laboratory conditions. Malacol Rev, 25: 1–12.

Etter, R. J., M. A. Rex, M. C. Chase and J. M. Quattro. 1999. A genetic dimension to deep-sea biodiversity. Deep Sea Res I, 46: 1095–1099.

Fabioux, C., A. Huvet, C. Lelong et al. 2004. Oyster *vasa*-like gene as a marker of the germ line cell development in *Crassostrea gigas*. Biochem Biophys Res Comm, 320: 592–598.

Fabry, V. J., B. A. Seibel, R. A. Feely and J. C. Orr. 2008. Impacts of ocean acidification on marine fauna and ecosystem processes. ICES J Mar Sci, 65: 414–432.

Fankhauser, G. 1945. The effects of changes in chromosome number on amphibian development. Q Rev Biol, 20: 20–78.

FAO. 2014. The State of World Fisheries and Aquaculture: Opportunities and Challenges. Food and Agriculture Organisation, Rome. p 221.

Feely, R. A. and S. C. Doney. 2011. Ocean acidification: the other CO_2 problem. ASLO Web Lectures, 3: 1–59.

Feely, R. A., R. H. Byrne, J. G. Acker et al. 1988. Winter-summer variations of calcite and aragonite saturation in the northeast Pacific. Mar Chem, 25: 227–241.

Feely, R.A., C.L. Sabine, J.M. Hernandez-Ayon et al. 2004. Impact of anthropogenic CO_2 on the $CaCO_3$ system in the oceans. Science, 305: 362–366

Feely, R. A., C. L. Sabine, J. M. Hernandez-Ayon et al. 2008. Evidence for upwelling of corrosive acidified water onto the continental shelf. Science, 320: 1490–1492.

Feely, R. A., J. Orr, V. J. Fabry et al. 2009. Present and future changes in seawater chemistry due to ocean acidification. In: *Carbon Sequestration and its Role in the Global Carbon Cycle*. (eds) B. J. McPherson and E. T. Sundquist. Geophysical Monograph Series, American Geophysical Union, Washington, DC. 183: 175–188.

Feng, Z. F., Z. F. Zhang, M. Y. Shao and W. Zhu. 2011. Developmental expression pattern of the F_c *vasa*-like gene, gonadogenesis and development of germ cells in Chinese shrimp *Fenneropenaeus chinensis*. Aquaculture, 314: 202–209.

Féral, C. and S. Le Gall. 1982. Induction expérimentale par un polluant marin (le tributylétain), de l'activité neuroendocrine contrôlant la morphogenèse du pénis chez les femelles d'*Ocenebra erinacea* (Mollusque Prosobranche gonochorique). Comp Rend Acad Sci Paris, Série III 295: 627–630.

Feral, J. 1979. La regeneration des bras de la *Sepia officinalis* L. (Cephalopoda: Sepioidea). 2. Etude histologique of cytologique. Cah Biol Mar, 20: 29–42.

Féral, J. P. 1988. Wound healing after arm amputation in *Sepia officinalis* (Cephalopoda: Sepioidea). J Invert Pathol, 52: 380–388.

Finn, J. K. 2009. Systematics and biology of the argonauts or 'paper nautiluses' (Cephalopoda: Argonautidae). Ph.D. Thesis, La Trobe University, Bundoora, Australia.

Finn, J. K. and M. D. Norman. 2010. The *Argonaut* shell: gas-mediated buoyancy control in a pelagic octopus. Proc R Soc B, doi. 10. 1098/rspb. 2010. 0155.

Fioroni, P., J. Oehlmann and E. Stroben. 1991. The pseudohermaphroditism of prosobranchs; morphological aspects. Zool Anz, 226: 1–26.

Flore, G., A. Poli, A. Di Cosmo et al. 2004. Dopamine in the ink defence system of *Sepia officinalis*: biosynthesis, vesicular compartmentation in mature ink gland cells, nitric oxide (NO)/ cGMP-induced depletion and fate in secreted ink. Biochem J, 378: 785–791.

Floren, A. S. 2003. The Philippine shell industry with special focus on *Mactan*, Cebu. Report on Coastal Resource Management Project. p 50.

Forsythe, J. W and W. F. Van Heukelem. 1987. Growth. In: *Cephalopod Life Cycles*. (ed) P. R. Boyle. Academic Press, London. 2: 135–156.

Forsythe, J. W., R. H. DeRusha and R. T. Hanlon. 1984. Growth, reproduction and life span of *Sepia officinalis* (Cephalopoda: Mollusca) cultured through seven consecutive generations. J Zool, 233: 175–192.

Foster-Smith, R. L. 1975. The role of mucus in the mechanism of feeding in three filter-feeding bivalves. J Moll Stud, 41: 571–588.

Frank, P. W. 1969. Growth rates and longevity of some gastropod mollusks on the coral reef at Heron Island. Oecologia, 2: 232–250.

Franklin, H. M., Z. E. Squires and D. Stuart-Fox. 2012. The energetic cost of mating in a promiscuous cephalopod. Biol Lett, 8: 754–756.

Franklin, J. B. and R. P. Rajesh. 2005. A sleep-inducing peptide from the venom of the Indian cone snail *Conus araneosus*. Toxicon, 103: 39–49.

Franzen, A. 1955. Comparative morphological investigations into the spermiogenesis among Mollusca. Zool Bidr Uppsala, 30: 399–456.

Freeman, G. and J. W. Lundelius. 1982. The developmental genetics of dextrality and sinistrality in the gastropod *Lymnaea peregra*. Wilhelm Roux's Arch Dev Boil, 191: 69–83.

Fritts, M. W., A. K. Fritts, S. A. Carleton and R. B. Bringolf. 2013. Shifts in stable-isotope signatures confirm parasitic relationship of freshwater mussel glochidia attached to fish host. J Moll Stud, 79: 163–169.

Frosch, D. 1974. The subpedunculate lobe of the octopus brain: evidence for dual function. Brain Res, 75: 277–285.

Fugino, K., K. Arai, K. Iwadare et al. 1990. Induction of gynogenesis diploid by inhibiting second meiosis in the Pacific abalone. Nippon Suisan Gakkaishi, 56: 1755–1763.

Fukumori, H., S. Y. Chee and Y. Kano. 2013. Drilling predation on neritid egg capsules by the muricid snail *Reishia clavigera*. J Moll Stud, 79: 139–146.

Gadgil, M. 1972. Male dimorphism as a consequence of sexual selection. Am Nat, 106: 574–580.

Gaffney, P. M. and B. McGee. 1992. Multiple paternity in *Crepidula fornicata* (Linnaeus). Veliger, 35: 12–15.

Gagne, F., C. Blevise, C. Salazar et al. 2001. Evaluation of estrogenic effects of municipal effluents to the freshwater mussel. *Elliptio complanata*. Comp Biochem Physiol, 128C: 213–225.

Gagne, F., C. Blaise, E. Pelleties et al. 2003. Sex alternation in softshell clam (*Mya arenaria*) in an intertidal gonad of the Saint Lawrence River (Qubec, Canada). Comp Biochem Physiol, 134B: 189–198.

Galante-Oliveira, S., R. Marcal, M. Pacheco and C. M. Barrroso. 2012. *Nucella lapillus* ecotypes at the southern distributional limit in Europe: variation in shell morphology is not correlated with chromosome counts on the Portuguese Atlantic coast. J Moll Stud, 78: 147–150.

Gallager, S. M., R. D. Turner and C. J. Berg Jr. 1981. Physiological aspects of wood consumption, growth and reproduction in the shipworm *Lyrodus pedicullatus* Quatrefages (Bivalvia: Teredinidae). J Exp Mar Biol Ecol, 52: 63–77.

Gallardo, C. S. 1979. Developmental pattern and adaptations for reproduction in *Nucella crassilabrum* and other muricacean gastropods. Biol Bull, 157: 453–463.

Gallardo, C. S. and O. Garrido. 1987. Nutritive egg formation in the marine snails *Crepidula dilatata* and *Nucella crassilabrum*. Int J Invert Reprod, 11: 239–254.

Galtsoff, P. S. 1964. The American oyster *Crassostrea virginica* Gmelin. Fish Bull, Fish Wildlife Ser, 64: 1–480.

Gamarra-Luques, C., M. Giraud-Billoud and A. Castro-Vazquez. 2013. Reproductive organogenesis in the apple snail *Pomacea canaliculata* (Lamarck, 1822), with reference to the effects of xenobiotics. J Moll Stud, 79: 147–162.

Gannefors, C., M. Boer, G. Kattner et al. 2005. The Arctic sea butterfly *Limacina helicina*, lipids and life strategy. Mar Biol, 147: 169–175.

Gaudron, S. M., E. Demoyencourt and S. Duperron. 2012. Reproductive traits of the cold-seep symbiotic mussel *Idas modiolaeformis*: gametogenesis and larval biology. Biol Bull, 222: 6–16.

Gaylord, B., T. M. Hill, E. Sanford et al. 2011. Functional impacts of ocean acidification in an ecologically critical foundation species. J Exp Biol, 214: 2586–2594.

Gazeau, F., L. M. Parker and S. Comeau. 2007. Impacts of ocean acidification on marine shelled molluscs. Mar Biol, doi: 10.1007/s00227-013-2219-3.

Ge, J., Q. Li, H. Yu and L. Kong. 2015. Mendelian inheritance of golden shell color in the Pacific oyster *Crassostrea gigas*. Aquaculture, 441: 21–24.

Geiger, D. L. 2006. Marine Gastropoda. *In: The Mollusks, A Guide to their Study, Collection, and Preservation.* (eds) C. F. Sturm, T. A. Pearce and A. Valdes. Universal Publishers, Boca Raton, Florida. 24: 295–312.

Geller, J. B. 1990. Consequences of a morphological defense: growth, repair and reproduction by thin- and thick-shelled morphs of *Nucella emarginata* (Deshayes) (Gastropoda: Prosobranchia). J Exp Mar Biol Ecol, 144: 173–184.

Geyer, K. K., C. M. Rodriguez-Lopez, I. W. Chalmers et al. 2011. Cytosine methylation regulates oviposition in the pathogenic blood fluke *Schistosoma mansoni*. Nature Comm, 2: 424/ doi10.1038/ncomms1433.

Ghiselin, M. T. 1969. The evolution of hermaphroditism among animals. Q Rev Biol, 44: 189–208.

Ghiselli, F., L. Milani, P. L. Chang et al. 2012. De novo assembly of the Manila clam *Ruditapes phillippinarum* transcriptome provides new insights into expression bias, mitochondrial doubly uniparental inheritance and sex determination. Mol Biol Evol, 29: 771–786.

Ghode, G. S. and V. Kripa. 2001. *Polydora* infestation on *Crassostrea madrasensis*: a study on the infestation rate and eradication methods. J Mar Biol Ass India, 43: 110–119.

Giard, A. 1905. La Poecilogonie. Comp Rend Siances Six Cong Int Zool, Berne. pp 617–646.

Gibbon, M. C. and M. Castagna. 1984. Serotonin as an inducer of spawning in six bivalve species. Aquaculture, 40: 189–191.

Gibbs, P. E., M. J. Bebianno and M. R. Coelho. 1997. Evidence of the differential sensitivity of neogastropods to tributyltin (TBT) pollution, with notes on a species (*Columbella rustica*) lacking the imposex response. Environ Technol, 18: 1219–1224.

Gibson, G. D. and F. S. Chia. 1989. Development variability (pelagic and benthic) in *Haminoea callidegenita* (Opisthobranchia: Cephalapsidea) is influenced by egg mass jelly. Biol Bull, 176: 103–110.

Gibson, G. D. and F. S. Chia. 1991. Contrasting reproductive modes in two sympatric species of *Haminoea* (Opisthobranchia: Cephalaspidea). J Moll Stud, 57(Supplement Part 4): 49–60.

Gibson, G. D. and F. S. Chia. 1994. A metamorphic inducer in the opisthobarnch *Haminoea callidegenita*: partial purification and biological activity. Biol Bull, 187: 133–142.

Gibson, G. D. and F. S. Chia. 1995. Development variability in the poecilogonous opisthobranch *Haminoea callidegenita*: life-history traits and effects of environmental parameters. Mar Ecol Prog Ser, 121: 139–155.

Gilbert, S. F. 1988. *Developmental Biology*. Sinauer Associates, Sunderland, USA.

Glaubrecht, M. 1999. Systematic and the evolution of viviparity in tropical freshwater gastropods (Cerethioidea: Thiaridae sensu lato)—an overview. Cour Forsch Inst Seneckenberg, 215: 91–96.

Gofas, S. and A. Waren. 1998. Europe's smallest gastropod: habitat, distribution and relationships of *Retrotortina fuscata* (Omalogyridae). Cah Biol Mar, 39: 9–14.

Goffredi, S. K., L. A. Hurtado, S. Hallam and R. C. Vrijenhoek. 2003. Evolutionary relationship of deep-sea vent and cold seep clams (Mollusca: Vesicomyidae) of the "pacifica/lepta" species complex. Mar Biol, 142: 311–320.

Gooding, M. P. and G. A. LeBlanc. 2001. Biotransformation and disposition of testosterone in the eastern mudsnail *Ilyanassa obsoleta*. Gen Comp Endocrinol, 122: 172–80.

Gooding, M. P., V. S. Wilson, L. C. Folmar et al. 2003. The biocide tributyltin reduces the accumulation of testosterone as fatty acid esters in the mudsnail (*Ilyanassa obsoleta*). Environ Health Perspect, 111: 426–430.

Gosselin, L. A. and P. Y. Qian. 1997. Juvenile mortality in benthic marine invertebrates. Mar Ecol Prog Ser, 146: 265–282.

Goto, Y., M. Kajiwara, Y. Yanagisawa et al. 2012. Detection of vertebrate-type steroid hormones and their converting activities in the neogastropod *Thais clavigera* (Küster, 1858). J Moll Stud, 78: 197–204.

Gould, H. N. 1952. Studies on sex in the hermaphrodite mollusk *Crepidula plana*. IV. Internal and external factors influencing growth and sex development. J Exp Zool, 119: 93–163.

Gracio, M. A. A. 1988. A comparative laboratory study of *Bulinus globosus* uninfected and infected with *Schistosoma haematobium*. Malacol Rev, 21: 123–127.

Graf, D. L. and K. S. Cummings. 2007. Review of the systematic and global diversity of freshwater mussel species (Bivalvia: Unionoida). J Moll Stud, 73: 291–314.

Grahame, J. 1977. Reproductive effort and r- and K-selection in two species of *Lacuna* (Gastropoda: Prosobranchia). Mar Biol, 40: 212–224.

Grassle, J. F. 1985. Hydrothermal vent animals and distributions and biology. Science, 229: 713–717.

Grassle, J. F. and N. Maciolek. 1992. Deep-Sea species richness: regional and local diversity estimates from quantitative bottom samples. Am Nat, 139: 313–341.

Green, M. A., M. E. Jones, C. L. Boudreau et al. 2004. Dissolution mortality of juvenile on coastal marine deposits. Limnol Oceanogr, 49: 724–734.

Gregory, T. R. 2005. Animal genome size database. (http://www.genomesize.com)

Griffiths, C. L. and R. J. Griffths. 1987. Bivalvia. In: *Animal Energetics*. (eds) T. J. Pandian and F. J. Vernberg. Academic Press, San Diego. 2: 1–88.

Griffiths, R. J. 1977. Reproductive cycles in littoral populations of *Choromytilus meridionalis* (Kr.) and *Aulacomya ater* (Molina) with a quantitative assessment of gamete production in the former. J Exp Mar Biol Ecol, 30: 53–71.

Griffiths, R. J. 1981a. Aerial exposure and energy balance in littoral and sublittoral *Choromytilus meridionalis* (Kr.) (Bivalvia). J Exp Mar Biol Ecol, 52: 231–241.

Griffths, R. J. 1981b. Production and energy flow in relation to age and shore level in the bivalve *Choromytilus meridionalis* (Kr.). Estu Coast Shelf Sci, 13: 477–493.

Griffiths, R. J. and J. A. King. 1979. Energy expended on growth and gonad output in the ribbed mussel *Aulacomya alter*. Mar Biol, 53: 217–222.

Griffiths, R. J. and R. Buffenstein. 1981. Aerial exposure and energy input in the bivalve *Choromytilus meridionalis* (Kr.). J Exp Mar Biol Ecol, 52: 219–229.

Grubich, J. R. 2005. Disparity between the feeding performance and predicted muscle strength in the pharyngeal musculature of black drum *Pogonias cronis* (Sciaenidae Gmellin 1971), Environ Biol Fish, 74: 261–272.

Gruner, H. E. 1993. *Textbook Special Zoology, Vol 1: Invertebrates*. Part 3: Mollusca, Sipunculida, Echgiurida, Annelida, Onychophora, Tardigrada, Pentastomida, Gustav Fischer, Jena.

Guinez, R. 2005. A review on shell thickening in mussels. Rev Biol Mar Oceanogr, 40: 1–6.

Guo, X. and P. M. Gaffney. 1993. Artificial gynogenesis in the Pacific oyster *Crassostrea gigas*. II Allozyme inheritance and early growth. J Hered, 84: 311–315.

Guo, X. and S. K. Allen Jr. 1994a. Sex determination and polyploidy gigantism in the dwarf surf clam *Mulinia lateralis*. Genetics, 138: 1199–1206.

Guo, X. and S. K. Allen Jr. 1994b. Viable tetraploids in the Pacific oyster (*Crassostrea gigas* Thunberg) by inhibiting polar body I in eggs from triploids. Mol Mar Biol Biotechnol, 3: 42–50.

Guo, X., K. Cooper, W. K. Hershberger and K. K. Chew. 1992a. Genetic consequences of blocking polar body I with cytochalasin B in fertilized eggs of Pacific oyster *Crassostrea gigas*: I Ploidy of resultant embryos. Biol Bull, 183: 381–386.

Guo, X., K. Cooper, W. K. Hershberger and K. K. Chew. 1992b. Genetic consequences of blocking polar body I with cytochalasin B in fertilized eggs of Pacific oyster *Crassostrea gigas*: II Segregation of chromosomes. Biol Bull, 183: 387–393.

Guo, X., K. Cooper, W. K. Hershberger and K. K. Chew. 1993. Artificial gynogenesis with ultraviolet-irradiated sperm in the Pacific oyster *Crassostrea gigas*. I. Induction and survival. Aquaculture, 113: 201–214.

Guo, X., G. A. DeBrosse and S. K. Allen Jr. 1996. All-triploid Pacific oyster (*Crassostrea gigas* Thunberg) produced by mating tetraploids and diploids. Aquaculture, 142: 149–161.

Guo, X., D. Hedgecock, W. K. Hershberger et al. 1998. Genetic determinants of protandric sex in the Pacific oyster, *Crassostrea gigas* Thunberg. Evolution, 52: 394–402.

Gusselin, L. A. and P. Y. Quan. 1997. Juvenile mortality in benthic marine invertebrates. Mar Ecol Prog Ser, 146: 265–282.

Gutiérrez, J. L., C. G. Jones, D. L. Strayer and O. O. Iribarne. 2003. Mollusks as ecosystem engineers: the role of shell production in aquatic habitats. Oikos, 101: 79–90.

Haag, R. and J. L. Staton. 2003. Variation in fecundity and other reproductive traits in freshwater mussels. Freshwater Biol, 48: 2118–2130.

Hadfield, M. G. 1989. Latitudinal effects on juvenile size and fecundity in *Petaloconchus*. (Gastropoda). Bull Mar Sci, 45: 369–376.

Hadfield, M. G. and M. Switzer-Dunlap. 1984. Opisthobranchs. In: *The Mollusca: Reproduction*. (eds) A. S. Tompa, N. H. Vendonk and J. A. M. van den Biggelaar. Academic Press, Orlanda. pp 209–350.

Haga, T. and T. Kase. 2013. Progenetic dwarf males in the deep-sea wood-boring genus *Xylophaga* (Bivalvia: Pholadoidea). J Moll Stud,79: 90–94.

Haley, L. E. 1977. Sex determination in the American oyster. J Hered, 68: 114–116.

Haniffa, M. A. 1975. Ecophysiological studies in a chosen gastropod. Ph.D. Thesis. Madurai University, Madurai.

Haniffa, M. A. 1978a. Secondary productivity and energy flow in a tropical pond. Hydrobiologia, 59: 49–65.

Haniffa, M. A. 1978b. Energy loss in an aestivating population of the tropical snail *Pila globosa*. Hydrobiologia, 61: 169–182.

Haniffa, M. A. 1982. Effects of feeding level and body size and food utilization of the freshwater snail *Pila globsa*. Hydrobiologia, 97: 141–149.

Haniffa, M. A. and T. J. Pandian. 1974. Effects of body weight on feeding and radula size in the freshwater snail *Pila globosa*. Veliger, 164: 415–418.

Haniffa, M. A. and T. J. Pandian. 1978. Morphometry, primary productivity and energy flow in a tropical pond. Hydrobiologia, 59: 23–48.

Hanlon, R. T. 1983a. *Octopus briareus*. In: *Cephalopod Life Cycles: Species Accounts*. (ed) P. R. Boyle. Academic Press, London. 1: 251–267.

Hanlon, R. T. 1983b. *Octopus joubini*. In: *Cephalopod Life Cycles: Species Accounts*. (ed) P. R. Boyle. Academic Press, London. 1: 293–310.

Hanlon, R. T. 1987. Mariculture. In: *Cephalopod Life Cycles. Comparative Reviews*. (ed). P. R. Boyle. Academic Press, London. 2: 291–305.

Hanlon, R. T. and J. B. Messenger. 1996. *Cephalopod Behaviour*. Cambridge University Press, Cambridge. p 256.

Haque, Z. 1999. Role of the dopaminergic neuron (RPeD1) in the control of respiratory behaviour in *Lymnaea stagnalis*. MSc Thesis, University of Calgary.

Hardy, J. T. and S. A. Hardy. 1969. Ecology of *Tridacna* in Palau. Pacific Sci, 23: 467–472.

Harris, J. O., G. B. Maguire, S. J. Edwards and S. M. Hindrum. 1999. Effect of pH on growth rate, oxygen consumption rate, and histopathology of gill and kidney tissue for juvenile greenlip abalone, *Haliotis laevigata* Donovan and blacklip abalone, *Haliotis rubra* Leach. J Shellfish Res, 18: 611–619.

Hassan, M. M., J. G. Qin and X. Li. 2015. Sperm cryopreservation in oysters: a review of its current status and potentials for future application in aquaculture. Aquaculture, 438: 24–32.

Hazlett, B. A. 1981. The behavior ecology of hermit crabs. Ann Rev Ecol Syst, 12: 1–22.

He, M., W. Jiang and L. Huang. 2004. Studies on aneuploid pearl oyster (*Pinctada martensii* Dunker) produced by crossing triploid females and a diploid male following the inhibition of PBI. Aquaculture, 203: 117–124.

Heard, W. H. 1975. Sexuality and other aspects of reproduction in *Anodonta* (Pelecypoda: Unionidae). Malacologica, 15: 81–103.

Heard, W. H. 1979. Hermaphroditism in *Elliptio* (Pelecypoda: Unionidae). Malacol Rev, 12: 21–28.

Hedgecock, D., P. Gaffney, P. Goulletquer et al. 2005. The case for sequencing the oyster genome. J Shellfish Res, 24: 429–442.

Heller, J. 1993. Hermaphroditism in molluscs. Biol J Linn Soc, 48: 19–42.

Heller, J. and N. Farstey. 1990. Sexual and parthenogenetic populations of the freshwater snail *Melanoides tuberculata*. Israel J Zool, 37: 75–87.

Hellou, J., P. S. S. Yeats and F. Gagné. 2003. Chemical contaminants and biological indicators of mussel health during gametogenesis. Environ Toxicol Chem, 22: 2080–2087.

Hendricks, J. R. 2008. Sinistral snail shells in the sea: developmental causes and consequences. Lethaia, 42: 55–66.

Hettinger, A., E. Sanford, T. M. Hill et al. 2012. Persistent carry-over effects of planktonic exposure to ocean acidification in the Olympia oyster. Ecology, 93: 2758–2768.

Hibbert, C. J. 1977. Energy relations of the bivalve *Mercenaria mercenaria* on an intertidal mudflat. Mar Biol, 44: 77–84.

Hickmann, C. S. and S. S. Porter. 2007. Nocturnal swimming, aggregation at light traps and mass spawning of scissurellid gastropods (Mollusca: Vetigastropoda). Inverteb Biol, 126: 10–17.

Himes, J. E., J. A. Riffell, C. A. Zimmer and R. K. Zimmer. 2011. Sperm chemotoxis as revealed with live and synthetic eggs. Biol Bull, 220: 1–5.

Hiramatsu, N., T. Matsubara, T. Fujita et al. 2006. Multiple piscine vitellogenins: biomarkers of fish exposure to estrogenic endocrine disruptors in aquatic environments. Mar Biol, 149: 35–47.

Hirohashi, N. and Y. Iwata. 2013. The different types of sperm morphology and behavior within a single species. Why do sperm of squid sneaker males for a cluster? Comp Integ Biol, 7: doi.org/10.4161/c1b 26729.

Hirohashi, N., L. Alvarez, K. Shiba et al. 2013. Sperm of sneaker male squids exhibit chemotactic swarming to CO_2. Cur Biol, 23: 775–781.

Hixon, R. F. 1983. *Loligo opalescens*. In: *Cephalopod Life Cycles: Species Accounts*. (ed) P. R. Boyle. Academic Press, London. 1: 95–114.

Hoagland, K. E. 1978. Protandry and evolution of environmentally-mediated sex change-study of mollusca. Malacologia, 17: 365–391.

Hoagland, K. E. 1986. Patterns of encapsulation and brooding in the Calyptraeidae. Am Malacol Bull, 4: 173–183.

Hoagland, K. E. and R. Robertson. 1988. An assessment of poecilogony in marine invertebrates: phenomenon or fantasy. Biol Bull, 174: 109–125.

Hodgson, A. N. 1982. Studies on wound healing and an estimate of the rate of regeneration of the siphon of *Scrobicularia plana* (da Coasta). J Exp Mar Biol Ecol, 62: 117–128.

Hodgson, A. N. and J. Heller. 2000. Spermatozoan structure and spermiogenesis in four species of *Melanopsis* (Gastropoda, Prosobranchia, Cerethioidea) from Israel. Inverteb Reprod Dev, 37: 185–200.

Hoeh, W. R., D. T. Stewart, C. Saavedra et al. 1997. Phylogenetic evidence for role-reversals of gender-associated mitochondrial DNA in *Mytilus* (Bivalvia: Mytilidae). Mol Biol Evol, 14: 959–967.

Hoffmann, K. F. and D. W. Dunne. 2003. Characterization of the *Schistosoma* transcriptome opens up the world of helminth genomics. Genom Biol, 5: 203–203.

Hogg, N. A. S. and J. Wijdenes. 1979. A study of gonadal organogenesis and the factors influencing regeneration following surigical castration in *Deroceras reticulatum* (Pulmonata: Limacidae). Cell Tiss Res, 198: 295–307.

Horiguchi, T., Z. Li, S. Uno et al. 2003. Contamination of organotin compounds and imposex in molluscs from Vancouver, Canada. Mar Environ Res, 57: 75–88.

Horiguchi, T., M. Kojima, N. Takiguchi et al. 2005. Continuing observation of disturbed reproductive cycle and ovarian spermatogenesis in the giant abalone, *Haliotis madaka* from an organotin-contaminated site of Japan. Mar Pollut Bull, 51: 817–822.

Horvath, A., A. Bubalo, A. Cucevic et al. 2012. Cryopreservation of sperm and larvae of the European flat oyster (*Ostrea edulis*). J Appl Ichthyol, 28: 948–951.

Horvath, T. G. and G. Lamberti. 1997. Drifting macrophytes as a mechanism for zebra mussel (*Dreissena polymorpha*) invasion of lake outlet streams. Am Midl Nat, 138: 29–36.

Hoving, H. J. T. and V. Laptikhovsky. 2007. Getting under the skin: autonomous implantation of squid spermatophores. Biol Bull, 212: 177–179.

Hoving, H. J. T., V. Laptikhovsky, U. Piatkowski and B. Onsoy. 2008. Reproduction in *Heteroteuthis dispar* (Ruppell, 1844) (Mollusca: Cephalopoda): a sepiolid reproductive adaptation to an oceanic lifestyle. Mar Biol, 154: 219–230.

Huelsken, T. 2011. First evidence of drilling predation by *Comber sordidus* (Swaimson, 1821) (Gastropoda: Naticidae) on soldier crabs (Crustacea : Mictyridae). Moll Res, 31: 125–132.

Hughes, R. N. 1970. An energy budget for a tidal-flat population of the bivalve *Scrobicularia plana* (Da Costa). J Anim Ecol, 39: 357–381.

Hughes, R. N. 1971a. Ecological energetics of *Nerita* (Archaeogastropoda, Neritacea) populations on Barbados, West Indies. Mar Biol, 11: 12–22.

Hughes, R. N. 1971b. Ecological energetics of the keyhole limpet *Fissurella barbadensis* Gmelin. J Exp Mar Biol Ecol, 6: 167–178.

Hughes, R. N. and D. J. Roberts. 1980. Reproductive effort of winkles (*Littorina* spp.) with contrasted methods of reproduction. Oecolgia, 47: 130–136.

Hunter, R. D. 1975. Growth, fecundity, and bioenergetics in three populations of *Lymnaea palustris* in upstate New York. Ecology, 56: 50–63.

Hyman, L. H. 1967. *The Invertebrates Mollusca* 1. McGraw Hill, New York. p 702.

Idler, D. R. and P. Wiseman. 1972. Molluscan sterols: a review. J Fish Res Board Can, 29: 385–398.

Iguchi, T., Y. Katsu, T. Horiguchi et al. 2007. Endocrine disrupting organotin compounds are potent inducers of imposex in gastropods and adipogenesis in vertebrates. Mol Cell Toxicol, 3: 1–10.

Iijima, R., J. Kisugi and M. Yamazaki. 1994. Biopolymers from marine invertebrates. XIV. Antifungal properties Dolobellanin A, a putative self defense molecule of sea hare *Dolabella auricularia*. Biol Pharm Bull, 17: 1144–1146.

Iijima, R., J. Kisugi and M. Yamazaki. 1995. Antifungal activity of aplysianin E, a cytotoxic protein of sea hare (*Aplysia kurodai*) eggs. Dev Comp Immunol, 19: 13–19.

Iijima, R., J. Kisugi and M. Yamazaki. 2003. A novel antimicrobial peptide from the sea hare *Dolabella auricularia*. Dev Comp Immunol, 27: 305–311.

IPCC (Intergovernmental Panel on Climate Change). 2013. Climate Change 2013. The Physical Basis Contribution of Working Group I to the Fifth Assessment Report of the Intergovernmental Panel on Climate Change. (eds) T. F. Stocker, D. Qiu, G. K. Platter et al. Cambridge University Press, Cambridge.

Iwata, Y., H. Munehara and Y. Sakurai. 2005. Dependence of paternity rates on alternative reproductive behaviors in the squid *Loligo bleekeri*. Mar Ecol Prog Ser, 298: 219–228.

Iwata, Y., P. Shaw, E. Fujiwara et al. 2011. Why small males have big sperm: dimorphic squid sperm linked to alternative mating behaviours. BMC Evol Biol, 11: 236, doi: 10.1186/1471-2148-11-236.

Iwata, Y., Y. Sakurai and P. Shaw. 2015. Dimorphic germ transfer strategies and alternative mating tactics in lobgonid squid. J Moll Stud, 81: 147–151.

Jacob, J. 1954. Parthenogenesis and allopolyploidy in the melaniid snails (Gastropoda: Prosobranchia). Curr Sci, 48: 793–794.

Jacob, J. 1958a. XVI—Cytological studies of Melaniidae (Mollusca) with special reference to parthenogenesis and polyploidy. I. Oogenesis of the parthenogenetic species of *Melanoides* (Prosobranchia-Gastropoda). Trans R Soc Edinburgh, 63: 341–352.

Jacob, J. 1958b. XX—Cytological studies of Melaniidae (Mollusca) with special reference to parthenogenesis and polyploidy. II. A study of meiosis in the rare males of the polyploid race of *Melanoides tuberculatus* and *Melanoides lineatus*. Trans R Soc Edinburgh, 63: 433–444.

Jacobsen, H. P. and O. B. Stabell. 2004. Antipredator behaviour mediated by chemical cues: the role of conspecific alarm signalling and predator labelling in the avoidance response of a marine gastropod. Oikos, 104: 43–50.

Janeczko, A. and A. Skoczowski. 2005. Mammalian sex hormones in plants. Folia Histochem Cytobiol, 43: 71–9.

Janer, G. and C. Porte. 2007. Sex steroids and potential mechanisms of non-genomic endocrine disruption in invertebrates. Ecotoxicology, 16: 145–160.

Janer, G., G. A. LeBlanc and C. Porte. 2005a. Androgen metabolism in invertebrates and its modulation by xenoandrogens. A comparative study. Ann NY Acad Sci, 1040: 354–356.

Janer, G., R. Lavado, R. Thibaut and C. Porte. 2005b. Effects of 17b-estradiol exposure in the mussel *Mytilus galloprovincialis*: a possible regulating role for ste+-roid acyltransferases. Aquat Toxicol, 75: 32–42.

Janer, G., R. M. Sternberg, G. A. LeBlanc and C. Porte. 2005c. Testosterone conjugating activities in invertebrates: are they target for endocrine disruption? Aquat Toxicol, 71: 273–282.

Jarne, P., B. Selay, C. Bellec et al. 1992. Analysis of mating systems in the schistosome vector hermaphrodite snail *Bulinus globosus* by DNA finger printing. Heredity, 68: 141–146.

Jarne, P., M. Vianey-Liaud and B. Delay. 1993. Selfing and outcrossing in hermaphrodite freshwater gastropods (Basommatophora): where, when and why. Biol J Linn Soc, 49: 99–125.

Jenner, C. and A. McCrary. 1968. Sexual dimorphism in erycinacean bivalves. Am Malacol Bull, 35: 43.

Jeppersen, L. L. 1976. The control of mating behavior in *Helix pomatia* L. (Gastropoda: Pulmonata). Anim Beh, 24: 275–290.

Jiang, W., G. Xu, Y. Linn and G. Li. 1991. Comparison of growth between triploid and diploid of *Pinctada martensii* (D). Trop Oceanol, 10: 1–7.

Jimbo, M., F. Nakanishi, R. Sakai et al. 2003. Characterization of L-amino acid oxidase and antimicrobial activity of aplysianin A, a sea hare-derived antitumor-antimicrobial protein. Fish Sci, 69: 1240–1246.

Johannes, R. E. 1963. A poison-secreting nudibranch (Mollusca: Opisthobranchia). Veliger, 5: 104–105.

Johnson, M. S. 1982. Polymorphism for direction of coil in *Patula suturalis*: behavioural isolation and positive frequency dependent selection. Heredity, 49: 145–151.

Johnson, S. G., C. M. Lively and S. J. Schrag. 1995. Evolution and ecological correlates of uniparental reproduction in freshwater snails. Experientia, 51: 498–509.

Jokela, J. and C. M. Lively. 1995. Parasites, sex and early reproduction in a mixed population of freshwater snails. Evolution, 49: 1268–1271.

Jokela, J., C. M. Lively, M. F. Dybdahl and J. A. Fox. 1997. Evidence for cost of sex in the freshwater snail *Potamopyrgus antipodarum*. Evolution, 78: 452–460.

Jokinen, E. H. 1978. The aestivation pattern of a population of *Lymnaea elodes* (Say) (Gastropoda: Lymnaecidae). Am Midland Nat, 100: 43–53.

Joll, L. M. 1983. *Octopus tetricus*. In: *Cephalopod Life Cycles: Species Accounts*. (ed) P. R. Boyle. Academic Press, London. 1: 325–334.

Jones, A. G. 2005. Gerud 2.0: a computer programme for the reconstruction of parental genotypes half side progeny arrays with known or unknown parents. Mol Ecol Notes, 5: 708–711.

Jones, A. G., E. M. Adams and S. J. Arnold. 2002. Topping off a mechanism of first male sperm precedence in a vertebrate. Proc Natl Acad Sci USA, 99: 2078–2081.

Jones, J. B. 1992. Environmental impact of trawling on the seabed: a review. New Zealand J Mar Freshwater Res, 26: 59–67.

Jones, O. R. and J. Wang. 2010. COLONY: a program for parentage sibship inference from genotypes data. Mol Ecol Resour, 10: 551–555.

Jordaens, K., L. Dillen and T. Backeljau. 2009. Shell shape and mating behavior in pulmonate gastropods (Mollusca). Biol J Linn Soc, 96: 306–321.

Jouaux, A., C. Heude-Berthelin, P. Sourdaine et al. 2010. Gametogenic stages in triploid oysters *Crassostrea gigas*: irregular locking of gonad proliferation and subsequent reproductive effort. J Exp Mar Biol Ecol, 395: 162–170.

Jungbluth, J. H. and D. von Knorre. 1995. Red list of island molluscs [Snails (Gastropods) and mussels (Bivalves)] in Germany. Mitteil Deutsch Malakol Geselsch, 56/57: 1–17.

Kabat, A. R. 1985. The allometry of brooding in *Transennella tantilla* (Gould) (Mollusca: Bivalvia). J Exp Mar Biol Ecol, 91: 271–279.

Kajiwara, M., S. Kuraku, T. Kurokawa et al. 2006. Tissue preferential expression of estrogen receptor gene in the marine snail, *Thais clavigera*. Gen Comp Endocrinol, 148: 315–326.

Kamiya, H., K. Muramoto and M. Yamazaki. 1986. Aplysianin-A, an antibacterial and antineoplastic glycoprotein in the albumen gland of a sea hare, *Aplysia kurodai*. Experientia, 42: 1065–1067.

Kamiya, H., K. Muramoto, R. Goto and M. Yamazaki. 1988. Characterization of the antibacterial and antineoplastic glycoproteins in a sea hare *Aplysia juliana*. Nippon Suisan Gakkaishi, 54: 773–777.

Kanda, A., T. Takahashi, H. Satake and H. Minakata. 2006. Molecular and functional characterization of a novel gonadotropin-releasing-hormone receptor isolated from the common octopus (*Octopus vulgaris*). Biochem J, 395: 125–135.

Kappes, H. and P. Haase. 2012. Slow but steady dispersal of freshwater molluscs. Aquat Sci, 74: 1–14.

Kaustky, N. 1982. Quantitative studies on gonad cycle, fecundity, reproductive output and recruitment in a Baltic *Mytilus edulis* population. Mar Biol, 68: 143–160.

Kawano, T. and J. L. M. Leme. 1994. Chromosomes of three species of *Megalobulinus* (Gastropoda: Mesurethra: Megalobulinidae) from Brazil. Malacol Rev, 27: 47–52.

Keay, J., J. T. Bridgham and J. W. Thornton. 2006. The *Octopus vulgaris* estrogen receptor is a constitutive transcriptional activator: evolutionary and functional implications. Endocrinology, 147: 3861–3869.

Kenchington, E., B. MacDonald, B. Cao et al. 2002. Genetics of mother-dependent sex ratio in blue mussel (*Mytilus* spp) and implication for doubly uniparental inheritance of mitochondrial DNA. Genetics, 161: 1579–1588.

Kenchington, E. L., L. Hamilton, A. Cogswell and E. Zouros. 2009. Paternal mtDNA and maleness are co-inherited but not casually linked in mytilid mussels. PLoS ONE, 4: doi. org/10.1371/journal.pone.0006976

Kennish, M. J. and R. A. Lutz. 1992. The hydrothermal vent clam, *Calyptogena magnifica* (Boss and Turner, 1980): a review of existing literature. Rev Aquat Sci, 6: 29–66.

Kent, R. M. L. 1979. The infestations of *Polydora ciliata* on the flesh content of *Mytilus edulis* L. J Mar Biol Ass UK, 59: 289–297.

Kent, R. M. L. 1981. The effect of *Polydora ciliata* on the shell strength of *Mytilus edulis*. J Cons Perm Int Explor Mer, 39: 252–255.

Ketata, I., X. Danier, A. Haniza-Chaffai and C. Minier. 2008. Endocrine related reproductive effects in molluscs. Comp Biochem Physiol. 147C: 261–270.

Kicklighter, C. E. and C. D. Derby. 2006. Multiple components in ink of the sea hare *Aplysia californica* are aversive to the sea anemone *Anthopleura sola*. J Exp Mar Biol Ecol, 334: 256–268.

Kicklighter, C. E., S. Shabani, P. M. Johnson and C. D. Derby. 2005. Sea hares use novel antipredatory chemical defenses. Curr Biol, 15: 549–554.

Kingsley-Smith, P. R., C. A. Richardron and R. Seed. 2005. Growth and development of the veliger larvae and juveniles of *Polinices pulchellus* (Gastropoda: Naticidae). J Mar Biol Ass UK, 55: 171–174.

Kinnel, R. B., R. K. Dieter, J. Meinwald et al. 1979. Brasilenyne and *cis*-dihydrorhodophytin: antifeedant medium-ring haloether from a sea hare (*Aplysia brasiliana*). Proc Natl Acad Sci USA, 76: 3576–3579.

Kisugi, J., H. Kamiya and M. Yamazaki. 1987. Purification and characterization of aplysianin E, an antitumor factor from sea hare eggs. Cancer Res, 47: 5649–5653.

Kisugi, J., H. Ohye, H. Kamiya and M. Yamazaki. 1989. Biopolymers from marine invertebrates. X. Mode of action of an antibacterial glycoprotein, aplysianin E, from eggs of a sea hare. Chem Pharm Bull, 37: 3050–3053.

Kisugi, J., H. Ohye, H. Kamiya and M. Yamazaki. 1992. Biopolymers from marine invertebrates. XIII. Characterization of an antibacterial protein, dolabellanin A, from the albumen gland of the sea hare, *Dolabella auricularia*. Chem Pharm Bull, 40: 1537–1539.

Kiyomoto, M., A. Komaru, J. Scapra et al. 1996. Abnormal gametogenesis, male dominant sex ratio and Sertoli cell morphology in induced triploid mussels *Mytilus galloprovincialis*. Zool Sci, 13: 393–402.

Kjesbu, O. S., H. Murura, F. Saborida-Rey and P. R. Whithames. 2010. Method development and evaluation of stock potential of marine fish. Fish Res, 104: 1–7.

Knutzen, J. 1981. Effects of decreased pH on marine organisms. Mar Pollut Bull, 12: 25–29.

Kofoed, L. H. 1975. The feeding biology of *Hydrobia ventrosa* (Montagu). II. Allocation of the components of the carbon-budget and the significance of the secretion of dissolved organic material. J Exp Mar Biol Ecol, 19: 243–253.

Kohler, H. R., W. Kloas, M. Schulling et al. 2007. Sex steroid receptor evolution and signalling in aquatic invertebrates. Ecotoxicology, 16: 131–143.

Kohn, A. J. and F. B. Perron. 1994. *Life History and Bibliography: Patterns in Conus*. Oxford University Press, New York.

Komaru, A. and K. T. Wada. 1989. Gametogenesis and growth induced triploid scallops *Chlamys nobilis*. Nippon Suisan Gakkaishi, 55: 447–452.

Komaru, A. and K. T. Wada. 1994. Meiotic maturation and progeny of oocytes from triploid Japanese pearl oysters (*Pinctada fucata martensii*) fertilized with sperm from diploids. Aquaculture, 120: 61–70.

Komaru, A. and K. Konishi. 1999. Non-reductional spermatozoa in three shell color types of the freshwater clam *Corbicula fluminea* in Taiwan. Zool Sci, 16: 105–108.

Komaru, A., H. Matsuda, T. Yamakawa and K. T. Wada. 1990. Chromosome-behavior of meiosis-inhibited eggs with cytochalasin B in Japanese pearl oyster. Nippon Suisun Gakkaishi, 569: 1419–1422.

Komaru, A., K. Konishi, I. Nakayama et al. 1997. Hermaphroditic freshwater clams in the genus *Corbicula* produce non-reductional spermatozoa with somatic DNA content. Biol Bull, 193: 320–323

Komaru, A., T. Kawagishi and K. Konishi. 1998. Cytological evidence of spontaneous androgenesis in the freshwater clam *Corbicula leana* Prone. Dev Genes Ecol, 208: 46–50.

Kossak, U. 2006. How climate change translates into ecological change: impacts of warming and desalination on prey properties and predator-prey interactions in the Baltic Sea, Ph.D. Thesis, Christian-Albrechts-Universitat, Kiel. p 97.

Kroeker, K. J., R. L. Kordas, R. N. Crim and G. G. Singh. 2010. Meta-analysis reveals negative yet variable effects of ocean acidification on marine organisms. Ecol Lett, 13: 1419–1434.

Kroeker, K. J., E. Sanford, B. M. Jellism and B. Gaylord. 2014. Predicting the effects of ocean acidification on predator-prey interactions: a conceptual framework based on coastal molluscs. Biol Bull, 226: 211–222.

Kruczynski, W. L. 1972. The effects of the pea crab *Pinnotheres maculatus* Say on the ground of the bay scallop *Argopecten irradians concentricus* (Say). Chesapeake Sci, 13: 218–220.

Krug, J. E., R. A. Ellington, R. Burton and A. Valdes. 2007. A new poecilogonous species of sea slug (Opisthobranchia: Sacoglossa) from California: comparison with the planktotrophic congener *Alderia modesta* (Loven, 1884). J Moll Stud, 73: 29–38.

Krug, P. J. 1998. Poecilogony in an estuarine opisthobranch: planktotrophy, lecithotrophy, and mixed clutches in a population of the ascoglossan *Alderia modesta*. Mar Biol, 132: 483–494.

Krug, P. J. 2001. Bet-hedging dispersal strategy of a specialist marine herbivore: a settlement dimorphism among sibling larvae of *Alderia modesta*. Mar Ecol Prog Ser, 213: 177–192.

Krug, P. J. 2007. Poecilogony and larval ecology in the gastropod genus *Alderia*. Am Malacol Bull, 23: 99–111.

Krug, P. J. 2009. Not my "Type": larval dispersal dimorphisms and bet-hedging in opisthobranch life histories. Biol Bull, 216: 355–372.

Krug, P. J., J. A. Riffell and R. K. Zimmer. 2009. Endogenous signaling pathways and chemical communication between sperm and eggs. J Exp Biol, 212: 1092–1100.

Kuenzler, E. J. 1961. Structure and energy flow of a mussel population in a Georgia salt marsh. Limnol Oceanogr, 6: 191–204.

Kurihara, H., T. Asai, S. Kato and A. Ishimatsu. 2008. Effects of elevated pCO$_2$ on early development in the mussel *Mytilus galloprovincialis*. Aquat Biol, 4: 225–233.

Lafont, R. and M. Mathieu. 2007. Steroids in aquatic invertebrates. Ecotoxicology, 16: 109–130.

Laha, T., P. Pinlaor, J. Mulvenna et al. 2007. Gene discovery for the carcinogenic human liver fluke *Opisthorchis viverrini*. BMC Genomics, 8: 189. doi. 10. 1186/1471-2164-8. 189.

Lai, J. H., J. C. del Alamo, J. Rodriquez-Rodriquez and J. C. Lasheras. 2010. The mechanics of the adhesive locomotion of terrestrial gastropods. J Exp Biol, 21: 3920–3933.

Laidlaw, F. F. 1963. Notes on the genus *Dyakia* with a list of species. J Conchol, 25: 137–157.

Lalli, C. M. and F. E. Wells Jr. 1973. Brood protection in an epipelagic thecosomatus pteropod *Spiratella* (*Lamacina*) *inflata* (D'Orbigny). Bull Mar Sci, 23: 933–944.

Lallias, D., N. Taris, P. Bowdry et al. 2010. Variance in the reproductive success of flat oyster *Ostrea edulis* L. assessed by parentage analyses in natural and experimental conditions. Genet Res, 92: 175–187.

Lambert, W. J., C. D. Todd and J. P. Thorpe. 2000. Variation in growth rate and reproductive output in British populations of the dorid nudibranchs *Adalaria proxima*: consequences of restricted larval dispersal? Mar Biol, 137: 149–159.

Landes, A. and M. Zimmer. 2012. Acidification and warming affect both a calcifying predator and prey, but not their interaction. Mar Ecol Prog Ser, 450: 1–10.

Laptikhovsky, V. 2001. Fecundity, egg masses and hatchlings of *Benthoctopus* spp. (Octopodidae) in Falkland waters. J Mar Biol Ass UK, 81: 267–270.

Laptikhovsky, V. and A. Salman. 2003. On reproductive strategies of the epipelagic octopods of the superfamily Argonautoidea (Cephalopoda: Octopoda). Mar Biol, 142: 321–326.

Laptikhovsky, V., M. A. Collins and A. Arkhipin. 2013. First case of possible of iteroparity among coeloid cephalopods: the giant warty squid *Kondakovia longimana*. J Moll Stud, 79: 270–272.

Laptikhovsky, V. V. and C. M. Nigmatullin. 1993. Egg size, fecundity, and spawning in females of the genus *Illex* (Cephalopoda: Ommastrephidae). ICES J Mar Sci, 50: 393–403.

Lauckner, G. 1983a. Diseases of Mollusca: Bivalvia. In: *Diseases of Marine Animals*. (ed) O. Kinne. Biologische Anstalt Helgoland, Hamburg. 2: 477–962.

Lauckner, G. 1983b. Diseases of Mollusca: Amphineura. In: *Diseases of Marine Animals*. (ed) O. Kinne. Biologische Anstalt Hengoland, Hamburg. 2: 963–976.

Lauckner, G. 1983c. Diseases of Mollusca: Scaphopoda. In: *Diseases of Marine Animals*. (ed) O. Kinne. Biologische Anstalt Helgoland, Hamburg. 2: 979–984.

Laviolette, P. 1951. Regenerats germinaux multiples chez *Arion rufus* L. C R Acad Sci Paris, 233: 1139–1141.

Laviolette, P. 1954. Role de la gonade dan le determinisme humoral de la maturite glanduire du tractus genital chez quelques gastropods Arionidae et Limacidae. Bull Biol Fr Belg, 88: 310–312.

Laviolette, P. 1955. Etude cytologique et experimentale de la regeneration germinale après castration chez *Arion rufus* L. (gastropode pulmone). Ann Sci nat (Zool) IIe Series, 16: 427–535.

Laws, H. M. 1973. The chromosome of *Succinea australis* (Fernssac). J. Malacol Soc Aust, 2: 289–292.

Le Curieux-Belfond, O., S. Moslemi, M. Mathieu and G. E. Seralini. 2001. Androgen metabolism in oyster *Crassostrea gigas*: evidence for 17β HSD activities and characterization of an aromatase like activity, inhibited by pharmacological compounds and a marine pollutant. J Steroid Biochem Mol Biol, 78: 359–366.

Le Curieux-Belfond, O., B. Fievet, G. E. Seralini and M. Mathieu. 2005. Short-term bioaccumulation, circulation and metabolism of estradiol-17β in the oyster *Crassostrea gigas*. J Exp Mar Biol Ecol, 325: 125–133.

Le Gall, P. 1979. Interspecificite du facteur mitogene d'origine cerebral chezles mollusques. C.R. Acad Sci, Paris, 288: 1231–1233.

Le Gall, S. 1981. Experimental study of morphologenic factor controlling differentiating of external male genital tract of *Crepidula fornicata* (Protandrous Hermaphrodite Mollusk). Gen Comp Endocrinol, 43: 51–62.

Le Pennec, M. and P. G. Beninger. 1997. Ultrastructural characteristics of spermatogenesis in three species deep sea hydrothermal vent mytilids. Can J Zool, 75: 308–316.

Le Pennec, M., P. G. Beninger, G. Dorance and Y. -M. Paulet. 1991. Trophic sources and pathways to the developing gametes of *Pecten maximus* (Bivalvia: Pectonidae). J Mar Biol Ass UK, 71: 451–463.

Lei, S., X. Zhang, P. Zhang and Y. Ikeda. 2014. Biochemical composition of cuttlefish (*Sepia esculenta*) eggs during embryonic development. Mol Res, 34: 1–9.

Leighton, D. L. and C. A. Lewis. 1982. Experimental hybridization in abalones. Int Invert Reprod, 5: 273–282.

Leonard, G. H., M. D. Bertness and P. O. Yund. 1999. Crab predation, waterborne cues, and inducible defenses in the blue mussel, *Mytilus edulis*. Ecology, 80: 1–14.

Leuven, R. S. E. W., G. van der Velde, I. Baijens et al. 2009. The river Rhine: a global highway for dispersal of aquatic invasive species. Biol Invasions, 11: 1989–2008.

Lewis, S. L., D. E. Lyons, J. M. Meekins and J. M. Newcomb. 2011. Serotonin influences locomotion in the nudibranch mollusc *Melibe leonina*. Biol Bull, 220: 155–160.

Lindinger, M. I., D. J. Lauren and D. G. McDonald. 1984. Acid-base balance in the sea mussel *Mytilus edulis*. III. Effects of environmental hypercapnia on intra- and extracellular acid-base status. Mar Biol Lett, 5: 371–381.

Lipton, A. P., G. S. Rao and I. Jagadis. 2013. *The Indian Sacred Chank*. Central Marine Fisheries Research Institute, Kochi. p 250.

Liu, Y. and Y. Li. 2015. Successful oocyte cryopreservation in the blue mussel *Mytilus galloprovincialis*. Aquaculture, 438: 55–58.

Liuzzi, M. G. and D. G. Zelaya. 2013. Egg-hull ultrastructure of *Ischnochiton streamineus* (Sowerby, 1832), a South American brooding chiton (Chitoneina: Ischnochitonidae). J Moll Stud, 79: 372–377.

Lively, C. M. 1987. Evidence from a New Zealand snail for the maintenance for sex by parasitism. Nature, 328: 519–521.

Lively, C. M. 1992. Parthenogenesis in freshwater snail: reproductive assurance versus parasitic release. Evolution, 46: 907–913.

Lively, C. M. and J. Jokela. 2002. Temporal and spatial distributions of parasites and sex in a freshwater snail. Evol Ecol Res, 4: 219–226.

Lombardi, S. A., G. D. Chon, J. J. -W. Lee et al. 2013. Shell hardness and compressive strength of the eastern oyster *Crassostrea virginica* and the Asian oyster *Crassostrea ariakensis*. Biol Bull, 225: 175–183.

Lombardo, R. C., F. Takeshita, S. Abe and S. Goshima. 2012. Mate choice by males and paternity distribution in offspring of triple-mated females in *Neptunea arthritica* (Gastropoda: Buccinidae). J Mollus Stud, 78: 283–289.

Longley, R. D. 2011. Neurogenesis in the procerebrum of the snail *Helix aspersa*: a quantitative analysis. Biol Bull, 221: 215–226.

Longo, F. J. 1972. The effects of cytochalasin B on the events of fertilization in the surf clam, *Spisulu solidissima* I polar body formation. J Exp Zool, 182: 321–344.

Lord, A. 1986. Are the contents of egg capsules of the marine gastropod *Nucella lapillus* (L.) axenic? Am Malacol Bull, 4: 201–203.

Lorion, J., S. Halari, J. Nascimento et al. 2012. Evolutionary history of *Idas modiolaeformus* (Bivalvia, Mytilidae), a cold seep mussel bearing multiple symbionts. Cah Biol Mar, 53: 77–87.

Lu, M., T. Horiguchi, H. Shiraishi et al. 2002. Determination of testosterone in an individual shell of *Thais clavigera* by ELISA. Bunseki Kagaku, 51: 21–27.

Lubet, P. E. and M. Mathieu. 1978. Experimental studies on the control of the annual reproductive cycle in the pelecypod molluscs (*Mytilus edulis* L.) and *Crassostrea gigas* (Th). Gen Comp Endocrinol, 34: 109.

Lubet, P. and M. Mathieu. 1990. Les régulations endocriniennes chez les mollusques bivalves. Ann Biol, 29: 235–252.

Lukowiak, K., Z. Haque, G. Spencer et al. 2003. Long-term memory survives nerve injury and the subsequent regeneration process. Learn Mem, 10: 44–54.

Lum-Kong, A. 1993. Oogenesis, fecundity and pattern of spawning in *Loligo forbesi* (Cephalopod; Loliginicidae). Malacol Rev, 26: 81–88.

Lupo di Prisco, C. and F. D. Fulgheri. 1975. Alternative pathways of steroid biosynthesis in gonads and hepatopancreas of *Aplysia depilans*. Comp Biochem Physiol, 50B: 191–195.

Lutzen, J. 1968. Unisexuality in the parasitic family Entoconchidae (Gastropoda: Prosobranchia). Malacologia, 7: 7–15.

Lutzen, J. 1972. Studies on parasitic gastropods from echinoderms. II. On Stilijer Broderip. Biol Skr, 19: 1–18.

Lutzen, J., A. Jespersen and M. P. Russel. 2015. The Pacific clam *Nutricola tantilla* (Bivalvia: Veneridae) has separate sexes and makes use of brood production and sperm storage. J Moll Stud, 81: 397–406.

Lyons, J. and P. Kanehl. 2002. Seasonal movements of smallmouth bass in streams. In: *Black Bass: Ecology, Conservation, and Management*. (eds) D. P. Philipp and M. S. Ridgeway. American Fisheries Society, Bethesda, Maryland, Symp. 31: 149–160.

MacGinitie, G. 1934. Egg-laying activities of the sea hare *Tethys californicus* (Cooper). Biol Bull, 67: 300–303.

Mackas, D. L. and M. D. Galbraith. 2012. Pteropod time-series from the NE Pacific. ICES J Mar Sci, 69: 448–459.

Makinen, T., M. Panova and C. Andre. 2007. High levels of multiple paternity in *Littorina saxatilis*: hedging the bets? J Hered, 98: 705–711.

Malaquias, M. A. E., L. Bentes, K. Erzini and T. C. Borges. 2006. Molluscan diversity caught by trawling fisheries: a case study in southern Portugal. Fish Manag Ecol, 13: 39–45.

Malleswar, V. N. S., R. Basavaraj and S. Kirupanidhi. 2013. Behavioral and physiological challenges in *Pila globosa* (Indian apple snail) during aestivation. Int J S Res, 1: 54–55.

Mallia, J. V., P. Muthiah and P. C. Thomas. 2006. Growth of triploid oyster *Crassostrea madrasensis* (Preston). Aquacult Res, 37: 718–724.

Mallia, J. V., P. C. Thomas and P. Muthiah. 2007. Induction and evaluation of triploidy in edible oyster *Crassostrea madrasensis* (Preston)—an approach to enhance bivalve aquaculture. Ind J Fish, 54: 417–421.

Mallia, J. V., P. Muthiah and P. C. Thomas. 2009. Performance of triploid edible oyster *Crassostrea madrasensis* (Preston)—gonad development and biochemical composition. J Mar Biol Ass India, 51: 81–86.

Mangold, K. 1983. *Octopus vulgaris*. In: *Cephalopod Life Cycles: Species Accounts*. (ed) P. R. Boyle. Academic Press, London. 1: 335–364.

Mangold, K. 1987. Reproduction. In: *Cephalopod Life Cycles: Comparative Reviews*. (ed) P. R. Boyle. Academic Press, London. 2: 157–200.

Manriquez, P. H., M. E. Jara, M. L. Mardones et al. 2013. Ocean acidification disrupts prey responses to predator cues but not prey shell growth in Concholepas concholepas (Ioco). PLoS ONE, 8: e68643.

Maran, C., E. Centanni, F. Pellizzato and B. Pavoni. 2006. Organochlorine compounds (polychlorinated biphenyls and pesticides) and polycyclic aromatic hydrocarbons in populations of *Hexaplex trunculus* affected by imposex in the Lagoon of Venice, Italy. Environ Toxicol Chem, 25: 486–495.

Marche-Marchad, I. 1963. Un nouvel hrdraire a medusa non-pelagique *Monobrachium drachi* n sp (Limnomedusae) trouve dans la baie Goree (Senegal). c.r. hebd Séanc Acad Sci, Paris (Ser D), 257: 1347–1349.

Marian, J. E. A. R. 2012. A model to explain implantation in cephalopods (Mollusca: Cephalopoda) and a discussion on its evolutionary origins and significance. Biol J Linn Soc, 105: 711–726.

Markov, G. V., R. Tavares, C. Dauphin-Villemant et al. 2009. Independent elaboration of steroid hormone signaling pathways in metazoans. Proc Natl Aacd Sci USA, 106: 11913–11918.

Marshall, D. J. and A. Rajkumar. 2003. Imposex in the indigenous *Nassarius kraussianus* (Mollusca: Neogastropoda) from South African harbours. Mar Pollut Bull, 46: 1150–1155.

Marshall, D. J., J. H. Santos, K. M. Leung and W. H. Chak. 2008. Correlations between gastropod shell dissolution and water chemical properties in a tropical estuary. Mar Environ Res 66: 422–429.

Marshall, T. C., J. Slate, L. E. B. Kruuk and J. M. Pemberton. 1998. Statistical confidence for likelihood-based paternity inference in natural populations. Mol Ecol, 7: 639–655.

Martel, A. and D. Larrivee. 1986. Behaviour and timing of copulation and egg-laying in the neogastropod *Buccinum undatum*. J Exp Mar Biol Ecol, 96: 27–42.

Martinez, C. and A. Rivera. 1994. Role of monoamines in reproductive process of *Argopecten purpuratus*. Invert Reprod Dev, 25: 167–174.

Maschner, R. P. Jr. 2000. Studies on tooth strength of the Atlantic cow-nose ray *Rhinoptera bonasus*. Master's Thesis, California State Polytechnic University.

Mathieu, M. 1994. Endocrine control of carbohydrate metabolism in molluscs. In: *Perspective Aquaculture Research Trends*. (eds) K. G. Dovery, R. E. Peter and S. S. Tolie. Natl Res Council Canada, Ottawa, pp. 471–474.

Mathieu, M., I. Robbins and P. Lubet. 1991. The neuroendocrinology of *Mytilus edulis*. Aquaculture, 94: 213–223.

Matozzo, V., A. Chinellato, M. Munari et al. 2012. First evidence of immunomodulation in bivalves under seawater acidification and increased temperature. PLoS ONE, 7: doi. org/10.1371/journal.pone.0033820.

Matsumoto, T., M. Osada, Y. Osawa and K. Mori. 1997. Gonadal estrogen profile and immunohistochemical localization of steroidogenic enzymes in the oyster and scallop during sexual maturation. Comp Biochem Physiol, 118B: 811–817.

Matsuo, R. and E. Ito. 2011. Spontaneous regeneration of central nervous system in gastropod. Biol Bull, 221: 35–42.

Maxwell, W. L. 1974. Spermiogenesis of *Eledone cirrhosa* Lamarck (Cephalopoda, Octopoda). Proc R Soc, 186B: 181–190.

Maxwell, W. L. 1977. Free-etching studies of pulmonate spermatozoa. Veliger, 20: 71–74.

Maxwell, W. L. 1983. Mollusca. In: *Reproductive Biology of Invertebrates*. (eds) K. G. Adiyodi and R. G. Adiyodi. John Wiley & Sons, New York. 2: 275–320.

McConnaughey, T. A. 1995. Ion transport and the generation of biomineral supersaturation. Bull Inst Oceanogr Monaco, 14: 1–18.

McDonald, K. A., R. Collin and M. P. Lesoway. 2014. Poecilogony in the caenogastropod *Calyptraea lichen* (Mollusca: Gastropoda). Invert Biol, 133: 213–220.

McManus, C. 2002. *Right Hand Left Hand: The Origins of Asymmetry in Brain, Bodies, Atoms and Cultures*. Weidenfeld and Nicolson, London.

Meechonkit, P., S. Asuvaponey, P. W. Jumromn et al. 2012. Sexual differences in serotonin distribution and induction of synchronous release by serotonin in the freshwater mussel *Hyriopsis bialatus*. J Moll Stud, 78: 297–303.

Meenakshi, V. R. 1956. Physiology of hibernation of the apple-snail *Pila virens* (Lamarck). Curr Sci 25: 321–322.

Meenakshi, V. R. 1964. Aestivation in the Indian apple snail *Pila*—I. Adaptation in natural and experimental conditions. Comp Biochem Physiol, 11: 379–386.

Melo, V. M. M., A. B. G. Duarte, A. F. F. U. Carvalho et al. 2000. Purification of a novel antibacterial and haemagglutinating protein from the purple gland of the sea hare, *Aplysia dactylomela* Rang, 1828. Toxicon, 38: 1415–1427.

Melzner, F., P. Stange, K. Trubenbach et al. 2011. Food supply and seawater CO_2 impact calcification and internal shell dissolution in the blue mussel *Mytilus edulis*. PLoS ONE, 6(9): e24223.

Menge, J. L. 1974. Prey selection and foraging period of the predaceous rocky intertidal snail *Acanthina punctulata*. Oecologia, 17: 293–316.

Metalunan, S., P. Calosi, S. D. Rundle et al. 2013. Effects of ocean acidification and elevated temperature on shell plasticity and energetic basis in the intertidal gastropods. Mar Ecol Prog Ser, 293: 109–118.

Meuleman, E. A. 1971. Host-parasite interrelationships between the freshwater pulmonate *Biomphalaria pfeifferi* and the trematode *Schistosoma mansoni*. Netherl J Zool, 22: 355–427.

Michaelidis, B., C. Ouzounis, A. Paleras and H. O. Portner. 2005. Effects of long-term moderate hypercapnia on acid-base balance and growth rate in marine mussels *Mytilus galloprovincialis*. Mar Ecol Prog Ser, 293: 109–118.

Mikhailov, A. T., M. Torrado and J. Mendez. 1995. Sex differentiation of reproductive tissue in bivalve molluscs: identification of male associated polypeptide in the mantle of *Mytilus galloprovincialis* Lmk. Int J Dev Biol, 39: 545–548.

Miles, C. D. 1961. Regeneration and lesions in pulmonate gastropods. Ph.D. Thesis, University of Arizona.

Milione, M., P. Saucedo and P. Southgate. 2011. Sexual development, sex ratio and morphometrics of *Pteria penguin* (Bivavia: Pteridae) in north-eastern Australia. Moll Res, 31: 30–36.

Minoretti, N., P. Stoll and B. Baur. 2013. Heritability of sperm length and adult shell size in the land snail *Arianta arbustorium* (Linnaeus, 1758). J Moll Stud, 79: 218–224.

Moffet, S. 1995. Neural regeneration in gastropod molluscs. Prog Neurobiol, 46: 289–330.

Moltschwaniwskyj, N. A. 1995. Multiple spawning in the tropical squid *Photololigo* sp: what is the cost of somatic growth? Mar Biol, 124: 127–135.

Moomjian, L., S. Nystrom and D. Rittschof. 2002. Behavioral responses of sexually active mud snails: kairomones and pheromones. J Chem Ecol, 29: 497–501.

Moore, J. D. and E. R. Trueman. 1971. Swimming of the scallop, *Chlamys opercularis* (L.). J Exp Mar Biol Ecol, 6: 179–185.

Moran, A. L. 1999. Intracapsular feeding by embryos of the gastropod genus *Littorina*. Biol Bull, 196: 229–244.

Morcillo, Y. and C. Porte. 1999. Evidence of endocrine disruption in the imposex-affected gastropod *Bolinus brandaris*. Environ Res, 81: 349–354.

Mori, K., T. Muramatsu and Y. Nakamura. 1969. Effect of steroid on oyster-III. Sex reversal from male to female in *Crassostrea gigas* by estradiol-17β. Bull Jap Soc Sci Fish, 35: 1072–1076.

Morley, N. J. 2006. Parasitism as a source of potential distortion in studies on endocrine disrupting chemical in molluscs. Mar Pollut Bull, 52: 1330–1332.

Morse, D. E. 1984. Biochemical and genetic engineering for improved production of abalones and other valuable molluscs. Aquaculture, 39: 263–282.

Morton, B. 1972. Some aspects of functional morphology and biology of *Pesudopythina subsinuata* (Bivalvia: Leptonacea) commensal on stomatopod crustaceans. J Zool Lond, 166: 79–89.

Morton, B. 1976. Secondary brooding of temporary dwarf males? *Ephippodonta* [*Ephippodontina*] *oedipus* sp. nov, (Bivalvia: Leptonacea). J Conchol Lond, 29: 31–39.

Morton, B. 1979. Population-dynamics and expression of sexuality in *Balcis shaplandi* and *Mucronalia fulvescens* (Mollusca, Gastropoda, Aglossa) parasitic upon *Archaster typicus* (Echinodermata, Asteroidea). Malacologia, 18: 327–346.

Morton, B. 1981. The biology and functional morphology of *Chlamydoconcha arcutti* with a discussion on the taxonomic status of Chlamydoconchacea (Mollusca: Bivalvia). J Zool Lond, 195: 81–121.

Morton, B. 1991. A molluscan fouling community dominated by *Septifer virgatus* (Bivalvia: Mytilidae) in Hong Kong. Malacol Rev, 24: 115–118.

Morton, J. E., A. D. Boney and E. D. S. Corner. 1957. The adaptation of *Lasaea rubra* (Montagu), a small intertidal lamellibranch. J Mar Biol Ass UK, 36: 383–405.

Mulvey, M. and R. C. Virjenhoek. 1981. Multiple paternity in the hermaphroditic snail *Biomphlaria obstructa*. J Hered, 72: 308–312.

Munday, P. L., D. L. Dixson, J. M. Donelson et al. 2009. Ocean acidification impairs olfactory discrimination and homing ability of a marine fish. Proc Natl Acad Sci USA, 106: 1848–1852.

Munuswamy, N. and T. Subramonian. 1985. Oogenesis and shell gland activity in a freshwater fairy shrimp *Streptocephalus dichotomus* Baird (Crustacea: Anostraca). Cytobios, 44: 137–147.

Murray, J. and B. Clarke. 1980. The genus *Partula* on Morrea: speciation in progress. Proc R Soc, 211B: 83–117.

Muthuvelu, S., P. Murugesan, M. Muniasamy et al. 2013. Changes in benthic macro faunal assemblages in relation to bottom trawling in Cuddalore and Parangipettai coastal waters, Southeast Coast of India. Ocean Sci J, 48: 183–195.

Myers, M. J., C. P. Meyer and V. H. Resh. 2000. Neritid and thiarid gastropods from French Polynesian streams: how reproduction (sexual, parthenogenetic) and dispersal (active, passive) affect population structure. Freshwater Biol, 44: 535–545.

Nagarajan, R., S. E. G. Lea and J. D. Goss-Custard. 2006. Seasonal variations in mussel *Mytilus edulis* L. shell thickness and strength and their ecological implications. J Exp Mar Biol Ecol, 339: 241–250.

Nair, N. B. and M. Saraswathy. 1971. The biology of wood-boring teredinid molluscs. Adv Mar Biol, 9: 335–509.

Naraoka, T., H. Uchisawa, H. Mori et al. 2003. Purification characterization and molecular cloning of tyrosinase from the cephalopod mollusk *Illex argentinus*. Eur J Biochem, 270: 4026–4038.

Nasution, S. 2003. Intra-capsular development in marine gastropod *Buccinum undatum* (Linnaeous 1758). J Natur Indonesia, 5: 124–128.

Nasution, S., D. Roberts, K. Farnsworth et al. 2010. Maternal effect on offspring size and packaging constraints in the whelk. J Zool, 281: 112–117.

Natarajan, P., K. Ramadass, D. Sivalingam and P. Thillairajan. 1988. Ornamental shell industry of Ramanathapuram coast. CMFRI Bull, 42: 106–110.

Natarajan, R. 1958. Studies on the egg masses and larval development of some prosobranchs from the Gulf of Mannar and the Palk Bay. Proc Ind Acad Sci, 46B: 170–228.

Natarajan, R. 1959. Chromosomes of the slug *Oncidium esculentum* Cuv. (Mollusca: Pulmonata). J Zool Soc India, 11: 30–33.

Navas, A., C. Carvalho and J. Equaordo. 2010. *Aestivation: Progress in Molecular and Subcellular Aspects*. Springer Berlin, p 268.

Nesis, K. N. 1983. *Dosidicus gigas*. In: *Cephalopod Life Cycles: Species Accounts*. (ed) P. R. Boyle. Academic Press, London. pp 215–232.

Newell, R. C., V. I. Pye and M. Ahsanulla. 1971. Factors affecting feeding rate of the winkle *Littorina littorea*. Mar Biol, 9: 138–144.

Newell, R. L. E., V. S. Kennedy and K. S. Shaw. 2007. Comparative vulnerability to predators and unduced defense responses of eastern oysters *Crassostrea virginica* and non-native *Crassostrea ariakensis* oysters in Chesapeake Bay. Mar Biol, 152: 449–460.

Ng, T. P. T., M. S. Davies, R. Stafford and G. A. Williams. 2011. Mucus trail following as a mate searching strategy in mangrove littornid snails. Anim Beh, 82: 459–465.

Nienhuis, S., A. R. Palmer and C. D. G. Harley. 2010. Elevated CO_2 affects shell dissolution rate but not calcification rate in a marine snail. Proc R Soc Lond B, 277: 2553–2558.

Nishikawa, J., S. Mamiya, T. Kanayama et al. 2004. Involvement of the retinoid X receptor in the development of imposex caused by organotins in gastropods. Environ Sci Technol, 38: 6271–6276.

Noisetter, F., T. Cometet, E. Legrand et al. 2014. Does encapsulation protect embryos from the effects of ocean acidification? The example of *Crepidula fornicata*. PLoS ONE, 9(3): E93021.

Nolen, T. G. and P. M. Johnson. 2001. Defensive inking in *Aplysia* spp: multiple episodes of ink secretion and the adaptive use of a limited chemical resource. J Exp Biol, 204: 1257–1268.

Nolen, T. G., P. M. Johnson, C. E. Kicklighter and T. Capo. 1995. Ink secretion by the marine snail *Aplysia californica* enhances its ability to escape from a natural predator. J Comp Physiol A, 176: 239–254.

Nordsieck, H. 1963. Zur Anatomie und Systematik der Clausilien. Arch Molluskenkund, 92: 81–115.

Oberdorster, E., J. Romano and P. McClellan-Green. 2005. The neuropeptide APGWamide as a penis morphogenic factor (PMF) in gastropod molluscs. Integ Comp Biol, 45: 28–32.

Ockelmann, K. W. and K. Muss. 1978. The biology, ecology and behavior of the bivalve *Mysella bidentata* (Montagu). Ophelia, 17: 1–93.

O'Connor, W. A. and S. J. O'Connor. 2011. Early ontogeny of the pipi *Donax (Plebidonax) deltoids* (Donacidae:Bivalvia). Moll Res, 31: 53–56.

O'Donnell, M. J., M. N. George and E. Carrington. 2013. Mussel byssus attachment weakened by ocean acidification. Nature Clim Change, 3: 587–590.

O'Dor, R. K. 1983. *Illex illecebrosus*. In: *Cephalopod Life Cycles: Species Accounts*. (ed) P. R. Boyle. Academic Press, London. 1: 175–200.

O'Dor, R. K. and E. G. Macalaster. 1983. *Bathypolypus arcticus*. In: *Cephalopod Life Cycles: Species Accounts*. (ed) P. R. Boyle. Academic Press, London. 1: 401–410.

O'Dor, R. K. and M. J. Wells. 1987. Energy and nutrient flow. In: *Cephalopod Life Cycles: Species Accounts*. (ed) P. R. Boyle. Academic Press, London. 2: 109–134.

Oehlmann, J., U. Schulte-Oehlmann, M. Tillmann and B. Markert. 2000. Effects of endocrine disruptors on prosobranch snails (Mollusca: Gastropoda) in the laboratory. Part I: Bisphenol A and octylphenol as xeno-estrogens. Ecotoxicology, 9: 383–397.

Oehlmann, J., P. Di Benedetto, M. Tillmann et al. 2007. Endocrine disruption in prosobranch molluscs: evidence and ecological relevance. Ecotoxicology, 16: 29–43.

Oetken, M., J. Bachmann, U. Schulte-Oehlmann and J. Oehlmann. 2004. Evidence for endocrine disruption in invertebrates. Int Rev Cytol, 236: 1–44.

O'Foighil, D. 1985. Form, function and origin of temperory dwarf males in *Pseudopythina rugifer* (Carpenter, 1964) (Bivalvia, Galeommatacea). Veliger, 27: 245–252.

O'Foighil, D. 1987. Cytological evidence for self-fertilisation in the brooding bivalve *Lasaea subviridis*. Int J Invert Reprod Dev, 12: 83–90.

O'Foighil, D. and D. J. Eernisse. 1988. Geographically widespread, non-hybridizing sympatric strains of the hermaphroditic brooding clam *Lasaea* in the north eastern Pacific Ocean. Biol Bull, 175: 218–229.

O'Foighil, D. and C. Thiriot-Quevreux. 1991. Ploidy and pronuclear interaction in northeastern Pacific *Lasaea* clones (Mollusca: Bivalvia). Biol Bull, 181: 222–231.

O'Foighil, D. and M. J. Smith. 1995. Evolution of asexuality in the cosmopolitan marine clam *Lasaea*. Evolution, 49: 140–150.

Ohman, M. D. and B. E. Lavaniegos. 2009. Multi-decadal variations in calcareous holozooplankton in the California current system: thecosome pteropods, heteropods and foraminifera. Geophysic Lett, 36: doi.10.1029/2009/gi039901.

Ojeda, J. A. and O. R. Chaparro. 2004. Morphological, gravimetric and biochemical changes in *Crepidula fecunda* (Gatropoda: Calyptraeidae) egg capsule walls during development. Mar Biol, 144: 263–269.

Okutani, T. 1983. *Todarodes pacificus*. In: *Cephalopod Life Cycles: Species Accounts*. (ed) P. R. Boyle. Academic Press, London. 1: 201–214.

Okutani, T. and J. A. McGowan. 1969. Systematics, distribution and abundance of epiplanktonic squid (Cephalopoda: Decapoda) larvae of the California current, April 1954–March 1957. Bull Scripps Inst Oceanogr, 14: 1–98.

Oliveira, R. F. 2006. Neuroendocrine mechanism of alternative reproductive tactics in fish. In: *Fish Physiology Behaviour and Fish Physiology*. (eds) K. Sloman, S. Bashine and R. Wilson Elsevier, Amsterdam. 24: 297–357.

Omar, II. M. El-Din. 2013. The biological and medical significance of poisonous animals. J Biol Ear Sci, 3: M25–M41.

Orr, J. C., V. J. Fabry, O. Aumont et al. 2005. Anthropogenic ocean acidification over the twenty first century and its impact on calcifying organisms. Nature, 437: 681–686.

Orstan, A. and F. Welter-Schultes. 2002. A dextral specimen of *Albinaria cretensis* (Pulmonata: Clausiliidae). Triton, 5: 25–28.

Paalvast, P. and G. van der Velde. 2013. What is the main food source of the shipworm (*Teredo navalis*?). A stable isotope approach. J Sea Res, 80: 58–60.

Page, H. M., A. Fiala-Medioni, J. Childress and C. R. Fisher. 1991. Experimental evidence for filter-feeding by the hydrothermal vent mussel, *Bathymodiolus thermophiles*. Deep Sea Res, 38A: 1455–1461.

Paine, R. T. 1971. Energy flow in a natural population of the herbivorous gastropod *Tegula funebralis*. Limnol Oceanogr, 16: 86–98.

Painter, S. D., B. Glouch, R. W. Garden et al. 1988. Characterization of *Aplysia* attractin: the first water-borne peptide pheromone in invertebrate. Biol Bull, 194: 120–131.

Pakes, D. 2002. Changes in strength of selection exerted on prey during the ontogeny of an experimentally introduced predator. M.Sc Thesis, University of Guelph, Guelph, Ontario.

Palmer, A. R. 1981. Do carbonate skeletons limit the rate of body growth? Nature, 292: 150–152.

Palmer, A. R. 1983. Relative cost of producing skeletal organic matrix versus calcification evidence from marine gastropods. Mar Biol, 75: 287–292.

Palmer, A. R. 1990. Effect of crab effluent and scent of damaged conspecifics on feeding, growth, and shell morphology of the Atlantic dogwhelk *Nucella lapillus* (L.). Dev Hydrobiol, 56: 155–182.

Palmer, A. R. 1992. Calcification in marine molluscs: How costly is it? Proc Natl Acad Sci USA, 89: 1379–1382.

Palmer, A. R. 1996. From symmetry to asymmetry. Phylogenetic patterns of asymmetry variation in animals and their evolutionary significance. Proc Natl Acad Sci USA, 93: 14279–14286.

Pandian, T. J. 1969. Yolk utilization in the gastropod *Crepidula fornicata*. Mar Biol, 3: 117–121.

Pandian, T. J. 1975. Mechanism of heterotrophy. In: *Marine Ecology*. (ed) O. Kinne. John Wiley, London, Vol 3 Part 1. pp 61–249.

Pandian, T. J. 1987. Fish. In: *Animal Energetics*. (eds) T. J. Pandian and F. J. Vernberg. Academic Press, San Diego. 2: 357–465.

Pandian, T. J. 1994. Arthropoda-Crustacea. In: *Reproductive Biology of Invertebrates*. (eds) K. G. Adiyodi and R. G. Adiyodi. Oxford and IBH Publishers New Delhi, Vol 6 Part B. pp 39–166.

Pandian, T. J. 2002. Biodiversity: status and endeavors of India. ANJAC J Sci, 1: 21–32.

Pandian, T. J. 2010. *Sexuality in Fishes*. Science Publishers/CRC Press, USA. p 208.

Pandian, T. J. 2011. *Sex Determination in Fish*. Science Publishers/CRC Press, USA. p 270.

Pandian, T. J. 2012. *Genetic Sex Differentiation in Fish*. CRC Press, Boca Raton, USA. p 214.

Pandian, T. J. 2013. *Endocrine Sex Differentiation in Fish*. CRC Press, Boca Raton, USA. p 303.

Pandian, T. J. 2015. *Environmental Sex Determination in Fish*. CRC Press, Boca Raton, USA. p 299.

Pandian, T. J. 2016. *Reproduction and Development in Crustacea*. CRC Press, Boca Raton, USA. p 301.

Pandian, T. J. and C. Balasundaram. 1980. Contribution to the reproductive biology and aquaculture of *Macrobrachium nobilii*, Proc Symp Invert Reprod, Madras University, Madras, 1: 183–193.

Park, G. -M. 1994. Cytotaxonomic studies of freshwater gastropods in Korea. Malacol Rev, 27: 23–41.

Park, G. -M. and J. B. Burch. 1995. Karyotype analyses of six species of North American freshwater mussels (Bivalvia: Unionidae). Malacol Rev, 28: 43–61.

Park, G. -M., J. S. Lee, H. B. Song and O. K. Kwon. 1988. Cytological studies of *Cipangopaludina chinensis melleata* (Mesogastropoda: Viviparidae) in Korea. Korean J Malacol, 4: 41–49.

Park, G. -M., O. K. Kwon and P. R. Chung. 1992. Cytotaxonomic studies of two freshwater snail species of the family Lymnaeidae in Korea. J Sci Technol Kangwon Natl Univ, 31: 171–177.

Parker, L. M., P. M. O'Conor, W. A. Borysko et al. 2012. Adult exposure influences off spring response to ocean acidification in oysters. Glob Change Biol, 18: 82–92.

Parry, G. D. 1982. Reproductive effort in four species of intertidal limpets. Mar Biol, 67: 267–282.

Pascual, M. S., O. O. Iribarne, E. A. Zampatti and A. H. Bocca. 1989. Female-male interaction in the breeding system of the puelche oyster *Ostrea puelchana*. J Exp Mar Biol Ecol, 132: 209–219.

Passamonti, M. and V. Scali. 2001. Gender-associated mitochondrial DNA heterogamety in the venerid clam *Tapes philippinarum* (Mollusca, Bivalvia). Curr Genet, 39: 117–124.

Paterson, I. G., V. Partridge and J. Buckland-Nicks. 2001. Multiple paternity in *Littorina obtusata* (Gastropoda, Littorinidae) revealed by microsatellite analyses. Biol Bull, 200: 261–267.

Patterson, C. M. 1969. Chromosomes of molluscs. In: *Proc Symp Mollusks II. Mar Biol Ass India*. Kochi. pp 635–686.

Paul, V. J. and K. L. Van Alstyne. 1988. Use of ingested algal diterpenoids by *Elysia halimedae* Macnae (Opisthobranchia: Ascoglossa) as anti-predator defenses. J Exp Mar Biol Ecol, 119: 15–29.

Paul, V. J. and S. C. Pennings. 1991. Diet-derived chemical defenses in the sea hare *Stylocheilus longicauda* (Quoy et Gaimard 1824). J Exp Mar Biol Ecol, 151: 227–243.

Pavlova, G. A. 2001. Effects of serotonin, dopamine and ergometrine on locomotion in the pulmonate mollusk *Helix lucorum*. J Exp Biol, 204: 1025–1033.

Pazos, A. J. and M. Mathieu. 1999. Effects of five natural gonadotropin-releasing hormones on cell suspensions of marine bivalve gonad: stimulation of gonial DNA synthesis. Gen Comp Endocrinol, 113: 112–120.

Peake, J. F. 1973. Species isolation in sympatric populations of the genus *Diplommatina* (Gastropoda: Prosobranchia, Cyclophoridae, Diplommatinae). Malacologia, 14: 303–312.

Pearson, E. J. and T. C. Cheng. 1985. Studies in parasitic castration: occurrence of a gametogenesis-inhibiting factor in extract of *Zoogonus lasius* (Trematoda). J Invert Pathol, 46: 239–246.

Pechenik, J. A. 1986. The encapsulation of eggs and embryos by mollusks-an overview. Am Malacol Bull, 4: 165–172.

Pechenik, J. A., S. C. Chang and A. Lord. 1984. Encapsulated development of the marine prosobranch gastropod *Nucella lapillus*. Mar Biol, 78: 223–229.

Pechenik, J. A., I. D. Marsden and O. Pechenik. 2003. Effects of temperature, salinity, and air exposure on development of the estuarine pulmonate gastropod *Amphibola crenata*. J Exp Mar Biol Ecol, 292: 159–176.

Peck, M. R., P. Labadie, C. Minier and E. M. I Iill. 2007. Profiles of environmental and endogenous estrogens in the zebra mussel *Dreissena polymorpha*. Chemosphere, 69: 1–8.

Peek, A. S., R. A. Feldman, R. A. Lutz and R. C. Vrijenhoek. 1998. Co-speciation of chemoautotrophic bacteria and deep sea clams. Proc Natl Acad Sci USA, 95: 9962–9966.

Penchaszadeh, P. E., A. Averbuj and M. Cledón. 2001. Imposex in gastropods from Argentina (south-western Atlantic). Mar Pollut Bull, 42: 790–791.

Pennings, S. C., V. J. Paul, D. C. Dunbar et al. 1999. Unpalatable compounds in the marine gastropod *Dolabella auricularia*: distribution and effect of diet. J Chem Ecol, 25: 735–755.

Perdue, J. A. 1983. The relationship between the gametogenic cycles of the Pacific oyster *C. gigas* and the summer mortality phenomenon in strains selectively bred oysters. Ph.D. Thesis, University of Washington, Seattle. p 205.

Perdue, J. A. and G. Erickson. 1984. A comparison of the gametogenic cycle between the Pacific oyster *Crassostrea gigas* and Sumione oyster *Crassostrea rivularis* in Washington state. Aquaculture, 37: 231–237.

Perron, F. E. 1981. The partitioning of reproductive energy between ova and protective capsules in marine gastropods of the genus *Conus*. Am Nat, 118: 110–118.

Perron, F. E. 1982. Inter- and intraspecific patterns of reproductive effort in four species of cone shells (*Conus* spp.). Mar Biol, 68: 161–167.

Perron, F. E. and G. C. Corpuz. 1982. Costs of parental care in the gastropod *Conus pennaceus*: age-specific changes and physical constraints. Oecologia, 55: 319–324.

Peterson, C. H. and M. L. Quiammen. 1982. Siphon nipping, its importance to small fishes and its impact on growth of the bivalve *Protothaca staminea* (Conrad). J Exp Mar Biol Ecol, 63: 349–368.

Phillips, N. E. 2007. High variability in egg size and energetic content among intertidal mussels. Biol Bull, 212: 12–19.

Picken, G. B. 1980. Reproductive adaptations of Antartic benthic invertebrates. Biol J Linn Soc, 14: 67–75.

Picken, G. B. and D. Allan. 1983. Unique spawning behavior by the Antarctic limpet. *Nucella* (*Patinigera*) *concinna* (Strebel, 1908). J Exp Mar Biol Ecol, 71: 283–287.

Poore, G. C. B. 1973. Ecology of New Zealand abalones *Haliotis species* (Mollusca: Gastropoda). New Zealand J Mar Freshwater Res, 7: 67–84.

Portner, H. O. and A. Reipschläger. 1996. Ocean disposal of anthropogenic CO_2: physiological effects on tolerant and intolerant animals. In: *Ocean Storage of CO_2 Environmental Impact*. (eds) B. Ormerod and M. Angel. Massachusetts Institute of Technology and International Energy Agency, Greenhouse Gas R&D Programme, Cheltenham/Boston. pp 57–81.

Pouvreau, S., C. Bacher and M. Heral. 2000a. Ecophysiological model of growth and reproduction of the black pearl oyster *Pinctada margaritifera*: potential applications for pearl farming in French Polynesia. Aquaculture, 186: 117–144.

Pouvreau, S., A. Gangnery, F. Tiapari et al. 2000b. Gametogenic cycle and reproductive effort in the tropical blacklip pearl oyster *Pinctada margaritifera* (Bivalvia: Pteriidae) cultivated in Takapoto atoll (French Polynesia). Aquat Living Resour, 13: 37–48.

Protasio, A. V., I. J. Tsai, A. Babbage et al. 2012. A systematically improved high quality genome and transcriptome of the human blood fluke *Schistosoma mansoni*. PLoS Negl Trop Dis, 6(1). doi: 10.1371/journal.pntd.0001455.

Przeslawski, R. 2004. A review of the effects of environmental stress on embryonic development within intertidal gastropod egg masses. Moll Res, 24: 43–63.

Przeslawski, R. 2011. Notes on th egg capsule and variable embryonic development of *Nerita melanotragus* (Gastropoda: Neritidae). Moll Res, 31: 1152–158.

Puilliandre, N., T. F. Duda, C. Meyer et al. 2015. One, four or 100 genera? A new classification of the cone snails. J Moll Stud, 81: 1–23.

Puinean, A. M. and J. M. Rotchell. 2006a. *Vitellogenin* gene expression as a biomarker of endocrine disruption in the invertebrate, *Mytilus edulis*. Mar Environ Res, 62: S211–S214.

Puinean, A. M., P. Labadi, E. M. Hill et al. 2006b. Laboratory exposure to 17β estradiol fails to induce vitellogenesis and estrogen receptor gene expression in the marine invertebrate *Mytilus edulis*. Aquat Toxicol, 79: 376–383.

Quale, D. B. 1961. Denman Island disease. Fish Res Bd Can, Ms Rep Ser (Biol) No, 713: 1–91.

Quinn, G. P. 1988. Ecology of the intertidal pulmonate limpet *Siphonaria diemenensis* Quoy et Gaimard. II. Reproductive pattern and energetics. J Exp Mar Biol Ecol, 117: 137–156.

Quinteiro, J., T. Baibai, L. Oukhattar et al. 2011. Multiple paternity in the common octopus *Octopus vulgaris* (Cuvier, 1797) as revealed by microsatellite DNA analysis. Moll Res, 31: 15–20.

Rajaganapathi, J., S. P. Thyagarajan and J. K. P. Edward. 2000. Study on cephalopod's ink for anti-retroviral activity. Indian J Exp Biol, 38: 519–520.

Rajaganapathi, J., K. Kathiresan and T. P. Singh. 2002. Purification of anti-HIV protein from purple fluid of the sea hare *Bursatella leachii* de Blainville. Mar Biotechnol, 4: 447–453.

Ramasamy, M. S. and A. Murugan. 2002. Imposex in muricid gastropod *Thais biserialis* (Mollusca: Neogastropoda) from function harbour southeast of India. Ind J Mar Sci, 31: 243–245.

Range, P., M. Chickaro, R. Ben-Hamadou et al. 2011. Calcification, growth and mortality of juvenile clam *Ruditapes decussatus* under measured CO_2 and reduced pH: variable responses to ocean acidification at local scale? J Exp Mar Biol Ecol, 396: 177–184.

Rasmussen, E. 1973. Systematics and ecology of the Isefjord marine fauna (Denmark). Ophelia, 11: 1–495.

Rawlings, T. A. 1996. Shields against ultraviolet-radiation: an additional protective role for the capsule of benthic marine gastropods. Mar Ecol Prog Ser, 136: 81–95.

Rawson, P. D. and T. J. Hilbish. 1995. Evolutionary relationships among male and female mitochondrial DNA lineages in the *Mytilus edulis* species complex. Mol Biol Evol, 12: 893–901.

Raymond, M. and F. Rousset. 1995. Genepop (version 1.2): population genetics software for exact tests and ecumenicism. J Hered, 86: 248–249.

Reed, A. J., J. P. Morris, K. Linse and S. Thatje. 2013. Plasticity in shell morphology and growth among deep-sea protobranch bivalves of the genus *Yoldiella* (Yoldidae) from contrasting Southern Ocean regions. Deep Sea Res Part 1, 81: 14–24.

Reel, K. R. and F. A. Fuhrman. 1981. An acetylcholine antagonist from the mucous secretion of the dorid nudibranch, *Doriopsilla albopunctata*. Comp Biochem Physiol, 68C: 49–53.

Reis, J. B., A. L. Cohen and D. C. McCorkle. 2009. Marine calcifiers exhibit mixed responses to CO_2-induced ocean acidification. Geology, 37: 1131–1134.

Reitzel, A. M. and A. M. Tarrant. 2010. Correlated evolution of androgen receptor and aromatase revisited. Mol Biol Evol, 27: 2211–2215.

Ricciardi, A., R. Serrouya and F. G. Whoriskey. 1995. Aerial exposure tolerance of zebra and quagga mussels (Bivalvia: Dreissenidae): implications for overland dispersal. Can J Fish Aquat Sci, 52: 470–477.

Richard, A. 1970. Differenciation sexualle des cephalopods en culture *in vitro*. Anne Biol, 9: 409–415.

Richards, C. S. 1967. Estivation of *Biomphalaria glabrata* (Basommatophora: Planorbidae). Associated characteristics' and relation to infection with *Schistosoma mansoni*. Am J Trop Med Hygiene, 16: 797–802.

Riffell, J. A., P. J. Krug and R. K. Zimmer. 2002. The effects of sperm density and gamete contact time on the fertilization success of blacklip (*Haliotis rubra*; Leach, 1814) and greenlip (*H. laevigata*; Donovan, 1808) abalone. J Exp Biol, 205: 1439–1450.

Riffell, J. A., P. J. Krug and R. K. Zimmer. 2004. The ecological and evolutionary consequences of sperm chemoattraction. Proc Natl Acad Sci USA, 101: 4501–4506.

Rivest, B. R. 1983. Development and the influence of nurse egg allotment on hatching size in *Searlesia dira* (Reeve, 1846) (Prosobranchia: Buccinidae). J Exp Mar Biol Ecol, 69: 217–241.

Robertson, R. 1993. Snail handedness: the coiling directions of gastropods. Natl Geogr Res Explor, 9: 104–119.

Rocha, F. and A. Guerra. 1996. Signs of an extended and intermittent terminal spawning in the squids *Loligo vulgaris* Lamarck and *Loligo forbesi* Steenstrup (Cephalopoda: Loliginidae). J Exp Mar Biol Ecol, 207: 177–189.

Rocha, F., A. Guerra and A. F. Gonzalez. 2001. A review of reproductive strategies in cephalopods. Biol Rev, 76: 291–304.

Rodhouse, P. G. 1978. Energy transformations by the oyster *Ostrea edulis* L. in a temperate estuary. J Exp Mar Biol Ecol, 34: 1–22.

Rodhouse, P. G. 1998. Physiological progenesis in cephalopod molluscs. Biol Bull, 195: 17–20.

Rodriques, M., M. E. Garci, J. S. Troncoso and A. Guerra. 2011. Spawning strategy in Atlantic boblaid squid *Sepiola atlantica* (Cephalopoda: Sepiolidae). Helgol Mar Res, 65: 43–49.

Roger, L. M., A. J. Richardson, A. D. McKinnon et al. 2012. Comparison of the shell structure of two tropical Thecosomata (*Creseis acicula* and *Diacavohnia longirostris*) from 1963 to 2009: potential implications of declining aragonite saturation. ICES J Mar Sci, 69: 465–474.

Ronis, M. J. J. and A. Z. Mason. 1996. The metabolism of testosterone by the periwinkle (*Littorina littorea*) *in vitro* and *in vivo*: effects of tributyltin. Mar Environ Res, 42: 161–166.

Rosa, R. and B. A. Seibel. 2008. Synergistic effects of climate-related variables suggest future physiological impairment in a top oceanic predator. Proc Natl Acad Sci USA, 105: 20776–20780.

Rose, R. A. 1985. The spawn and embryonic development of colour variants of *Dendrodoris nigra* Stempson (Mollusca: Nudibranchia). J Malacol Soc Aust, 7: 75–88.

Rudolph, P. H. 1983. Copulatory activity and sperm production in *Bulinus globosus* (Gastropoda: Planorbidae). J Mol Stud, 49: 125–132.

Rug, M. and A. Ruppel. 2000. Toxic activities of the plant *Jatropha curcas* against intermediate snail hosts and larvae of schistosomes. Trop Med Int Health, 5: 423–430.

Ruiz-Jones, G. J. and M. G. Hadfield. 2011. Loss of sensory elements in the apical sensory organ during metamorphosis in the nudibranch *Phestilla sibogae*. Biol Bull, 220: 39–46.

Rundle, S. D., J. I. Spicer, R. A. Coleman et al. 2004. Environmental calcium modifies induced defences in snails. Proc R Soc, 271B: S67–S70.

Runham, N. W. 1978. Reproduction and its control in *Deroceras reticulatum*. Malacologia, 17: 341–350.

Runham, N. W. 1988. Mollusca. In: *Reproductive Biology of Invertebrates*. (eds) K. G. Adiyodi and R. G. Adiyodi. Oxford & IBH Publishers, New Delhi, Vol 3, pp 112 188.

Runham, N. W. 1993. In: *Reproductive Biology of Invertebrates*. (eds) K. G. Adiyodi and R. G. Adiyodi. Oxford & IBH Publishers, New Delhi, Vol 6 Part A. pp 311–383.

Russel-Hunter, W. D. and A. G. Eversole. 1976. Evidence for tissue degrowth in starved freshwater pulmonate snails (*Helisoma trivolvis*) from carbon and nitrogen analyses. Comp Biochem Physiol, 54A: 447–453.

Russell-Hunter, W. D. and R. F. McMahon. 1976. Evidence for functional protandry in a freshwater basommatophoran limpet, *Laevapex fuscus*. Trans Am Microscop Soc, 95: 174–182.

Saavedra, C. and J. B. Peña. 2005. Nucleotide diversity and pleistocene population expansion in Atlantic and Mediterranean scallops (*Pecten maximus* and *P. jacobaeus*) as revealed by the mitochondrial 16S ribosomal RNA gene. J Exp Mar Biol Ecol, 323: 138–150.

Saavedra, C. and E. Bachere. 2006. Bivalve genomics. Aquaculture, 256: 1–14.

Saavedra, C., M. -G. Reyero and E. Zourus. 1997. Male-dependent doubly uniparental inheritance of mitochondrial DNA and female-dependent sex ratio in the mussel *Mytilus galloprovincialis*. Genetics, 145: 1073–1082.

Sabine, C. L. and R. A. Feely. 2007. The oceanic sink for carbon dioxide. In: *Greenhouse Gas Sink*. (eds) D. Reay, N. Hewitt, J. Grace and K. Smith. CABI Publishing, Oxfordshire, UK. pp 31–49.

Sabine, C. L., R. A. Feely, N. Gruber et al. 2004. The oceanic sink for anthropogenic CO_2. Science, 305: 367–371.

Salerno, J. L., S. A. Macko, S. J. Hallam et al. 2005. Characterization of symbiotic populations in life history stages of mussels from chemosymbiotic environment. Biol Bull, 208: 145–155.

Sanford, E., B. Gaylord, A. Hettinger et al. 2014. Ocean acidification increases the vulnerability of native oysters to predation by invasive snails. Proc R Soc B, 281(1778): 20132681.

Santos, M. M., L. Filipe, C. Castro et al. 2005. New insights into the mechanism of imposex induction in the dogwhelk *Nucella lapillus*. Comp Biochem Physiol, 141C: 101–109.

Sarma, S. S. S., S. Nandini and R. D. Gulati. 2005. Life history strategies of cladocerans: comparisons of tropical and temperature taxa. Hydrobiologia, 542: 315–333.

Sastry, A. N. 1979. Pelecypoda (including Ostreidae). In: *Reproduction of Marine Invertebrates.* (eds) A. C. Giese and J. S. Pearse. Academic Press, London. 5: 113–292.

Sato, N., M. -A. Yoshida, E. Fujiwara and T. Kasugeu. 2013. High speed camera observations of copulatory behavior in *Idiosepius paradoxus*: function of the dimorphic heterocotyli. J Moll Stud, 79: 183–186.

Saville, A. 1987. Comparison between cephalopods and fish of those aspect of the biology related to stock management. In: *Cephalopod Life Cycles.* (ed) P. R. Boyle. Academic Press, London. 2: 277–290.

Sbilordo, S. H., O. Y. Martin and G. Ribi. 2012. Chromosome inheritance and reproductive barriers in backcrosses between two hybridizing *Viviparus* snail species. J Moll Stud, 78: 357–363.

Schaefer, K. 1996. Review of data on cephalapside reproduction with special reference to the genus *Haminaea* (Gastropoda: Opisthobranchia). Ophelia, 45: 17–37.

Scheibenstock, A., D. Krygier, Z. Haque et al. 2002. The soma of RPeD1 must be present for long-term memory formation of associative learning in *Lymnaea.* J. Neurophysiol, 88: 1584–1591.

Scheltema, R. S. 1966. Trans-Atlantic dispersal of veliger larvae from shoal water benthic mollusca. Sec Int Oceanogr Cong (Mosco) Abstract No, 375: 320.

Scheltema, R. S. 1971. Larval dispersal as a means of genetic exchange between geographically separated populations of shallow-water benthic marine gastropods. Biol Bull, 140: 284–322.

Schilthuizen, M. and A. Davison. 2005. The convoluted evolution of snail chirality. Naturwissenschaften, 92: 504–515.

Schilthuizen, M. and M. Hasse. 2010. Disintegrating true shape differences and experimental bias: are dextral and sinistral shells exact mirror images? J Zool, 282: 191–200.

Schmidt, G. A. 1932. Dimorphisme embryonaire de *Lineus guserensis* Ruber de la cote Mourmanne et de Roscoft et ses relations avec les forms adultes. Ann Inst Oceanogr (Paris), 12: 65–103.

Scott, A. P. 2012. Do mollusks use vertebrate steroids as reproductive hormones? I. Critical appraisal of the evidence for the presence, biosynthesis and uptake of steroids. Steroids, 77: 1450–1468.

Scott, A. P. 2013. Do mollusks use vertebrate steroids as reproductive hormones? II. Critical review of the evidence that steroids have biological role. Steroids, 78: 268–281.

Scott, A. P. and T. Ellis. 2007. Measurement of fish steroids in water—a review. Gen Comp Endocrinol, 153: 392–400.

Sepúlveda, R. D., C. G. Jara and C. S. Gallardo. 2012. Morphological analysis of two sympatric ecotypes and predator-induced phenotypic plasticity in *Acanthina monodon* (Gastropoda: Muricidae). J Moll Stud, 78: 173–178.

Seuront, L. and N. Spilmont. 2015. The smell of sex: water borne and air-borne sex pheromones in the intertidal gastropod *Littorina littorea.* J Moll Stud, 81: 96–103.

Shabani, S., S. Yaldiz, L. Vu and C. D. Derby. 2007. Acidity enhances the effectiveness of active chemical defensive secretions of sea hares, *Aplysia californica*, against spiny lobsters, *Panulirus interruptus.* J Comp Physiol, 193: 1195–1204.

Shane, B. S. 1994. *Introduction to Ecotoxicology.* CRC Press, Boca Raton, FL.

Shaw, M. A. and G. L. Mackie. 1990. Effects of calcium and pH on the reproductive success of *Amnicola limosa* (Gastropoda). Can J Fish Aquat Res, 47: 1694–1699.

Shen, H., L. Wang, X. Davi and Z. Shi. 2010. Karyotypes in *Onchidium struma* (Gastropoda: Pulmonata: Systellommatophora). Malacol Res, 30: 113–116.

Shepherd, S. A. 1986. Studies on southern Australian abalone (genus *Haliotis*). VII. Aggregative behaviour of *H. laevigata* in relation to spawning. Mar Biol, 90: 231–236.

Shevtsov, G. A. 1973. Results of tragging of the Pacific squid *Todarodes pacificus* Steenstrup in the Kuril-Hokkaido region. Can Fish Mar Serv. Translation Ser, 3301 (1974).

Shi, H. H., C. J. Huang, S. X. Zhu et al. 2005. Generalized system of imposex and reproductive failure in female gastropods of coastal waters of mainland China. Mar Ecol Prog Ser, 304: 179–189.

Shibazaki, Y., M. Shimizu and R. Kuroda. 2004. Body handedness is directed by genetically determined cytoskeletal dynamics in the early embryo. Curr Biol, 14: 1462–1467.

Shirayama, Y. and H. Thornton. 2005. Effect of increased atmospheric CO_2 on shallow water marine benthos. J Geophys Res Oceans, doi:10.1029/2004jc002618.

Siegenthaler, U., E. Monnin, K. Kawamura et al. 2005. Supporting evidence from the EPICA Dronning Maud Land ice core for atmospheric CO_2 changes during the past millennium. Tellus, 57B: 51–57.

Silva, L., L. Meireles, S. Davila et al. 2013. Life history of *Bulimulus tenuissimus* (O'Orbigny, 1835) (Gastropoda, Pulmonata, Bulimullidae). Effect of isolation in reproductive strategy and in resource allocation over their life time. Moll Res, 33: 75–79.

Singh, G., J. D. Block and S. Rebach. 2000. Measurement of the crushing force of the crab claw. Crustaceana, 73: 633–637.

Skibinski, D. O. F., C. Gailacher and C. M. Beynon. 1994. Sex limited mitochondrial DNA transmission in the marine mussel *Mytilus edulis*. Genetics, 138: 801–809.

Smith, K. E. and S. Thatje. 2013. Nurse egg consumption and intracapsular development in the common whelk *Buccinum undatum* (Linnaeus, 1758). Helgol Mar Res, 67: 109–120.

Smolensky, N., M. R. Romers and P. J. Krug. 2009. Evidence for costs of mating and self-fertilization in a simultaneous hermaphrodite with hypodermic insemination, the opisthobranch *Alderia*. Biol Bull, 216: 188–199.

Soffker, M. and C. R. Tyler. 2012. Endocrine disrupting chemicals and sexual behaviors in fish: a critical review on effects and possible consequences. Crit Rev Toxicol, 42: 653–668.

Soldatenko, E. and A. Petrov. 2012. Mating behavior and copulatory mechanics in six species of Planorbidae (Gastropoda: Pulmonata). J Moll Stud, 78: 185–196.

Spencer, G. E., M. H. Kazmi, N. I. Syed and K. Lukowiak. 2002. Changes in the activity of a CPG neuron after the reinforcement of an operantly conditioned behavior in *Lymnaea*. J. Neurophysiol, 88: 1915–1923.

Spight, T. M. 1976. Ecology of hatching size for marine snails. Oecologia, 24: 283–294.

Spight, T. M. and J. Emlen. 1976. Clutch sizes of two marine snails with changing food supply. Ecology, 57: 1162–1178.

Squires, Z. E., B. B. M. Wong, M. D. Norman and D. Stuart-Fox. 2012. Multiple fitness benefits of polyandry in a cephalopod. PLoS ONE, 7(5): e37074. doi. 10. 1371/journal. pone.0037074.

Squires, Z. E., M. D. Norman and D. Stuart-Fox. 2013. Mating behavior and general spawning pattern of the southern dumpling squid *Euprymna tasmanica* (Sepiolidae): a laboratory study. J Moll Stud, 79: 263–269.

St Mary, C. M. 1993. Novel sexual patterns in two simultaneously hermaphroditic gobies. *Lythrypnus dalli* and *Lythrypnus zebra*. Copeia, 1993: 304–313.

St Mary, C. M. 2000. Sex allocation in *Lythrypnus* (Gobiidae) variations on a hermaphroditic theme. Environ Biol Fish, 58: 321–333.

Stadler, T., M. Frye, M. Neiman and C. M. Lively. 2005. Mitochondrial haplotypes and the New Zealand origin of clonal European *Potamopyrgus*, an invasive aquatic snail. Mol Ecol, 14: 2465–2473.

Stanczyk, F. Z., J. S. Lee and R. J. Santen et al. 2007. Standardization of steroid hormone assay: why, how and when? Cancer Epidem Biomark Prev, 16(9): 1713–9.

Stanley, J. G., H. Hidu and S. K. Allen Jr. 1984. Growth of American oyster increased by polyploidy induced by blocking meiosis I but not meiosis II. Aquaculture, 37: 147–155.

Steer, M. A., N. A. Moltschaniwskyj, D. S. Nichols and M. Miller. 2004. The role of temperature and maternal ration in embryo survival: using the dumpling squid *Euprymna tasmanica* as a model. J Exp Mar Biol Ecol, 307: 73–89.

Stenyakina, A., L. J. Walters, E. A. Hoffman and C. Calestani. 2010. Food availability and sex reversal in *Mytella charruana* an introduced bivalve in the southeastern United States. Mol Reprod Dev, 77: 222–230.

Stephano, J. L. and M. Gould. 1988. Avoiding polyspermy in the oyster *Crassostrea gigas*. Aquaculture, 73: 295–307.

Sternberg, R. M., A. K. Hotchkiss and G. A. Leblanc. 2008. Synchronized expression of retinoid X receptor mRNA with reproductive tract recrudescence in an imposex-susceptible mollusc. Environ Sci Technol, 42: 1345–1351.

Sticke, W. B. 1973. The reproductive physiology of the intertidal prosobranch *Thais lamellosa* (Gmelin). 1. Seasonal changes in the rate of oxygen consumption and body component indexes. Biol Bull, 144: 511–524.

Stone, J. and M. Björklund. 2002. Delayed prezygotic isolating mechanisms: evolution with a twist. Proc R Soc, 269B: 861–865.

Strathmann, M. E. and R. R. Strathmann. 2006. A vermetid gastropod with complex intracapsular cannibalism of nurse eggs and sibling larvae and a high potential for invasion. Pacific Sci, 60: 97–108.

Strong, E. E., O. Gargominy, W. F. Ponder and P. Bouchet. 2008. Global diversity of gastropod (Gastropoda: Mollusca) in freshwater. Hydrobiologia, 595: 149–166.

Sunil Kumar, P. and P. A. Thomas. 2011. Sponge infestation on *Perna indica* Kuriakose and Nair 1976 in experimental culture system. Ind J Geo-Mar Sci, 40: 731–733.

Sweeting, R. A. 1981. Hermaphrodite roach in the River Lee. Thames Water, Lea Division.

Taborsky, M. 2001. The evolution of bourgeois and cooperative reproductive behaviours in fish. J Hered, 72: 100–110.

Takahama, H., T. Kinoshita, M. Sato and F. Sasaki. 1991. Fine structure of the spermatophores and their ejaculated forms, sperm reservoirs, of the Japanese common squid, *Todarodes pacificus*. J Morphol, 207: 241–251.

Talmage, S. C. and C. J. Gobler. 2010. Effects of past, present and future ocean carbon dioxide concentrations on the growth and survival of larval shellfish. Proc Natl Acad Sci USA, 107: 17246–17251.

Talmage, S. C. and C. J. Gobler. 2011. Effects of elevated temperature and carbon dioxide on the growth and survival of larvae and juveniles of three species of northwest Atlantic bivalves. PLoS ONE, 5: doi.org/10.1371/journal.pone.0026941.

Tardy, J. and M. Dufrenne. 1976. Effets du groupement et du volume disponible sur la sexualitat du mollusque nudibranche *Eubranchia doriae* (Trinichese 1879). Haliotis, 7: 66–68.

Tervit, H. R., S. L. Adams, R. D. Roberts et al. 2005. Successful cryopreservation of Pacific oysters (*Crassostrea gigas*) oocytes. Cryobiology, 51: 142–155.

The *Schistosoma japonicum* Sequencing and Functional Analysis Consortium. 2009. The *Schistosoma japonicum* genome reveals features of host-parasite interplay. Nature, 460: 345–351.

Thiriot-Quevreux, C. 1994. Advances in cytogenetics of aquatic organisms. In: *Genetics and Evolution of Aquatic Organisms*. (ed) A. R. Beaumont. Chapman and Hall, London. pp 36–388.

Thiriot-Quiévreux, C. 2003. Advances in chromosomal studies of gastropod molluscs. J Moll Stud, 69: 187–202.

Thomas K., F. Dahboul, D. Alain and T. Monsinjon. 2014. The gametogenic cycle and oestradiol levels in the zebra mussel *Dreissena polymorpha*: a 1-year study. J Moll Stud, 81: 58–65.

Thomas, P. A., K. K. Appukuttan, K. Ramadoss and S. G. Vincent. 1983. Calcibiocavitological investigations. Mar Fish Infor Ser, 49: 1–13.

Thomas, P. A., K. Ramadoss and S. G. Vincent. 1993. Invasion of *Pione margaritifera* Dendy and *Cliona lobata* Hancock on the molluscan beds along the Indian coast. J Mar Biol Ass India, 35: 145–156.

Thomas, P. C., J. V. Mallia and P. Muthiah. 2004. Triploidy induction and confirmation edible oyster *Crassostrea madrasensis*. J Mar Biol Ass India, 46: 224–228.

Thomas, R. F. and L. Opresko. 1973. Observations on *Octopus joubini*: four laboratory reared generations. Nautilus, 87: 61–65.

Thomas, S., A. P. Dineshbabu and G. Sasikumar. 2014. Gastropod resource distribution and seasonal variation in trawling grounds of Konkan Malabar region, eastern Arabian Sea. Ind J Geo-Mar Sci, 43: 384–392.

Thompson, T. E. 1973. Euthyneuran and other molluscan spermatozoa. Malacologia, 14: 167-206+addendum, 443–444.

Thomsen, J., M. A. Gutowska, J. Saphorster et al. 2010. Calcifying invertebrates succeed in a naturally CO_2 rich costal habitat but are threatened by high levels of future acidification. Biogeogenesis, 7: 3879–3891.

Thomsen, J., I. Casties, C. Pansch et al. 2013. Food availability outweighs ocean acidification effects on juvenile *Mytilus edulis*: laboratory and field experiments. Glob Change Biol, 19: 1017–1027.

Thornton, J. W. 2001. Evolution of vertebrate steroid receptors from an ancestral estrogen receptor by ligand exploitation and serial genome expansions. Proc Natl Aacd Sci USA, 98: 5671–5676.

Thornton, J. W., E. Need and D. Crews. 2003. Resurrecting the ancestral steroid receptor: ancient origin of estrogen signaling. Science, 301: 1714–1717.

Thorson, G. 1950. Reproductive cycle and larval ecology of marine bottom invertebrates. Biol Rev, 140: 1139–1147.

Tillmann, M., U. Schulte-Oehlmann, M. Duft et al. 2001. Effects of endocrine disruptors on prosobranch snails (Mollusca: Gastropoda) in the laboratory. Part III: Cyproterone acetate and vinclozoline as antiandrogens. Ecotoxicology, 10: 373–388.

Todd, C. D. 1979. Reproductive energetic of two species of dorid nudibranchs with planktotrophic and lecithotrophic larval stages. Mar Biol, 53: 57–68.

Trask, J. L. and C. L. Van Dover. 1999. Site-specific and ontogenetic variations in nutrition of mussel (*Bathymodiolus* sp.) from the Lucky Strike hydrothermal vent field, Mid-Atlantic Ridge. Limnol Oceanogr, 44: 334–343.

Trevallion, A. 1971. Studies on *Tellina tenuis* Da Costa. III. Aspects of general biology and energy flow. J Exp Mar Biol Ecol, 7: 95–122.

Trickey, J. S., J. Vannar and N. G. Wilson. 2013. Reproductive variance in planar spawning *Chromodoris* species (Mollusca: Nudibranchia). Moll Res, 33: 265–271.

Trivers, R. L. 1972. Paternal investment and sexual selection. In: *Sexual Selection and Descent of Man*. (ed) B. Campbell. Aldine, Chicago. pp 136–179.

Trueb, H. and G. Ribi. 1997. High fecundity of hybrids between the sympatric snail species *Viviparus ater* and *V. contectus* (Gastropoda: Prosobranchia). Heredity, 79: 418–423.

Trueman, E. R. 1983. Locomotun in molluscs. In: *The Mollusca. Physiology*. (ed) A. S. M. Saleuddin and K. M. Wilbur. Academic Press, Vol 4, Part. 1: 155–198.

Trussell, G. C. 1996. Phenotypic plasticity in an intertidal snail: the role of a common crab predator. Evolution, 50: 448–454.

Trussell, G. C. 2000. Phenotypic clines, plasticity and morphological trade-offs in an intertidal snail. Evolution, 54: 151–166.

Trussell, G. C. and M. O. Nicklin. 2002. Cue sensitivity, inducible defense, and trade-offs in a marine snail. Ecology, 83: 1635–1647.

Trussell, G. C., A. S. Johnson, S. G. Rudolph and E. S. Gilfilian. 1993. Resistance to dislodgement: habitat and size-specific differences in morphology and tenacity in an intertidal snail. Mar Ecol Prog Ser, 100: 135–144.

Tsuchiya, M. 1980. Biodeposit production by the mussel *Mytilus edulis* L. on the rocky shore. J Exp Mar Biol Ecol, 47: 203–222.

Turekian, K. K., K. Cochran, D. P. Kharkar et al. 1975. Slow growth rate of a deep-sea clam determined by ^{228}Ra chronology. Proc Natl Acad Sci USA, 72: 2829–2832.

Turner, L. M. and N. G. Wilson. 2012. The *Chelidonura tsurugensis* complex: do reproductive decisions maintain colour polymorphism? J Moll Stud, 78: 166–172.

Tutman, P., S. K. Sifner, J. Dulcic et al. 2008. A note on the distribution and biology of *Ocythoe tuberculata* (Cephalopoda: Ocythoidae) in the Adriatic Sea. Vie et Milieu, 58: 215–221.

Urban-Rich, J., M. Dagg and J. Peterson. 2001. Copepod grazing on phytoplankton in the Pacific sector of the Antartic. Polar front Deep Sea Res II, 48: 4223–4246.

Vacquier, V. D. 1998. Evolution of gamete recognition proteins. Science, 281: 1995–1998.

Valladares, A., G. Manriquez and B. A. Suarez-Isla. 2010. Shell shape variation in populations of *Mytilus chilensis* (Hupe 1854) from southern Chile: a geometric morphometric approach. Mar Biol, 157: 2731–2738.

Van Mol, J. J. 1967. Eude morphologique et physiologique du ganglion cerebraoide des Gastropodes Pulomones (Molluscques). Mem Acad R Belg (Classe Ser), 37: 1–168.

Van der Spoel, S. 1979. Strobilization in a pteropod (Gastropoda: Opisthobranchia). Malacologia, 18: 27–30.

Van Heukelem, W. F. 1983. *Octopus maya*. In: *Cephalopod Life Cycles: Species Accounts*. (ed) P. R. Boyle. Academic Press, London. 1: 311–324.

Vendetti, J. E., C. D. Trorobridge and P. J. Krug. 2012. Poecilogony and population genetic structure in *Elysia pusilla* (Heterobranchia, Sacoglossa) and reproductive data for five sacoglossans that express dimorphisms in larval development. Integ Comp Bull, 52: 138–150.

Venkatesan, V. and K. S. Mohamed. 2015. Gastropod classification and taxonomy. Summer school on recent advances in marine biodiversity conservation and management, Central Marine Fisheries Research Institute, Kochi. 38–41.

Vermeij, G. J. 1975. Evolution and distribution of left-handed and planispiral coiling in snails. Nature, 254: 419–420.

Vermeij, G. J. 1993. *A National History of Shells*. Princeton University Press, Princeton, N.J.

Vianey-Liaud, M. 1972. Etude du controle de la maturite des tractus genitaux et de la ponte par castration chirurgicale chez le planorbe *Australorbis glabratus* Say (Pulmone Basommatophore). Bull Soc Zool Fr, 97: 675–690.

Vianey-Liaud, M. 1976. Influence de l'isolement ef la taille sur la fecondite du planorbe *Australorbis glabratus* (Gastropode: Pulmone). Bull Biol France Belgique, 111: 5–9.

Vianey-Liaud, M. 1989. Growth and fecundity in a black-pigmented and an albino strain of *Biomphalaria glabrata* (Gastropoda: Pulmonata). Malacol Rev, 22: 25–32.

Vianey-Liaud, M. 1995. Bias in the production of heterozygous pigmented embryos from successively mated *Biomphalaria glabrata* (Gastropoda: Pulmonata). Malacol Rev, 28: 97–106.

Vianey-Liaud, M. and G. Dussart. 2002. Aspects of pairing and reproduction in the hermaphroditic freshwater snail *Biomphalaria glabrata* (Gastropoda: Pulmonata). J Moll Stud, 68: 243–248.

Vianey-Liaud, M., J. Dupouy, F. Lancastre and H. Nassi. 1987. Genetical exchange between one *Biomphalaria glabrata* (Gastropoda: Planorbidae) and varying number of partners. Mem Inst Oswaldo Cruz, 82: 457–460.

Vianey-Liaud, M., H. Nassi, F. Lancastre and J. Dupouy. 1989. Duration of pairing and use of allosperm *Biomphalaria glabrata* (Gastropoda: Planorbidae). Mem Inst Oswaldo Cruz, 84: 41–45.

Vianey-Liaud, M., J. Dupouy, F. Lancastre and H. Nassi. 1991. Constant use of allosperm of female-acting snails after successive cross-fertilization *Biomphalaria glabrata* (Gastropoda: Planorbidae). Malacol Rev, 24: 73–78.

Vianey-Liaud, M., D. Jolly and G. Dussart. 1996. Sperm competition in the simultaneous hermaphrodite freshwater snail *Biomphalaria glabrata* (Gastropoda: Planorbidae). J Mol Stud, 62: 451–457.

Vidal, E. A. G., R. Villanueva, J. P. Andre et al. 2014. Cephalopod culture: current status of main biological models and research priorities. Adv Mar Biol, 67: 1–98.

Vinarski, M. V. 2007. An interesting case of predominantly sinistral population of *Lymnaea stagnalis* (L.) (Gastropoda: Pulmonata: Lymnaeidae). Malacol Bohemoslov, 6: 17–21.

Vitturi, R. and E. Catalano. 1988. The male XO sex-determining mechanism in *Theodoxus meridionalis* (Linnaeus, 1758) (Mollusca: Prosobranchia). Cytologia, 53: 131–138.

Vitturi, R., M. S. Colomba, V. Caputo and A. Pondolfa. 1998. XY chromosome sex systems in the neogastropods *Fasciolaria lignaria* and *Pisania striata* (Mollusca: Prosobranchia). J Hered, 89: 538–543.

Vivekanandan, E., V. V. Singh and J. K. Kizhokuran. 2013. Carbon foodprint by marine fishing boards of India. Curr Sci, 104: 361–366.

Voelz, N. J., J. V. McArthur and R. B. Rader. 1998. Upstream mobility of the Asiatic clam *Corbicula fluminea*: identifying potential dispersal agents. J Freshw Ecol, 13: 39–45.

von Brand, T. and B. Mehlman. 1953. Relations between pre-and post-anaerobic oxygen consumption and oxygen tensions in some freshwater snails. Biol Bull, 104: 301–312.

von Brand, T., M. O. Nolan and E. G. Mann. 1948. Observations on the respiration of *Australorbis glabratus* and some other aquatic snails. Biol Bull, 95: 199–213.

Voogt, P. A. 1969. Investigations of the capacity of synthesizing 3b-sterols in mollusca. IV: The biosynthesis of 3b-sterols in some mesogastropods. Comp Biochem Physiol, 31: 37–46.

Walker, D., A. J. Power, M. Sweeney-Reeves and J. C. Avise. 2007. Multiple paternity and female sperm usage along egg-case strings of the knobbed whelk *Busycon carica* (Mollusca: Melogenidae). Mar Biol, 151: 53–61.

Walne, P. R. 1964. Observations on the fertility of the oyster (*Ostrea edulis*). J Mar Biol Ass UK, 44: 293–310.

Wang, C. and R. P. Croll. 2004. Effects of sex steroids on gonadal development and gender determination in the sea scallop *Placopecten magellanicus*. Aquaculture, 238: 483–498.

Wang, C. and R. P. Croll. 2006. Effects of steroids on spawning in the sea scallop *Placopecten magellanicus*. Aquaculture, 256: 423–432.

Wang, C. and R. P. Croll. 2007. Estrogen receptors in the sea scallop: characterization and possible involvement in reproductive regulation. Comp Biochem Physiol B, doi:10.1016/j.cbpb.2007.06.008.

Wang, R. C. and Z. P. Wang. 2008. *Science of Marine Shellfish Culture*. Ocean University of China Press, Qingdao. p 613.

Ward, P. D. 1983. *Nautilus macromphalus*. In: *Cephalopod Life Cycles: Species Accounts*. (ed) P. R. Boyle. Academic Press, London, 1: 11–28.

Wassnig, M. and P. C. Southgate. 2012. Embryonic and larval development of *Pteria penguin* (Röding, 1798) (Bivalvia: Pteriidae). J Moll Stud, 78: 134–141.

Watanabe, S., Y. Kirino and A. Gelperin. 2008. Neural and molecular mechanisms of microrecognition in *Limax*. Learn Mem, 15: 633–642.

Watson, S. A., L. Peck, P. A. Tyler et al. 2012. Marine invertebrate skeleton size varies with latitude, temperature and carbonate saturation: implications for global change and acidification. Glob Change Biol, 18: 3026–3038.

Watson, S. A., S. Lefevre, M. I. McCornick et al. 2014. Marine mollusc predator-escape behavior altered by near-future carbon dioxide levels. Proc R Soc, 281B: doi.org./10.1098/rspb.2013.2377.

Webbe, G. 1962. The transmission of *Schistosoma haematobium* in an area of Lake Province, Tanganyika. Bull Wld Heth Org, 27: 59–85.

Welborn, J. R. and D. T. Manahan. 1990. Direct measurements of sugar uptake from sea water into the molluscan larvae. Mar Ecol Prog Ser, 65: 233–239.

Wells, F. E. Jr. 1976. Growth rate of four species of enthecosomatous pteropods occurring off Barbados, West Indies. Nautilus, 90: 114–116.

Wells, F. E. and J. Wells. 1977. Cephalopoda: Octopoda. In: *Reproduction of Marine Invertebrates*. (eds) A. C. Giese and J. S. Pearse. Academic Press, New York, 4: 291–336.

Wells, M. J. and J. Wells. 1972a. Optic glands and the state of the testes in *Octopus*. Mar Behav Physiol, 1: 71–83.

Wells, M. J. and J. Wells. 1972b. Sexual displays and mating of *Octopus vulgaris* (Cuvier) and *O. cyanea* (Gray) an attempt to alter performance by manipulating the glandular condition of the animals. Anim Behav, 20: 293–308.

Wells, M. J. and J. Wells. 1975. Optic gland implants and their effects on the gonads of octopus. J Exp Biol, 62: 579–588.

West, H. H., J. F. Harrigan and S. K. Pierce. 1984. Hybridization of two populations of a marine opisthobranch with different developmental patterns. Veliger, 26: 199–206.

Wethington, A. R. and R. T. Dillon. 1991. Sperm storage and evidence for multiple insemination in a natural population of the freshwater snail *Physa*. Am Malacol Bull, 9: 9–102.

Whitehead, D. L. 1977. Steroids enhance shell regeneration in an aquatic gastropod (*Biomphalaria glabrata*). Comp Biochem Physiol, 58C: 137–141.

Wickett, M. E., K. Caldeira, and P. B. Duffy. 2003. Effect of horizontal grid resolution on simulations of oceanic CFC-11 uptake and direct injection of anthropogenic CO_2. J Geophys Res, 108(C6): 3189, doi:10.1029/2001JC001130.

Wilke, T., G. M. Davies, A. Falniowski et al. 2001. Molecular systematic of Hydrobiidae (Mollusca: Gastropoda: Rissooidea): testing monophyly and phylogenic relationships. Proc Acad Natl Sci Philadelphia, 151: 1–21.

Wise, J., M. G. Harasewyen and R. T. Dillon Jr. 2004. Population divergence in the sinistral whelks of North America with special reference to the east Florida ecotone. Mar Biol, 145: 1167–1179.

Wodinsky, J. 1977. Hormonal inhibition of feeding and death in *Octopus*: control by optic gland secretion. Science, 198: 948–951.

Wolfe, K., A. M. Smith, P. Trimpy and M. Byrne. 2012. Vulnerability of the paper nautilus (*Argonauta nodosa*) shell to a climate-change ocean: potential for extinction of dissolution. Biol Bull, 223: 236–244.

Won, S. -J., A. Novillo, N. Custodia et al. 2005. The freshwater mussel (*Elliptio complanata*) as a sentinel species: vitellogenin and steroid receptors. Integ Comp Biol, 45: 75–80.

Won, Y. J., S. Hallam, G. D. O'Mullam et al. 2003. Environmental acquisition of thiotrophic endosymbionts by deep-sea mussels of the genus *Bathymodiolus*. Appl Environ Microbiol, 79: 6785–6792.

Wood, A. R., G. Turmer, D. O. F. Skinski and A. R. Beaumont. 2003. Distribution of doubly uniparental inheritance of mitochondrial DNA in hybrid mussels (*Mytilus edulis* x *M. galloprovincialis*). Heredity, 91: 354–360.

Wood, H. L., J. I. Spicer and S. Widdicombe. 2008a. Ocean acidification may increase calcification rates, but at a cost. Proc R Soc B, 275: 1767–1773.

Wood, L. R., R. A. Griffiths and K. Groh. 2008b. Interactions between freshwater mussels and newts: a novel form of parasitism? Amphibia-Reptilia, 29: 457–462.

Wootton, A. N., C. Herring, J. A. Spry and P. S. Goldfarb. 1995. Evidence for the existence of cytochrome P450 gene families (CYP1A, 3A, 4A, 11A) and modulation of gene expression (CYP1A) in the mussel *Mytilus* spp. Mar Environ Res, 39: 21–26.

Wright, J. M., L. M. Parker, W. A. O'Connor et al. 2014. Populations of Pacific oysters *Crassostrea gigas* respond variably to elevated CO_2 and predation by *Morula marginalba*. Biol Bull, 226: 269–281.

Wright, J. R. and R. G. Hartnoll. 1981. An energy budget for a population of the limpet *Patella vulgata*. J Mar Biol Ass UK, 61: 627–646.

Wright, R. T., R. B. Coffin, C. P. Ersing and D. Pearson. 1982. Field and laboratory measurements of bivalve filtration of natural marine bacterioplankton. Limnol Oceanogr, 27: 91–98.

Xu, W., J. Zhang, S. Du et al. 2014. Sex differences in alarm response and predation risk in the freshwater snail *Pomacea canaliculata*. J Moll Stud, 80: 117–122.

Xue, D., T. Zhang and J. -X. Liu. 2014. Microsatellite evidence for high frequency of multiple paternity in the marine gastropods *Rapana venosa*. PLoS ONE, 9(1): e86508, doi: 10.1371/journal.pone.0086508.

Yakovlev, Y. and V. Malakhov. 1985. The anatomy of dwarf males of *Zachsia zenkewitschi* (Bivalvia: Teredinidae). Asian Mar Biol, 2: 47–56 .

Yamamoto, S., Y. Sugawara, T. Nomura and T. Oshino. 1988. Induced triploidy Pacific oyster *Crassostrea gigas* and performance of triploid larvae. Tohoku J Agri Res, 39: 47–59.

Yamazaki, M., J. Kisugi and H. Kamiya. 1989c. Biopolymers from marine invertebrates. XI. Characterization of an antineoplastic glycoprotein, dolabellanin A, from the albumen gland of a sea hare, *Dolabella auricularia*. Chem Pharm Bull, 37: 3343–3346.

Yan, H., Q. Li, W. Him et al. 2011. Seasonal changes of estradiol-17β and testosterone concentrations in the gonad of the razor clam *Sinonovacula constricta* (Lamark, 1818). J Moll Stud, 77: 116–122.

Yang, H. and X. Guo. 2004. Tetraploid induction by meiosis induction in the dwarf clam *Mulinia lateralis* (Say 1822): effects of cytochalasin B duration. Aquacult Res, 35: 1187–1194.

Yang, H. and X. Guo. 2006. Polyploid induction by heat shock-induced meiosis and mitosis inhibition in the dwarf surf clam *Mulinia lateralis* Say, Aquaculture, 252: 171–182.

Yang, H., F. Zhang and X. Guo. 2000. Triploid and tetraploid zhikong scallop *Chlamys farrei* Jones et Preston produced by inhibiting polar body I. Mar Biotechnol, 2: 466–475.

Yang, H., P. M. Johnson, K. C. Ko et al. 2005. Cloning, characterization and expression of escapin, a broadly antimicrobial FAD-containing L-amino acid oxidase from ink of the sea hare *Aplysia californica*. J Exp Biol, 208: 3609–3622.

Yang, H., E. Hu, R. Cuevas-Uribe et al. 2012. High-throughput sperm cryopreservation of eastern oyster *Crassostrea virginica*. Aquaculture, 344: 223–230.

Yankson, K. and J. Moyse. 1991. Cryopreservation of the spermatozoa of *Crassostrea tulipa* and the three other oysters. Aquaculture, 97: 259–267.

Young, K. G., J. P. Chang and J. I. Goldberd. 1999. Gonadotropin-releasing hormone neuronal system of the freshwater snails *Helisoma trivolvis* and *Lymnaea stagnalis*: possible involvement in reproduction. J Comp Neurol, 404: 427–437.

Young, N. D., R. S. Hall, A. R. Jex. 2010. Elucidating the transcriptome of *Fasciola hepatica*—a key to fundamental and biotechnological discoveries for a neglected parasite. Biotechnol Adv, 28: 222–231.

Yusa, Y. 1994. Size-related egg production in a simultaneous hermaphrodite, the sea hare *Aplysia kurodai* Baba (Mollusca: Opisthobranchia). Publ Seto Mar Biol Lab, 36: 249–254.

Yusa, Y. 2004a. Inheritance of colour polymorphism and the pattern of sperm competition in the apple snail *Pomacea canaliculata* (Gastropoda: Ampularidae). J Moll Stud, 70: 43–46.

Yusa, Y. 2004b. Brood sex ratio in the apple snail *Pomacea canaliculata* (Gastropoda: Ampullariidae) is determined genetically and not by environmental factors. J Moll Stud, 70: 269–275.

Yusa, Y. 2007. Causes for variations in sex ratio and modes of sex determination in the Mollusca: an overview. Am Malacol Bull, 23: 89–98.

Yusa, Y. and Y. Suzuki. 2003. A snail with unbiased population sex ratios but highly biased brood sex ratios. Proc R Soc, 270B: 283–288.

Zabala, S., G. N. Hermida and J. Gimenez. 2009. Ultrastructure of euspermatozoa and paraspermatozoa in the volutid snail *Adelomelon ancilla* (Mollusca: Gastropoda) Helgol Mar Res, 63: 181–188.

Zabala, S., G. Hermidia and J. Gimenez. 2012. Spermatogenesis in the marine snail *Adelomelon ancilla* (Volutidae) from Patagonia. J Moll Stud, 78: 52–65.

Zeebe, R. E., J. C. Zachos, K. Caldeira and T. Tyrrell. 2008. Oceans: carbon emissions and acidification. Science, 321: 51–52.

Zelaya, D. G., J. A. Pechenik and C. S. Gallardo. 2012. *Crepipatella dilatata* (Lamarck, 1822) (Calyptraeidae): an example of reproductive variability among gastropods. J Moll Stud, 78: 330–336.

Zhang, L., N. L. Wayne, N. M. Sherwood et al. 2000. Biological and immunological characterization of multiple GnRH in an opisthobranch mollusk, *Aplysia californica*. Gen Comp Endocrinol, 118: 77–89.

Zhang, L., J. A. Tello, W. Whang and P. Tsai. 2008. Molecular cloning, expression pattern and immunocytochemical localization of a gonadotropin releasing hormone-like molecule in the gastropod mollusk *Aplysia californica*. Gen Comp Endocrinol, 156: 201–209.

Zouros, E. 2001. The exceptional mitochondrial DNA system of the mussel family Mytilidae. Genes Genet Syst, 75: 313–318.

Zouros, E., A. O. Ball, C. Saavedra and K. R. Freeman. 1994a. Mitochondrial DNA Inheritance. Nature, 368: 818.

Zouros, E., B. A. Oherhanser, C. Saavedra and K. R. Freeman. 1994b. An unusual type of mitochondrial DNA inheritance in the blue mussel *Mytilus*. Proc Natl Acad Sci USA, 91: 7463–7467.

Author Index

Species Index

A

Abra ovata, 52
Acanthiana monodon, 50
A.punctulata, 55
Acanthinucella spirata, 114
Acanthochiton, 6, 46
Acanthopagrus schlegeli, 137
Aceton, 138
Acheilognathus rhombeus, 33
A. tabiara tabiara, 33
Acmae rubella, 27, 124
Adalaria proxima, 28, 30–31, 133
Adelomelon ancilla, 24–25
Aeolida, 6
Agriolimax reticulatus, 26
Albinaria cretensis, 65
Alderia, 122
A.willowi, 111, 118–119, 122, 127, 221–222
Amblema dolomeiu, 34
A. plicata, 33, 127, 129, 131
Ammonicera minortalis, 22
Amnicola limosa, 128, 206
Amphibola, 29
Amphiodromus inversus, 66
Ampulla priamus, 39
Anadra divulli, 39
Ancylus fluviatilis, 73–74, 86, 145
Anisus vortex, 85
Anodonta, 89, 124
A. californiensis, 88–89
A. corpulenta, 89
A. couperiana, 84, 88–89
A. gibbosa, 88–89
A. hallenbeckii, 88–89
A. imbecilis, 84, 88–89
A. peggyae, 88–89
A. wahlamartensis, 88–89
A. woodiana, 33
Aporocotlyle simplex, 152
Aplysia spp, 6, 28, 68–69, 72, 224
A.brasiliana, 69
A. californica, 29, 69–70, 78, 85–86, 104,
    107–109, 129, 133, 182, 191, 224

A. californicus, 2
A. dactylomela, 18–19, 68–69
A. depilans, 182
A. fasciata, 70
A. juliana, 8, 19, 68
A. kurodai, 68, 116, 125, 129
A. punctata, 18–19
A. vaccaria, 22, 220
Archaeospira arnata, 65
Archidoris, 137–138
Architeuthis sp, 21–22, 221
Arctica islandica, 38–39, 159–160
Arianta arbustorum, 25, 45
Argonauta, 6
A. argo, 18
A. hians, 19
A. nodosa, 46, 204–205
Argopecten irradians, 53–54, 159–160, 173, 207,
    209–210, 216
A.purpuratus, 186
Arion sp, 138–139
A. alter, 140
A. lusitanicus, 219
A. rufus, 136, 139, 143
Artacama proboscidea, 152
Artemia, 228
A. salina, 173
Astatotilapia burtoni, 101
Aulacomya alter, 20, 56, 59
Austropeplea ollula, 157

B

Bacillus sphericus, 227
Balcis shaplandi, 90
Bankia gouldi, 92
Basommatrophora, 152
Bathymodiolus sp, 12
B. azoricus, 12–13
B. childressi, 12
B. elongatus, 12
B. heckerae, 12
B. puteoserpentis, 12
B. thermophilus, 12, 33

Subject Index

A

Accessory Boring Organ, 54–55
Agametic cloning, 134, 137
Alarm signals, 70
Anaerobiosis, 145, 149–150
Androdioecy, 83, 98
Aragonite, 45, 201, 204
Attractive chemicals,
    Attractin, 78
    L-tryptophan, 78–79
    Temptin, 78
Aufwuchs, 14–15

B

Ballast water, 33
Biogenic habitats, 7, 39
Brood protection, 123–124
Bursa cupulatrix, 74, 83, 106

C

Capsular fluid, 112–113
Central pattern generator, 140
Chemical triploids, 163
Chemoautotrophic bivalves, 2
Compressive force (N), 55–56
Conchin, 45, 59
Conotoxin, 3
Crab cue, 62, 213–214

D

Dark meat, 51
Dead Sea salt, 160
Defensive chemicals,
    Achain, 68
    Anti-feedant, 68
    Aplysianin, 68
    Dolabellarianin, 68
    Escapin, 68
    Inking, 69–70

Desperate larvae, 122
Dimorphism,
    Dextral/sinistral, 44–45, 65–66
    Eggs, 119–120, 222
    Phally, 87, 97–98
    Shell thickness, 60–62, 218
    Sperm, 25–26, 101–103, 223
Dwarf male, 46, 97

E

Extra Capsular Yolk, 112

F

Fecundity,
    Absolute, 7, 126–127, 132–133
    Batch, 125
    Determinate, 127
    Indeterminate, 127
    Potential, 127
    Realized, 127
    Relative, 127
    Seasonal, 126
Fertilization,
    Monospermic, 77, 206
    Polyspermic, 162, 205–206
    success, 76, 78, 103, 161

G

Global warning, 200

H

Hypotheses,
    Bed-hedging, 100, 122, 128
    Inhibited T excretion, 197
    Neuropeptides, 197
    Palmer, 42
    Red Queen, 82
    Reproductive assurance, 83
    Retinoid X receptor, 197

Milton Keynes UK
Ingram Content Group UK Ltd.
UKHW040448071024
449327UK00020B/1078